Monte

Neural Mechanisms of Startle Behavior

The vertical leap of a nine-banded armadillo, *Dasypus novemcinctus,* surprised by the flash of a photographer's lighting system. Startle responses are readily elicited by unexpected disturbances. In mammals the natural role of this behavior is not well understood. Presumably it functions in predator avoidance, as demonstrated for many invertebrates and teleost fishes. Because of the startle behavior, an attacking predator might miss its target, or itself be alarmed by the sudden movement of the prey. By Bianca Lavies, © 1982 National Geographic Society.

Neural Mechanisms of Startle Behavior

Edited by

Robert C. Eaton
University of Colorado
Boulder, Colorado

Plenum Press • New York and London

Library of Congress Cataloging in Publication Data

Main entry under title:

Neural mechanisms of startle behavior.

Includes bibliographies and index.

1. Startle reaction. 2. Neural circuitry. I. Eaton, Robert C. [DNLM: 1. Nervous System—Physiology. 2. Startle Reaction—physiology. WL 106 N493]

QP372.6.N48 1984 156′.232 84-13375

ISBN 0-306-41556-9

A Division of Plenum Publishing Corporation
233 Spring Street, New York, N.Y. 10013

Printed in the United States of America

This book is dedicated to
C.H.E. and J.P.B.

Contributors

MICHAEL V. L. BENNETT, Division of Cellular Neurobiology, Department of Neuroscience, Albert Einstein College of Medicine, Bronx, New York 10461

THEODORE HOLMES BULLOCK, Neurobiology Unit, Scripps Institution of Oceanography and Department of Neurosciences, School of Medicine, University of California at San Diego, La Jolla, California 92093

MICHAEL DAVIS, Department Psychiatry, Connecticut Mental Health Center, Yale University, New Haven, Connecticut 06508

CHARLES D. DREWES, Zoology Department, Iowa State University, Ames, Iowa 50011

ROBERT C. EATON, Behavioral Biology Group, Department of Biology, E.P.O., University of Colorado, Boulder, Colorado 80309

JOHN T. HACKETT, Department of Physiology, School of Medicine, University of Virginia, Charlottesville, Virginia 22908

HOWARD S. HOFFMAN, Department of Psychology, Bryn Mawr College, Bryn Mawr, Pennsylvania 19010

DAVID G. KING, Department of Zoology, Southern Illinois University, Carbondale, Illinois 62901

FRANKLIN B. KRASNE, Department of Psychology, University of California at Los Angeles, Los Angeles, California 90024

GEORGE O. MACKIE, Biology Department, University of Victoria, Victoria, British Columbia, Canada V8W 2Y2

MICHAEL O'SHEA, Departments of Pharmacology and Physiology, University of Chicago Medical School, Chicago, Illinois 60637. *Present address*: Département de Biologie Animale, Université de Genève, CH-1211, Genève 4, Switzerland

KEIR G. PEARSON, Department of Physiology, University of Alberta, Edmonton, Alberta, Canada, T6G 2H7

ROY E. RITZMANN, Department of Biology, Case Western Reserve University, Cleveland, Ohio 44106

LAWRENCE SALKOFF, Department of Neurobiology, Washington University School of Medicine, St. Louis, Missouri, 63110.

JOHN B. THOMAS, Department of Biological Sciences, Stanford University, Stanford, California 94305

JEFFREY J. WINE, Department of Psychology, Stanford University, Stanford, California 94305

ROBERT J. WYMAN, Department of Biology, Yale University, New Haven, Connecticut 06511

Preface

In the past fifteen years there has been considerable interest in neural circuits that initiate behavior patterns. For many types of behaviors, this involves decision-making circuits whose primary elements are neither purely sensory nor motor, but represent a higher order of neural processing. Of the large number of studies on such systems, analyses of startle circuits compose a major portion, and have been carried out on systems found throughout the animal kingdom. Startle has been an important model because of the reliability of the behavioral act for laboratory study and the accessibility of the underlying neural circuitry. However, probably because of the breadth of the subject, this material has never been reviewed in a comprehensive way that presents the elements common to startle circuits in the different animal systems in which they occur.

This book presents a diversity of approaches based on a broad background of animal groups ranging from the earliest nervous systems in cnidarians to the most recently evolved and advanced in mammals. The behaviors themselves are all short latency, fast motor acts, when considered on the time scale of the organism, and involve avoidance or evasion, although in some cases we do not yet completely understand their natural role. These behaviors occur in response to stimuli that have sudden or unexpected onset. Classically, these behaviors are called "startle responses," although in recent years this usage has given way to the terminology "escape response" when the function has become understood to be used in avoiding predators or threatening objects.

In this book, each chapter describes various approaches that have been taken in the study of startle systems. Each uses the different systems and exploits their most favorable features in individual ways. For example, Wyman and colleagues show us remarkable progress in the understanding of neurogenetic mechanisms by studying the giant fiber system of *Drosophila*. Ritzmann demonstrates how single giant fibers in the cockroach can trigger different behavioral responses depending on sensory input. For the crayfish, Krasne and Wine provide a description of an escape network that is understood in exquisite detail from input to output. As can be seen from these examples, the theme that unites this book is the attempt to explain the behavioral act in terms of the neural mechanisms. Thus, although there is diversity in experimental approach, this book is primarily oriented toward neural networks and behavior.

It is the intent of this book to serve as a dialogue of ideas for those interested in the controlling mechanisms for the initiation of behavior patterns. As Bullock's introductory chapter tells us, the presence of specialized, fast-conducting giant fibers for initiating escape is common across many phyla and, with the exception of the mammals, the systems described in this book utilize these neurons. In mammals these cells are absent, but even here the critical cells, as shown in the chapter by Davis, are of large size. The electrophysiological dissection of the mammalian startle circuits is in its beginnings, but as the chapter by Hoffman shows, sensory aspects of behavioral responsiveness have been thoroughly investigated here. This approach could serve as a model for investigating these mechanisms on a cellular basis in invertebrate preparations.

What emerges from this book is the remarkable evolutionary convergence of mechanisms underlying startle in these phylogenetically diverse systems. We hope that these mechanisms will suggest general principles for the understanding of less specialized motor acts.

Robert C. Eaton

Boulder, Colorado

Contents

Chapter 3
Escape Reflexes in Earthworms and Other Annelids
Charles D. Drewes

Chapter 4
The Cockroach Escape Response
Roy E. Ritzmann

CHAPTER 9

Methodological Factors in the Behavioral Analysis of Startle: The Use of Reflex Modification Procedures and the Assessment of Threshold

HOWARD S. HOFFMAN

Chapter 10
The Mammalian Startle Response
Michael Davis

Chapter 11
Escapism: Some Startling Revelations
Michael V. L. Bennett

Neural Mechanisms of Startle Behavior

1

Comparative Neuroethology of Startle, Rapid Escape, and Giant Fiber-Mediated Responses

THEODORE HOLMES BULLOCK

1. Introduction

In this chapter I will explore the relationship between startle responses, rapid escape responses, and behaviors that are mediated by giant fibers. Are these independent, overlapping, or coextensive sets? It might be thought on superficial perusal of this book that it seems to support a view of these categories as being nearly universal, even if not quite coextensive. In their 1939 book *The Startle Pattern*, Landis and Hunt implied that this form of response seems confined to mammals. I will argue that all three categories are very ancient and widely distributed but by no means universal, and that they are in fact partially overlapping sets. This will require a brief survey of the animal kingdom with respect to criteria for each of the three categories of response. Hopefully, we will arrive at preferred usages for the terms and clear the air for conclusions about congruence of the three sets. This is no quibble over words, but an essential step toward the goal of contributing some well worked out examples of neuroethology—the adequate accounting for behavior in neural terms. The selection of papers in this book on a special class of behaviors

THEODORE HOLMES BULLOCK • Neurobiology Unit, Scripps Institution of Oceanography and Department of Neurosciences, School of Medicine, University of California at San Diego, La Jolla, California 92093.

is relevant not only to neuroethology but, as I will point out, to general neurobiology as well.

2. The Distribution of Startle and Escape Responses

Surveying the phyla for startle responses, rapid escape behavior, and giant fiber mediation is a task of quite unequal kind. Whereas a good body of data, though by no means sufficient, bears on the presence or absence of giant fiber responses (Figure 1), there appears to be little hard data on the two behaviorally defined categories. Rapid escape is such a fuzzy category that we need not attempt to make it precise, or compare animals systematically. Obviously it must include the sudden withdrawal of the ciliate protozoan *Stentor,* of many tube worms such as *Phoronis* (Phylum Phoronida, Figure 1) and *Sabella* (Annelida; class Polychaeta), of many burrowing animals from annelids such as *Lumbricus* to rodents such as prairie dogs, along with a wide variety of other avoidance responses: darting, jumping, coiling, scurrying, and taking flight. We cannot categorically exclude all avoidance movements that appear to us less than rapid, since the time scale that matters is that of the natural predators.

The only usefulness of the category in our context is to permit propositions such as the following, which I will assert for the purposes of this survey.

1. Not all rapid escape responses are startle responses; a fly may take off or a frog may jump during the slow approach of a threatening stimulus.
2. Not all rapid escape responses are giant fiber-mediated; heterotrich ciliates, crabs, octopuses, lizards, and mice exhibit conspicuous sudden escape, although they lack giant systems.
3. Not all startle responses lead to escape; they may be much too limited in amplitude to achieve translation of the body.

Startle seems potentially more amenable to definition than rapid escape. A startle response is an abrupt response, often of relatively short latency, to a sudden stimulus that we believe to be both unexpected and alarming (i.e., of high valence). The movement may or may not be a large one, that is, it may translate the whole body or move only limited parts of the body. The problem remains how fast is abrupt, given a spectrum of cases. We can simplify the experimental test by imposing quasinatural stimuli with artificial, virtually perfect abruptness and require that the response, even if the latency is not short, rise rapidly on the time scale of the species. Borderline cases are to be expected. However, it seems

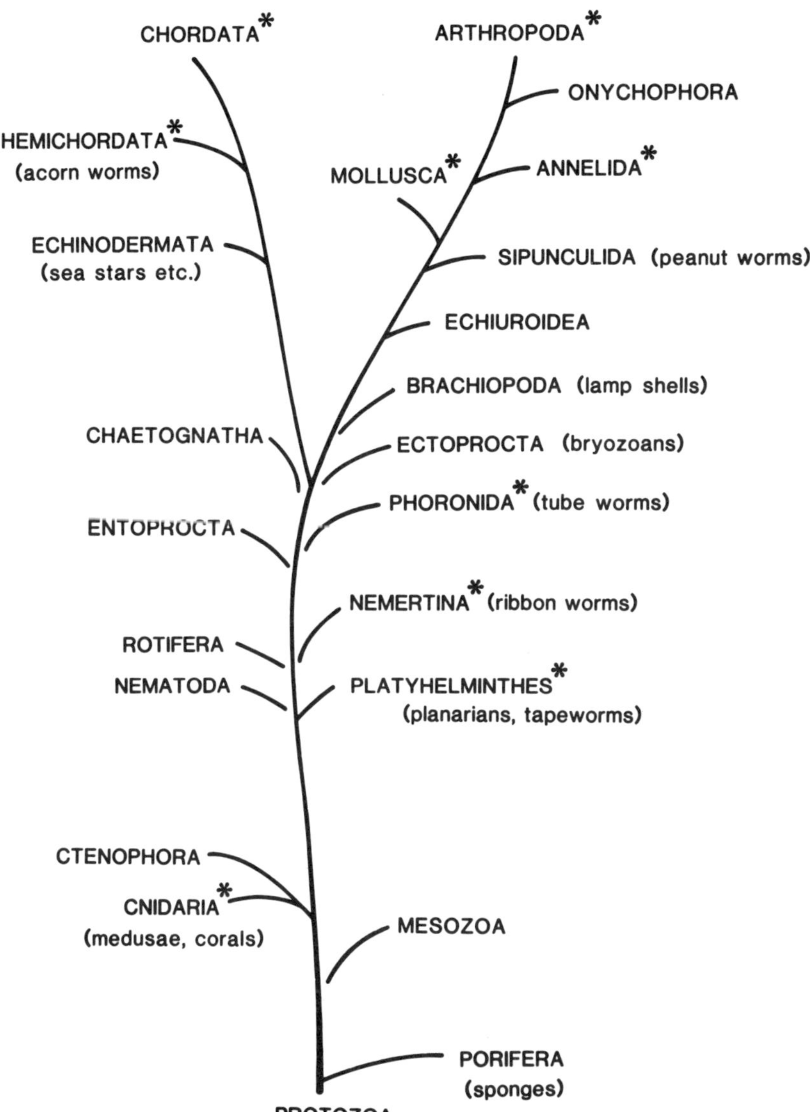

Figure 1. Evolutionary relationships of major animal groups. Phyla are written in capital letters and common name examples of less familiar groups are in parentheses. Phyla with asterisks have some members, although not necessarily all, with giant fibers. Scheme derived from Barnes (1968) after Hyman (1940).

clear from common experience that good startle responses may involve only a limited musculature and fail to translate the body; eye blink, facial twitch, crouch, or an isometric twitch of the limbs may constitute the whole startle response. The latency does not have to be minimal. The response depends not only on the abruptness of the stimulus but on some state of nonexpectancy that may be difficult to specify or control. In the absence of careful study of many groups, I will guess that it is common for animals without giant fiber systems to exhibit startle (see Chapter 10, this volume). Certainly I have observed a sudden twitch of a more or less extensive musculature in sloths (*Bradypus*) when the branch or substratum is struck sharply, although the twitch amplitude is very small.

3. The Distribution of Giant Fiber Systems and Responses They Mediate

Giant nerve fibers and giant fiber systems have been the subjects of a major thrust in cellular neurophysiology since the discovery of the giant axon in the squid and the giant system of which it is a part (Young, 1936). Even earlier, cytological and embryological studies owed much to the neuron of Mauthner (1859) in fish. Giant cells and fibers have contributed much to our fundamental knowledge of neurobiology (Hodgkin, 1964; Faber and Korn, 1978). Naturally, the question of their role in behavior was raised long ago, but among many functions proposed, the only one that has appeared to share some general applicability across diverse taxa is the mediation of startle responses (Bullock, 1948, 1953*a,b*). This is still a proposition under review. The present book turns the question around and asks what can be said of the neural basis of those forms of behavior that deserve to be called startle responses? From the findings reviewed in its chapters we can say that giant fibers are not essential to the mediation of startle responses.

Some questions remain to be answered: *Do giant fibers mediate only a part, that is, the initial phase of the startle response? Are giant fiber responses sometimes used for other abrupt behavior than startle responses? How widespread and diverse are giant fiber systems?* A short survey of the known giant systems may help to give perspective in a book that can treat only some examples in detail.

Once upon a time we used to say that giant fibers are not merely the largest caliber axons in the animal, but fibers that are discontinuously larger than the next largest fibers in that species (Bullock and Horridge, 1965). The absolute diameter was not considered crucial. The term was reserved for cases where there are no intergrades. However, the cockroach people "wanted in"; there must have been some kind of glamor to putting "giant" in the title of one's paper. They therefore undermined that definition because the roach's giant fibers are at the end of a continuous fiber diameter spectrum. The Mueller fiber people haven't been pushy, so that perhaps it is still true that giant fibers everywhere or nearly everywhere, except in cockroaches, are a distinct group far out in the diameter spectrum without intergrades.

It is important to notice that we are not talking about giant cell bodies; those are quite differently distributed and correlate poorly with giant axons. One of the basic neurobiological problems offered by giant axon systems is the meaning of the great range among them of the ratio of axoplasm to somatoplasm (axon volume to cell body volume) reported to vary from a few hundred to at least 5000.

Giant fibers occur in some hydrozoan jellyfish (Cnidaria, the most primitive phylum with a nervous system; see Chapter 2, this volume). They are apparently lacking in most platyhelminths, nematodes, rotifers, the bryozoans (Ectoprocta), and brachiopods (Figure 1; phyla with asterisks have members with giant fibers). They are clearly present in the Phoronida, physiologically studied by D.M. Wilson in 1959 in my laboratory (Wilson and Bullock, 1959). Among the mollusks only decapod cephalopods exhibit giant fibers. They are well developed in many families of polychaete annelids, which therefore offer a rich reservoir of diversity in anatomy, function, and behavioral role. However, there are scores of families of polychaetes that lack giant fibers. They are general but not universal among the oligochaetes (Annelida), but absent in leeches (Annelida), sipunculoids, echiuroids, and onychophorans. Giant fibers occur in many arthropods with an elongated abdomen, especially among the lower crustaceans, as well as the stomatopods and higher macrurous forms (shrimp, crayfish, and lobsterlike decapods), the scorpions, and a few orders of insects. In this phylum we have another great reservoir of diversity of form and function. They are lacking in echinoderms and prochordates except for enteropneustans (Hemichordata) which do have them. They are common but not universal among the varied taxa of lower vertebrates, from amphioxus and lampreys to teleosts, but lacking in adult anurans, reptiles, birds, and mammals.

Obviously many of the fastest animals lack giant systems. I think of the Sally Lightfoot (*Grapsus grapsus*), a high tide rock crab, and of the ghost crabs (*Ocypode*) I have seen flitting across the sand of a tropical beach, moving their legs faster than a hummingbird's wings, although 12°C cooler, and dodging the Governor's cat, which is itself a picture of quick movements, also achieved without giant fibers. Therefore agility and speed, whether in escape or attack, do not mean a giant system is responsible.

As in the cat, so also in many groups of vertebrates and invertebrates large caliber axons not only mediate escape; sometimes they mediate attack. Prey capture is not startle, but it may be startling and can occur in otherwise sluggish species: an alligator snaps, a stargazer (*Astroscopus*) gapes, a jumping spider pounces, a praying mantis snatches.

Giant fiber responses are only a subset of startle responses; startle can be recognized in species without giant fibers. Many ciliates, for example, *Euplotes* and *Stentor,* react extremely abruptly; bryozoans (Ectoprocta) snap their avicularia, a resting butterfly takes wing, a toad jumps, a kitten springs vertically, a human gives a jumpy twitch to certain startling stimuli.

Giant fibers, where they occur, sometimes mediate only an initial phase of the startle response, let alone the whole of the subsequent escape response, if there is one. Even the initial phase may not require, but may only be quantitatively accelerated by the giant fiber mediation, as Eaton and Hackett (Chapter 8, this volume) show for some teleosts. However, we must not be too quick to generalize from those fish. In other cases, such as serpulid polychaetes, the giant fiber is probably indispensable for the abrupt startle response.

4. The General Neurobiological Significance of Studies on Fast Systems

More than 40 years of a rich literature on the axonal membrane of the most medial of the third-order giant axons of squids, its biophysics, biochemistry, and ultrastructure, constitute a major part of our knowledge of cell and membrane biology. Giant systems in squids, earthworms, crayfish, and goldfish among others have provided a central core of our understanding of synaptic transmission including chemical, electrotonic, and mixed junctions, and excitation and inhibition in all combinations.

Deserving special recognition is another domain of basic problems at the next higher level, that of neuronal integrative mechanisms. By this I mean the variables available to neurons, at their somas, dendrites, pacemaker regions, branch points, axon terminals and synapses, with which they can determine output as a complex function of input, variables such as facilitation, accomodation, rebound, iteration, and many more (Bullock *et al.,* 1977). This field owes much to favorable preparations involving giant fiber systems and their inputs and outputs.

Yet another domain encompasses ontogeny, regeneration, and plasticity, which have greatly benefited from giant fiber studies. Finally, the domain of system organization including sensory–motor integration, convergence of inputs, specificity, and adaptiveness of connections and of dynamics of coupling functions is an integrative level for which giant fibers are advantageous, and which gets closer to the behavioral emphasis of this volume.

Broad categories such as those I have just enumerated are not vivid! To conjure up some specific mental images and make more real the contributions and the promise of further insight from exploiting favorable species with fast fiber systems, let me parade a few instances, more or less familiar depending on the reader's background. My purpose is to illustrate what we owe to and what we can look forward to in the study of giant fibers, especially their relevance to general neuroethology.

Facilitation of conduction velocity in axons had been denied, but was readily seen in earthworm giant and other axons (Bullock, 1951) in a critical range of intervals. In this volume, Drewes (Chapter 3) shows a remarkable use of this facilitation of axonal conduction in explaining behavioral discrimination.

An old speculation is still viable that electrical fields in the tissue over dimensions of a few micrometers to tens of micrometers are not only epiphenomenal results of neural activity of synchronized masses, but are sometimes causes (Gerard, 1971; Bremer, 1944; Bullock, 1945*a*, 1947, 1953*a*). That is to say, they exert some influence, at least modulatory or predisposing. This idea still has not caught on and to be sure it is hard to prove or to exclude! But it is potentially significant, and some of the newest observations seem to make it more than likely. I would list as one of the prime opportunities for new work, the testing of such places as the Mauthner's axon cap in teleosts and the neuropil around the larger giant fibers in polychaetes for local field effects, by which I mean electrotonic influence not adequately attributable to specialized gap junctions. It will probably be an effect additive to specific classical electrotonic transmission and perhaps difficult to distinguish from it.

Earthworm giant fibers have been favorable material on which Drewes (Chapter 3, this volume) and his co-workers have worked to reveal details of the recovery of cellular function in regenerating axons; they demonstrate a complex kind of labile electrical integrative junction, presumably much like that described long ago (Bullock and Turner, 1950) in the intact axon at loci of partial anodal block. One sees delays of 5 msec or more, local initiations of impulses after conduction has failed, reflected and multiple spikes and other asymmetries, all showing changes over time subject to recent history and local state. These are probably properties of septate electrotonic and regenerating electrotonic contacts. They may be related in some way to the remarkable axotomy-induced changes in both structure and function studied by Faber and Zottoli (1981); however, I suspect these are distinct phenomena: one immediate and the other developing over many hours. The slow changes may be similar to those found by Cohen and Jacklet (1965) in insects where axotomy was followed by redistribution of ribosomes in the soma and by spike invasion into the soma where invasion normally cannot occur.

The still puzzling case of the fastest of all axons, the giant fibers of shrimp, calls for new work. Except for a few studies that have appeared only in abstract form (Hsu *et al.*, 1975*a*,*b*; Hsu, 1982), this remarkable giant fiber has not had recent attention. Already in 1941, Holmes *et al.* had shown that the velocity in a prawn, *Leander,* whose giant fiber is only 35 μm in diameter, is 20 m/sec—not extraordinary absolutely but as a ratio of velocity to diameter far higher than any other invertebrate axon. In 1961, Fan *et al.* reported in a Chinese language journal that velocities exceeding 200 m/sec were measured at sea temperatures in *Penaeus,* a species of shrimp (see also Hsu *et al.*, 1964). We were then getting few Chinese journals and I heard about this, with some misgivings, in Moscow in 1961. I wrote to my old friend T.P. Feng, the head of the Chinese Academy Institute of Physiology where the work was done, asking if this report was reliable. In those days we didn't know whether it might hurt someone in China to receive a letter from the United States, so I had a Russian friend address and mail the letter from Moscow. Feng replied in full, confirming the report and giving me confidence that there had not been mistakes, as can easily happen in measuring short stimulus–response times. Eventually my former associate, Kiyoshi Kusano, took it up; he published several papers (1965, 1966, 1971) confirming and extending the facts. Huang *et al.* (1963), Hao and Hsu (1965), and Hama (1966) reported on the ultrastructure.

Velocity in the shrimp giant axon reaches 210 m/sec at 22 °C. The fiber with its 10 μm sheath is 120 μm in diameter, but the axon may be only 10 μm in diameter, so that there is a large space under the sheath. There are no Ranvier nodes, but conduction is thought to be saltatory by way of functional nodes at points of exit of branches (in much the same way as suggested by Yasargil and his colleagues in 1982 for the Mauthner axon). New study is needed and will doubtless turn up novel explanations. At the same time we need to think about the neuroethological aspects. Is the evolutionary pressure to develop this elaborate new fiber such that a shrimp with a 10-cm central path reduced its conduction time from the 2 msec of a goldfish to 0.5 msec, saving 1.5 msec? If nature can achieve these high velocities, why hasn't she used them more often in other groups?

What is the cost? To me it seems likely that this will be one more among many cases where the familiar limits that nature achieves are not theoretical limits or the best she can do, but crossover points in cost–benefit ratio curves.

These few examples suffice to illustrate the range of contributions to general neurobiology to be expected from analyses of the neural basis of a form of behavior widespread among animal groups.

5. Some Historical Notes to the Contributions That Follow

The chapters that follow show remarkable advances on many fronts. For historical interest, I will comment on a few. Charles Drewes's (Chapter 3) results with recording earthworm giant fiber spikes from the skin in the unanesthetized, active animal make me think, "Great! Look what they've been getting since 1978 with the method introduced by Rushton and Barlow in 1943 and hardly used since then." I remember the introduction to Rushton's paper (1945*a*), which he sent to me in draft manuscript by convoyed sea mail across the North Atlantic. Regrettably, the editor cut out one passage as immaterial, but the draft said something like this: Under war-time conditions it is hard to obtain normal laboratory animals to study, so we turned to our victory gardens and dug up a mess of angleworms—or proper British words to that effect. The implication was: we apologize for working on these humble creatures, but the precedent we lean on is the short note published ten years ago by a team of young Turks named John C. Eccles, Rägnar Granit, and John Z. Young (Eccles *et al.*, 1932). Of course, Rushton and Barlow soon abandoned this animal, as had Eccles and co-workers. I tried to follow up on these excellent papers (Rushton, 1945*a,b,* 1946; Rushton and Barlow, 1943), since I was doing similar things (Bullock, 1945*b*) in New Haven with night crawlers from the lawn of the Hall of Graduate Studies of Yale.

What impressed me was the opportunity—the first, I think, in biology—to examine noninvasively the labile state of a cellular unit; that is, the conduction velocity of a giant fiber, day after day in the same animal and in individual after individual, knowing that one would be dealing with the same unit. I made a circular earthworm tunnel with electrodes in the floor and a quick change system for connecting the stimulator and two amplifiers to different pairs rapidly. It worked well but I got distracted and did not pursue it; now it is gratifying to see this remarkable opportunity exploited so well.

The work of Charles Kimmel (1982), as with the recent papers of Robert Eaton and his co-workers (1982) on the role of non-Mauthner large fibers of the medial longitudinal fasciculus or reticulospinal tracts in escape responses, is also specially and personally welcome. It promises to solve an old puzzle. When my student Ellis Berkowitz (1956) recorded Mauthner fiber spikes from the surface of the intact cord in several teleost species, he kept getting spikes from other fibers almost as fast. He tried to study the neuroethological role of Mauthner's axon by itself, but the other fibers kept coming in at nearly the same threshold, electrically and mechanically. It wasn't until D.M. Wilson's prethesis excursion, published in 1959, that good all-or-none single unit Mauthner spikes were recorded to physiological stimuli such as a light tap on the tail or the table, just as for earthworm giant fibers. Don Wilson and I got the idea that permitted this finding from an anatomical note by I.C. Smith (1955), who pointed out that the African lungfish, *Protopterus,* of all unlikely animals, has a relatively, not absolutely, outstanding Mauthner fiber. Don got the long expected answer in the first preparation. As in the earthworm, stimulation of the giant fiber alone was sufficient to cause a startle response-like twitch. So, for the first time we had the evidence that Mauthner's fiber functions to mediate the fast startle response twitch, as I had hypothesized by analogy with invertebrates, instead of locomotion and equilibrium, as was the prevailing view at the time.

In this book we will read the latest word on the differences in role between Mauthner's and the next largest fibers, and I expect we will learn that fish are not all alike in these respects, just as Zottoli (1978*a,b*) has shown us for cytological differences and for input connections.

A number of giant fiber preparations offer favorable opportunities to study the neural basis of plasticity as in habituation or in recalibration; giant fiber circuits are not too lowly or automatic to show plastic adaptability in their coupling functions, depending on the recent history of input and feedback. Habituation was studied many years ago in the earthworm giant fiber system, and to some degree in crayfish and cockroaches; it seemed likely to be mainly localized in the small fiber afferents to the giants. It remains a good question how much plasticity will be found in a Mauthner system; I would give odds there is a significant amount under the right conditions. Eaton and colleagues (1977) showed a marked change early in life from a time when repetitive firing is common and inhibition much less prominent, to the adult state when inhibition is prompt and strong and the Mauthner cell usually fires only once for each stimulus. Assessment of the normal range of adaptability is part of the opportunity ahead of us.

I believe giant systems are also opportune places to study the labile influence of cerebral, thalamic, hypothalamic, and even cerebellar descending modulation. Adrenergic and other widespread, stage-setting systems might be expected to affect at least the thresholds, if not the symmetry of directional sensitivity and possibly the distribution of the stochastic second stage of fast escape by nongiants.

I can't help remarking on the refreshing change it represents to be arguing that giant fiber systems deserve more attention. For years, during the 1930s and 1940s, Ladd Prosser was saying at national meetings that the time had come in invertebrate physiology to spend part of our time studying small fiber systems! For a long time this was difficult. Today, with some conspicuous lacunae, we can record from almost any cells we want, and most of the attention is focused on small or nongiant neurons. The collection of papers in this volume shows a new and healthy emphasis on following clues wherever they lead: toward giants, nongiant large caliber fibers or small cells and processes, even local circuits, as in the Mauthner's axon cap.

6. References

Barnes, R. D., 1968, *Invertebrate Zoology,* W. B. Saunders Co., Philadelphia.

Barnes, R. D., 1980, *Invertebrate Zoology,* 4th ed., Saunders College Publishing Co., Philadelphia.

Berkowitz, E. C., 1956, Functional properties of spinal pathways in the carp, *Cyprinus carpio* L., *J. Comp. Neurol.* **106**:269–290.

Bremer, F., 1944, L'activité "spontanée" des centres nerveuses, *Bull. Acad. Roy. Med. Belgique* (Ser. 6) **9**:148–173.

Bullock, T. H., 1945*a*, Problems in the comparative study of brain waves, *Yale J. Biol. Med.* **17**:657–679.

Bullock, T. H., 1945*b*, Functional organization of the giant fiber system of *Lumbricus, J. Neurophysiol.* **8**:55–72.

Bullock, T. H., 1947, Problems in invertebrate electrophysiology, *Physiol. Rev.* **27**:643–664.

Bullock, T. H., 1948, Physiological mapping of giant nerve fiber systems in polychaete annelids, *Physiol. Comp. Oecol.* **1**:1–14.

Bullock, T. H., 1951, Facilitation of conduction rate in single nerve fibers, *J. Physiol.* **114**:89–97.

Bullock, T. H., 1953*a*, A contribution from the study of cords of lower forms, in: *The Spinal Cord* (G. E. W. Wolstenholme, ed.), J. A. Churchill, London, pp. 3–10.

Bullock, T. H., 1953*b*, Properties of some natural and quasi-artificial synapses in polychaetes, *J. Comp. Neurol.* **98**:37–68.

Bullock, T. H., and Horridge, G. A., 1965, *Structure and Function in the Nervous Systems of Invertebrates,* W. H. Freeman and Co., San Francisco.

Bullock, T. H., and Turner, R. S., 1950, Events associated with conduction failure in nerve fibers, *J. Cell. Comp. Physiol.* **36**:59–82.

Bullock, T. H., Orkand, R. O., and Grinnell, A. D., 1977, *Introduction to Nervous Systems,* W. H. Freeman and Co., San Francisco.

Cohen, M. J., and Jacklet, J. W., 1965, Neurons of insects: RNA changes during injury and regeneration, *Science* **148**:1237–1239.

Eaton, R. C., Farley, R. D., Kimmel, C. B., and Schabtach, E., 1977, Functional development in the Mauthner cell system of embryos and larvae of the zebra fish, *J. Neurobiol.* **8**:151–172.

Eaton, R. C., Lavender, W. A., and Wieland, C. M., 1982, Alternative pathways initiate fast-start responses following lesions of the Mauthner neuron in goldfish, *J. Comp. Physiol.* **145**:485–496.

Eccles, J. C., Granit, R., and Young, J. Z., 1932, Impulses in the giant nerve fibres of earthworms, *J. Physiol.* **77**:23P–25P.

Faber, D. S., and Korn, H., 1978, *Neurobiology of the Mauthner Cell,* Raven Press, New York.

Faber, D. S., and Zottoli, S. J., 1981, Axotomy-induced changes in cell structure and membrane excitability are sustained in a vertebrate central neuron, *Brain Res.* **223**:436–443.

Fan, S.-F., Hsu, K., Chen, F.-S., and Hao, B., 1961, On the high conduction velocity of the giant nerve fiber of shrimp *Penaeus orientalis, Kexue Tongbao* **4**:51–52 (in Chinese).

Gerard, R. W., 1941, The interaction of neurones, *Ohio J. Sci.* **41**:160–172.

Hama, K., 1966, The fine structure of the Schwann cell sheath of the nerve fiber in the shrimp (*Penaeus japonicus*), *J. Cell. Biol.* **31**:624–632.

Hao, B., and Hsu, K., 1965, The birefringence properties of the myelin sheath of shrimp nerve fiber, *Acta Physiol. Sin.* **28**:373–377 (in Chinese with English summary).

Hodgkin, A., 1964, *The Conduction of the Nervous Impulse,* Liverpool University Press, Liverpool.

Holmes, W., Pumphrey, R. J., and Young, J. Z., 1941, The structure and conduction velocity of the medullated nerve fibres of prawns, *J. Exp. Biol.* **18**:50–54.

Hsu, K., 1982, Structural and funtional characteristics of nerve fibers of shrimp *Penaeus orientalis, IBRO News* **10**:10–11.

Hsu, K., Tan, T.-P., and Chen, F.-S., 1964, On the excitation and saltatory conduction in the giant fiber of shrimp (*Penaeus orientalis*) in: *Theses of the 14th National Congress of Chinese Association of Physiologists* p. 17 (in Chinese).

Hsu, K., Tan, T.-P., and Chen, F.-S., 1975*a,* Saltatory conduction in the myelinated giant fibre of the shrimp (*Penaeus orientalis*), *Kexue Tongbao* **20**:380–382 (in Chinese).

Hsu, K., Yang, Q.-Zh., and Tsou, S.-Hs., 1975*b,* On the apparent lack of resting membrane potential in the shrimp giant nerve fibre, *Kexue Tangbao* **20**:383–386 (in Chinese).

Huang, S.-K., Yeh, Y., and Hsu, K., 1963, A microscopic and electron microscopic investigation of the myelin sheath of the nerve fiber of *Penaeus orientalis, Acta Physiol. Sin.* **26**:39–42 (in Chinese with English summary).

Hyman, L. H., 1940, *The Invertebrates: Protozoa Through Ctenophora,* McGraw-Hill Book Co., Inc., New York.

Kimmel, C. B., 1982, Reticulospinal and vestibulospinal neurons in the young larva of a teleost fish, *Brachydanio rerio, Prog. Brain Res.* **57**:1–23.

Kusano, K., 1965, Electrical characteristics and fine structure of the *Kuruma*-shrimp nerve fibres (*Penaeus japonicus*), *Proc. Jpn. Acad.* **41**:952–957.

Kusano, K., 1966, Electrical activity and structural correlates of giant fibers in *Kuruma* shrimp (*Penaeus japonicus*), *J. Cell. Physiol.* **68**:361–384.

Kusano, K., 1971, Impulse conduction in the shrimp medullated giant fiber with special reference to the structure of functionally excitable areas, *J. Comp. Neurol.* **142**:481–494.

Landis, C. and Hunt, W. A., 1939, *The Startle Pattern,* Farrar and Rinehart, New York.

Mauthner, L., 1859, Untersuchungen über den Bau des Rückenmarkes der Fische, *Sitz ber. Kgl. Preuss. Wiss.* **34**:31–36.

Rushton, W. A. H., 1945*a,* Action potentials from the isolated nerve cord of the earthworm, *Proc. R. Soc. (London) Ser. B* **132**:423–437.

Rushton, W. A. H., 1945*b,* Motor response from giant fibres in the earthworm, *Nature* **156**:109–110.

Rushton, W. A. H., 1946, Reflex conduction in the giant fibres of the earthworm, *Proc. R. Soc. (London) Ser. B.* **133**:109–120.

Rushton, W. A. H., and Barlow, H. B., 1943, Single fibre response from an intact animal, *Nature (London)* **152**:597–598.

Smith, I. C., 1955, Giant nerve fibres in *Protopterus, J., Physiol. (London)* **129**:42P.

Wilson, D. M., 1959, Function of giant Mauthner's neurons in the lungfish, *Science* **129**:841–842.

Wilson, D. M., and Bullock, T. H., 1959, Electrical recording from giant fiber and muscle in phoronids, *Anat. Rec.* **132**:518–519.

Yasargil, G. M., Greeff, N. G., Luescher, H. R., Akert, K., and Sandri, C., 1982, The structural correlate of saltatory conduction along the Mauthner axon in the tench (*Tinca tinca L.*): Identification of nodal equivalents at the axon collaterals, *J. Comp. Neurol.* **212**:417–424.

Young, J. Z., 1936, The structure of nerve fibres and synapses in some invertebrates, *Cold Spring Harbor Symp. Quant. Biol.* **4**:1–6.

Zottoli, S. J., 1978*a*, Comparative morphology of the Mauthner cell in fish and amphibians, in: *Neurobiology of the Mauthner Cell* (D. S. Faber and H. Korn, eds.), Raven Press, New York, pp. 13–45.

Zottoli, S. J., 1978*b*, Comparison of Mauthner cell size in teleosts, *J. Comp. Neurol.* **178**:741–758.

2

Fast Pathways and Escape Behavior in Cnidaria

GEORGE O. MACKIE

1. Introduction

Although giant axons were described in a cnidarian a hundred years ago (Korotneff, 1884), this discovery was lost through an historical accident and it is only in recent years that escape responses mediated by outsize or giant axons comparable to those found in polychaetes, squid, and others have again come to light. The best examples are from the Class Hydrozoa, specifically one particular trachyline jellyfish, *Aglantha digitale* (Figure 1A), and a number of siphonophores of which the best known is *Nanomia cara*. The greater part of this chapter will deal with these examples. Table I presents the taxonomic relationships of these animals. Giant axons are known in the Class Scyphozoa, but they coordinate normal locomotion, not escape per se, and the same is true in a number of hydromedusae (see reviews by Passano, 1982; Spencer and Schwab, 1982). These cases will not be covered here.

As with other animals, cnidarians have evolved a wide variety of protective responses. These range from the simple contractions of tentacles and other parts when pinched to the complex, "programmed" types of activity such as the escape swimming shown by some sea anemones in the presence of predators (Ross, 1974). In fact, it is hard to think of any cnidarian that does not show some sort of protective retraction or avoidance response in the presence of potentially damaging stimuli. To review the whole field would serve little purpose and our terms of ref-

GEORGE O. MACKIE • Biology Department, University of Victoria, Victoria, British Columbia, Canada V8W 2Y2.

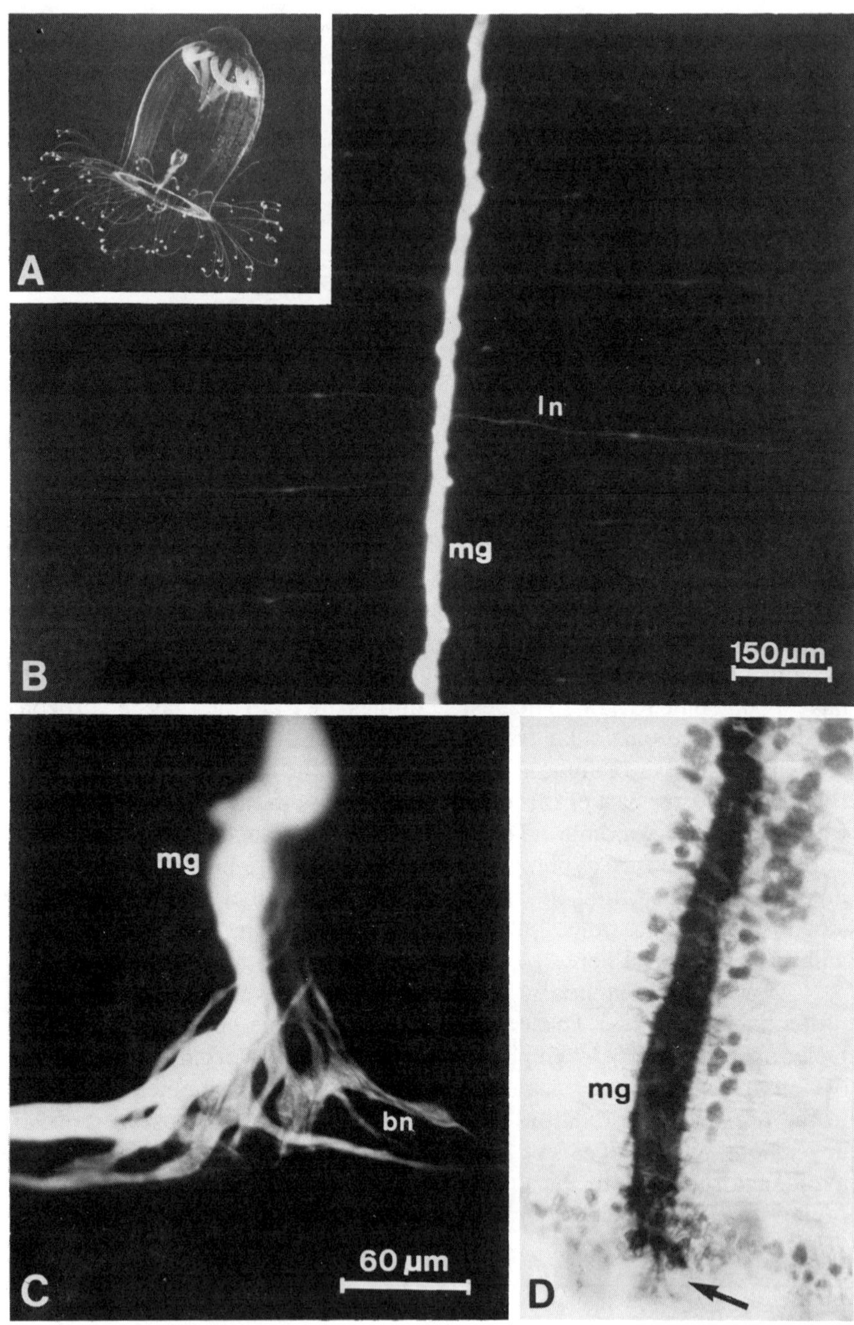
A
B
C
D
ln
mg
150 µm
mg
bn
60 µm
mg

erence here preclude such a review, since the emphasis is on fast pathways. We can, however, legitimately include a brief account of the protective retraction and closure of sea anemones; brief because there is still much uncertainty about the neural substrates of the response, and also because the main facts about sea anemone behavior are already well known from the pioneering studies of Pantin and co-workers (see Pantin, 1952) and from several more recent views (e.g., Josephson, 1974; Robson 1975; McFarlane, 1982).

2. Fast Pathways in Sea Anemones and Colonial Anthozoans

Sea anemones and the polyps of colonial anthozoans contract symmetrically and rapidly when stimulated. In *Metridium* and *Calamactis*, the major muscles involved are the retractors that run up the mesenteries. In *Calliactis* the main effector is the marginal sphincter. The response varies from species to species. The tentacles usually contract concurrently with the withdrawal of the body. In colonial forms, the response spreads for varying distances away from the site stimulated. The retractile movements show all the characteristics of a protective type of behavior.

The conducting system mediating protective retractions has been identified as the endodermal nerve net on anatomical and physiological grounds summarized by Josephson (1974). Recordings have not as yet been obtained from individual identifiable nerve elements in the net and no intracellular recordings are available, but there is little doubt that the nerve net is responsible. Fast spikes are recordable from the mesenteries and other regions (Pickens, 1969; Robson and Josephson, 1969) and are apparently conducted within a through conducting nerve net (TCNN), a distinct subsystem of the animal's nervous system (McFarlane, 1982).

Anatomically, the TCNN pathways in the mesenteries would appear to be the elongated bipolar cells described by various workers. These cells average 3.2 mm in length on the retractor face of the mesenteries of *Metridium* and average 1.6 μm in diameter (Batham *et al.*, 1960). Most are below 2.0 μm but a few lie in the 5.0–5.9 μm range. Possibly these

Figure 1. Anatomy of motor giant system in *Aglantha*. (A) Living specimen of *Aglantha* with extended tentacles, at rest. (B) Fluorescence photomicrograph of subumbrella whole mount. The motor giant (mg) has been injected with Lucifer Yellow. Dye has entered the lateral neurons (ln). (C) Fluorescence photomicrograph of motor giant at the point where it joins the inner nerve ring. Dye-coupled basal neurons (bn) enter the ring. (D) Motor giant, same region and same magnification as C, injected with horseradish peroxidase. The base of the axon is divided into a few short "horns" (arrow). (A is from the unpublished data of Claudia Mills, B–D are from Weber *et al.*, 1982).

Table I. Taxonomic Relationships of the Phylum Cnidaria[a]

Taxon			Description
Class 1: Hydrozoa or Hydromedusae			Small medusae and polyp forms (hydroids)
Order: Trachylina			Small oceanic medusae
		Genera: *Aglantha*	
		Rhopalonema	
Order: Siphonophora			Colonial forms made up of both medusalike and polyplike forms, attached to a common stem
	Suborder: Physonectae		With gas-filled float
		Genera: *Forskalia*	
		Nanomia	
		Physophora	
	Suborder: Calycophora		Without a float
		Genera: *Abylopsis*	
		Chelophyes	
		Hippopodius	
		Sulculeolaria	
Class 2: Scyphozoa or Scyphomedusae			Large medusae (jellyfishes)
Class 3: Anthozoa			Solitary polyps (sea anemones) and polyp colonies (corals)
Order: Actiniaria			Sea anemones
		Genera: *Calamactis*	
		Caliactis	
		Metridium	

[a]The most primitive animals having nerves. Only those groups covered in the text are included.

are the TCNN subset. However, the diameters grade continuously from <0.9 μm up to 5.9 μm, and there is no histologically distinct class of "giant axons" as usually understood (Bullock and Horridge, 1965).

Conduction velocities in the *Metridium* retractors lie in the range 70–120 cm/sec (Robson and Josephson, 1969). Values up to 120 cm/sec are also seen in *Calliactis* (Robson, 1961). These values are unexpectedly high for neurons less than 6.0 μm in diameter. In *Calamactis,* Pickens (1969) reports 90 cm/sec for neurons up to 12 μm in diameter, which would agree with expectations based on the evidence from hydrozoans.

Excitation in the TCNN system evokes muscle contractions. Some muscles give both fast and slow contractions. The former are associated with action potentials. Fast contractions sum and facilitate (Pantin, 1935).

In *Calliactis* the TCNN shows spontaneous activity (McFarlane, 1973) as well as responses to stimuli. Pacemakers are widely distributed through the system. In colonial forms, protective retractions are associated with activity in a system equivalent to the TCNN. Although impulses nearly always through-conduct across the whole colony, not all the zooids contract, or not to the same extent. The evidence and possible mechanisms are discussed by Shelton (1982).

3. The Escape System of *Aglantha*

Aglantha digitale (Figure 1A) is a small trachyline medusa, widely distributed in the oceans around the world. Giant axons were discovered in the subumbrella by Singla (1978), who suggested that they mediated escape swimming. Subsequent studies (Donaldson *et al.*, 1980; Mackie, 1980; Roberts and Mackie, 1980) confirmed that the animal can swim in two ways ("fast" and "slow") and that the giant axons are involved only in the fast response. Histological details were provided by Roberts and Mackie (1980) and by Weber and colleagues (1982). So far, *Aglantha* is the only jellyfish definitely known to possess an escape system, but one hundred years before Singla's discovery, a structure labelled as "empty space" was figured in *Rhopalonema* (another trachyline medusa) in a position corresponding to that of the ring giant of *Aglantha* (Hertwig and Hertwig, 1878), and it may be that *Aglantha* is not unique.

3.1. The Escape Response

Aglantha can swim slowly in a series of endogenously generated, rhythmic, weak pulsations or rapidly in response to external stimulation. In both responses, swimming is produced by contraction of the subumbrellar striated muscle, which constricts the bell, ejecting water posteriorly through the velar opening. Strain energy stored in the walls of the bell during contraction is released, restoring the bell to its previous shape and drawing in water.

The escape type of swimming can be evoked by touching the bell margin, pinching or tugging the tentacles, or giving electric shocks to the tentacles or margin. The response consists of one to three violent contractions of which the first propels the animal through a distance equivalent to about five body lengths (Figure 2). (In slow swimming, by contrast, a single contraction propels the animal only about one body length.) Escape swimming is accompanied by rapid, all-or-none retraction of the tentacles. At other times, the tentacles respond individually and gradedly. Total latency to peak thrust lies in the range 40–50 msec.

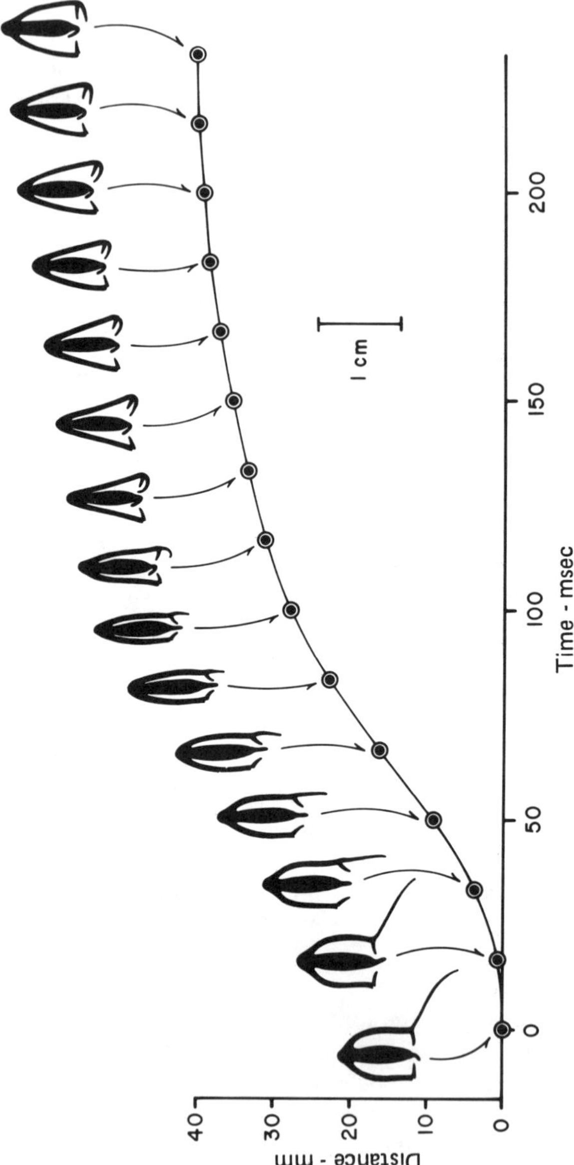

Figure 2. Longitudinal profiles during an escape swimming contraction and part of subsequent refilling stage in *Aglantha*. The accompanying graph depicts movement of the animal through the water (from Donaldson *et al.*, 1980).

Aglantha has no hydroid stage and spends its whole life in the highly competitive world of the mid-water and upper-water plankton, in the company of agile and fast moving crustaceans, fish larvae, chaetognaths, and others with which it performs diurnal vertical migrations. *Aglantha* has been observed from a submersible living in waters containing crustacean plankton in densities of thousands of individuals per cubic meter (Mackie and Mills, 1983). Escape responses following contact with euphausiids, amphipods, crab larvae, and copepods have been observed repeatedly from the submersible. The response probably helps to minimize damage to the medusa's delicate tissues. The escape response was also shown to prevent capture by a predator (*Aequorea*) in lab tests (Donaldson *et al.,* 1980).

3.2. Layout of Nerves and Muscles

A single giant axon (*ring giant*) runs around the bell margin on the outer side, next to the outer nerve ring (Figure 3). At its widest, it has a diameter of 22–35 μm. It differs from other giant axons in the same animal, indeed, from all other known axons, in having a large central vacuole running along it that restricts the cytoplasm to a narrow outer layer. The cytoplasm itself is densely granular, as seen in electron micrographs, in contrast to the rather featureless axoplasm of most axons. Sensory structures (comb pads) are present close to the ring giant, but it has not proved possible to demonstrate junctions between them and the giant. Sensory tranduction might thus be a property of the giant itself.

The tentacles are supplied with a network of small neurons and also with one relatively large axon. The tentacle muscles are striated, not smooth as in all other known medusae.

On the subumbrellar side, eight giant axons (*motor giants*) run out across the subumbrellar muscle sheet from a starting point in the nerve ring. *Lateral neurons* (Figure 1B) run out sideways from these units, forming numerous scattered junctions with the swimming muscles. At the nerve ring end, the motor giants are connected to a network of smaller neurons (*basal neurons*) that run laterally, mingling with inner nerve ring elements (Figure 1C). Each motor giant is an independent syncytial structure with many nuclei. Lucifer Yellow spreads evenly along it, as does horseradish peroxidase (Figure 1D), but only Lucifer Yellow spreads into the lateral neurons and into the basal plexus elements. This suggests that gap junctions are present between the dye-coupled components, and this has been confirmed by electron microscopy. No dye coupling or electrical coupling has been demonstrated between adjacent motor giants or their associated basal and lateral neurons. Figure 4 summarizes these features.

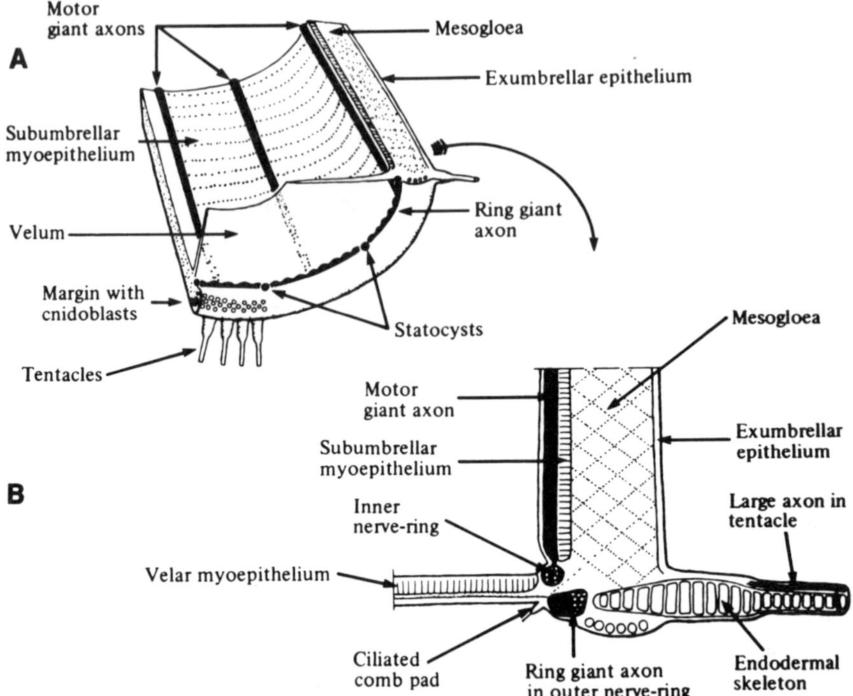

Figure 3. Anatomy of regions concerned with escape behavior in *Aglantha*. (A) Cut-away diagram showing location of giant axons. (B) Diagrammatic cross section through the margin omitting endodermal components (from Roberts and Mackie, 1980).

It is unclear exactly how the ring giant is connected to the motor giants. Direct extensions from one to the other have not been found, and it seems likely that the link is made indirectly through small neurons that cross between the two nerve rings and synapse on the motor giant or on its basal plexus elements. Synapses have been found in these places, but the presynaptic elements have not been traced back to their sources in the outer nerve complex.

3.3. Physiological Analysis

Spikes are conducted around the margin in the ring giant at velocities up to 2.6 m/sec. The axon tends to fire repetitively in short bursts (Figure 5A,B). These events are quite distinct from other, more slowly conducted potentials recordable extracellularly from the outer nerve ring. Likewise in the tentacles, extracellular recordings show both fast and slow systems,

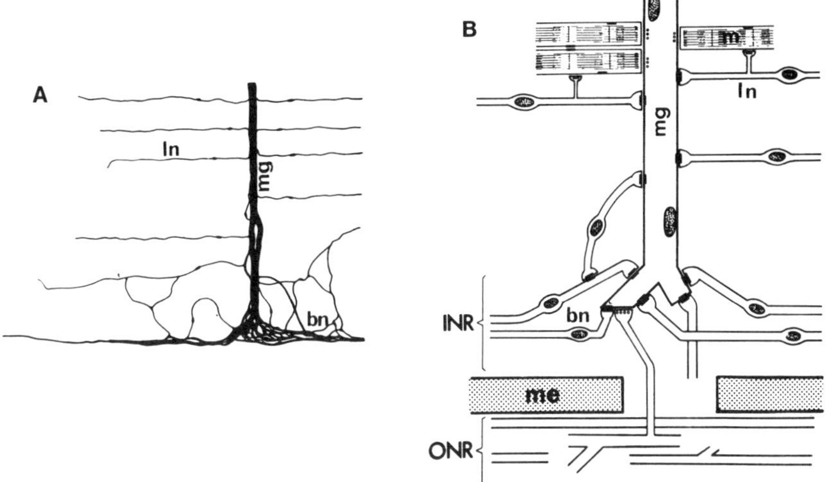

Figure 4. The motor giant and connected units in *Aglantha*. (A) Camera lucida drawing of a Lucifer Yellow preparation. (B) Diagram of the interconnections as determined by optical and electron microscopy. Abbreviations: bn, basal neurons; INR, inner nerve ring; ln, lateral neurons; m, swimming muscle; me, mesogloea; mg, motor giant; ONR, outer nerve ring (from Weber *et al.*, 1982).

conducting at velocities of 0.65 and 0.14 cm/sec, respectively, in the example shown in Figure 5C, and presumably representing the tentacle giant and the smaller tentacle elements, respectively. Events in the ring giant are conducted one-for-one along the tentacles (Figure 5B). Conduction is not blocked by elevated Mg^{2+}, and it is concluded that the large tentacle axons are electrically continuous with the ring giant: hence the rapid contraction of the tentacles during escape swimming.

By contrast, and perhaps surprisingly, spikes do not pass directly from the ring to the motor giants. Transmission is not one-for-one, is blocked by elevated Mg^{2+}, and involves a delay of 1.6–1.8 msec, measuring from the arrival of the spike in the ring giant to the start of the postsynaptic potential in the motor giant. It is concluded that transmission is chemical; the long delay may mean that more than one synapse is involved. Each motor giant is excited independently. In normal escape responses, the rapid conduction of signals around the ring giant would ensure that all eight motor giants are excited within a very short space of time.

Conduction up the motor giants takes place at velocities ranging from 1 m/sec in small animals to 4 m/sec in large ones. Spikes presumably pass

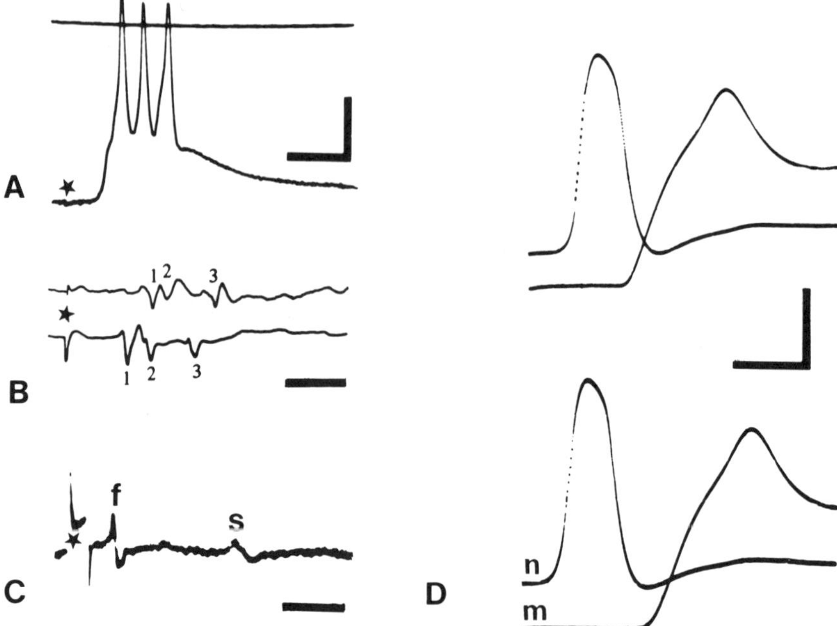

Figure 5. Electrical activity accompanying escape behavior in *Aglantha*. (A) Intracellular record of a typical triple firing from ring giant. Scale: 20 mV, 10 msec. (B) Extracellular records from a tentacle (above) and ring giant (below). Scale: 10 msec. (C) Extracellular record from one end of an isolated tentacle showing fast (f) and slow (s) propagated events following a shock at the other end. Scale: 5 msec. (D) Intracellular recordings from motor giant axon (n) and swimming muscle (m). The presynaptic and postsynaptic electrodes were 80 μm and 600 μm apart in the upper and lower records, respectively. Scale: 40 mV, 2 msec. Stars indicate shock artifacts in this and subsequent records (A and B from Roberts and Mackie, 1980; C from Mackie, 1980; and D from R. W. Meech, unpublished data).

directly to the lateral neurons via gap junctions and excite the muscle sheet at numerous points. The giant axon itself synapses directly with the muscle, and recordings with microelectrodes on either side of these junctions yield synaptic delays as low as 0.8 msec (Figure 5D, upper record). The delay is greater when the muscle electrode is moved further away from the motor giant (Figure 5D, lower record) presumably because of conduction time in the lateral neurons (R. W. Meech, unpublished data).

A scheme comparing the motor organization of *Aglantha* with that of a conventional medusa, *Sarsia,* is shown in Figure 6. *Aglantha* is completely without the radial muscles that bring about protective "crumpling" in *Sarsia*. It protects itself by its escape response. Excitable epithelia figure in the wiring of both species, but are not of importance in the present context. The scheme envisages a dual excitatory innervation of both the tentacles and the swimming muscle in *Aglantha* compared with the single innervation pattern found in *Sarsia*. Slow swimming in *Aglantha* is assumed to be controlled (as in other hydromedusae) by pacemaker nerves in the margin; impulses would spread out across the swimming muscle in some system other than the motor giant axons, producing subthreshold depolarizations, and hence weak contractions, in the muscle. Escape swimming would involve the giant axon system and its coupled side branches, and under these conditions it is assumed that the muscle would be excited to a point above threshold, and that spikes would then propagate through it bringing about rapid, total activation of the entire muscle sheet. Likewise in the tentacles, the double innervation would allow differentiation of slow, graded, and rapid all-or-none responses. It should be noted that neither in the tentacles nor in the swimming muscle is there any evidence for two sorts of muscle fiber. Differentiation of fast and slow responses would therefore seem to be a function of the innervation. The motor giants and the lateral/basal neurons make junctions with the swimming muscles that appear similar histologically but the synaptic "weighting" might not be the same in the two places.

Several features in this scheme are tentative, and recent work within our group casts doubt on certain points. It is not definite that the muscle sheet can propagate spikes without decrement. The muscle cells are elec-

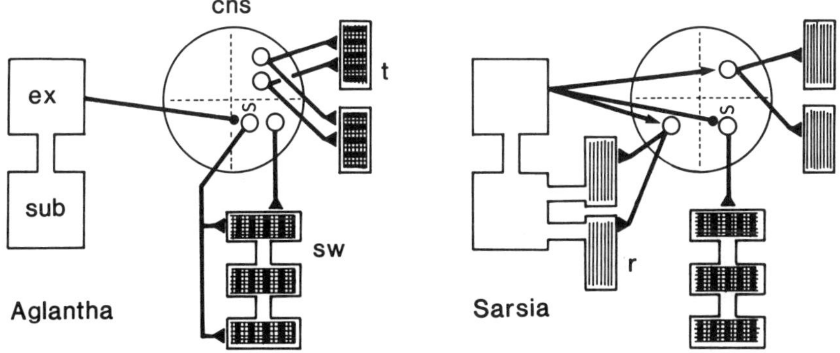

Figure 6. Comparison of motor organizations in *Aglantha* and *Sarsia*. Abbreviations: cns, central nervous system; ex, sub, excitable epithelia of exumbrella and subumbrella sides; r, radial muscle; sw, swimming muscle; t, muscle of tentacles. Smooth and striated muscles are differentiated graphically. ~ indicates an oscillator neuron system. Bridges connecting epithelial and muscle blocks represent electrically transmitting pathways (from Mackie, 1980).

trically coupled but full-scale spikes are not always seen; certain parts of the muscle field show attenuated spikes, at least under experimental conditions. Another point requiring clarification concerns the nervous pathway for slow swimming. Histologically there appears to be only one innervation consisting of the motor giants and the associated smaller neurons of the basal and lateral plexuses. The giants do not spike during slow swimming, which might suggest that the smaller, coupled units are also inactive. If so there would have to be a separate slow innervation, as yet unrecognized histologically. Alternatively, and more probably, the same small units might function as the final common pathway in both responses, with the giants active only during escape.

On the basis of present evidence, it seems likely that about 20% of the response time in escape swimming is taken up in the nervous system, the remainder being required for the production of muscle tension (Table II). These values may be compared with measurements for startle response latencies in goldfish where the time taken up in the nervous system again represents about 20% of the total response time (data summarized by Guthrie, 1980; see Chapter 8, this volume).

4. The Escape Systems of Siphonophores

Siphonophores are pelagic hydrozoan colonies consisting of a long stem from which are budded locomotory zooids (the nectophores, or swimming bells), feeding zooids (gastrozooids and palpons, with their tentacles), and reproductive zooids (gonophores). In physonectids such

Table II. *Aglantha* Escape Response[a]

	Estimated latencies (msec)
Conduction time around ring (5 mm at 2.6 m/sec)	2.0
Transmission delay to motor giant	1.6
Conduction up motor giant (15 mm at 4.0 m/sec)	4.0
Conduction in lateral nerves (0.5 mm at 0.3 m/sec)	1.6
Delay at neuromuscular junction	0.8
Total time to reach all parts of muscle sheet	10.0
Time to peak muscle tension	40.0
Total response time[b]	50.0

[a]Data from Roberts and Mackie, 1980; R. W. Meech, unpublished data.
[b]The sum of the total time to reach all parts of muscle sheet and the time to peak muscle tension.

as *Nanomia* (Figure 7), a float is present at the apex of the colony and the nectophores are arranged in rows facing out from the stem. In diphyids such as *Chelophyes* there is no float, and the two nectophores are arranged in line. In both cases, swimming is achieved by contractions of the swimming muscles lining the inner sides of the nectophores. Coordination of responses is mediated by pathways in the stem interconnecting the nectophores and other zooids. The stem and its appendages contract violently during escape behavior, and escape swimming occurs. *Nanomia* can perform escape swimming in both forward and reverse directions (Figure 7, inset), but *Chelophyes* can swim only forward. Stem organization as described below for *Nanomia* is much the same in the two species, but

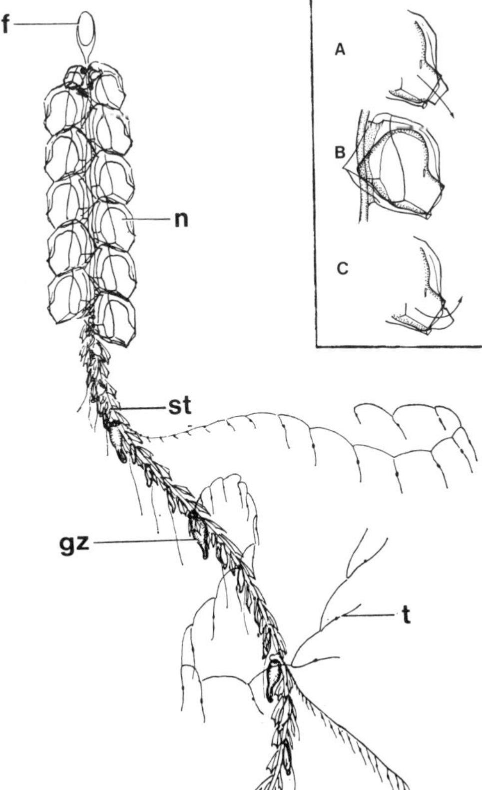

Figure 7. Upper portion of colony, at rest in *Nanomia*. Inset shows a nectophore during (A) forward and (C) reverse locomotion, and (B) at rest. Abbreviations: f, float; gz, gastrozooid; n, nectophore; st, stem; t, tentacle (from Mackie, 1964).

swimming is more complicated in *Nanomia*, being bidirectional. *Chelophyes* was studied chiefly in order to clarify details of nectophore–stem interactions that were and remain rather baffling in *Nanomia*. For this reason and because of the considerable differences between the two types of siphonophores, it seems best to treat *Nanomia* and *Chelophyes* sequentially here rather than to attempt a synthesis.

4.1. Nanomia

4.1.1. Responses to Stimulation and Early Evidence Concerning Pathways

If the stem is touched lightly, it will contract locally in the region of the stimulus. Stronger stimulation causes a rapid contraction of the whole stem plus escape swimming in the forward direction. Escape swimming is characterized by a single unified contraction of all the nectophores, and this may carry the animal a distance up to 10 cm in less than a second (exact measurements have not been made). Escape swimming differs from normal ("slow") swimming in the synchronous character of the response, and in the fact that it occurs only in response to stimulation. In slow swimming, which occurs spontaneously, the nectophores beat at their own individual frequencies, giving rise to a sort of zig-zag gliding movement that is in marked contrast to the much faster, arrow-straight darting of the escape response. When the colony is stimulated on its float, it performs *reverse* escape swimming. Again, the stem shortens and the nectophores all contract together, but deflector muscles at the corners of the bell mouths are concurrently activated and the water jet is ejected in the forwards direction (Figure 7, inset C). These muscles are not active during forward swimming.

The differences between escape and slow forward swimming responses were studied by Mackie (1964), who showed that escape swimming was mediated by a nerve tract that runs across the exumbrella of the nectophore connecting the locomotory centers with the stem. This tract was found to consist of a single giant axon 10 μm in diameter in *Nanomia* (Mackie, 1973). Cutting the axon isolated the nectophore functionally so far as participation in forward escape swimming is concerned. However, it can still join in reverse escape swimming, and perform uncoordinated (slow type) forward locomotion. Mackie (1964) showed that the reverse response was conducted across the nectophore by epithelial conduction in either the exumbrellar ectoderm or the subumbrellar endoderm. Later research on another siphonophore, *Hippopodius* (Bassot *et al.*, 1978), shows that the two epithelia are functionally coupled, and it is probable that in *Nanomia* both epithelia can relay the reverse escape response. The findings that forward and reverse swimming are mediated by distinct pathways in the nectophore led to the suggestion that two comparable pathways were again involved in the stem. Later, detailed study of the stem (Mackie, 1976, 1978) showed that this early concept of the stem pathways was wrong, but the conclusions regarding the nectophore pathways were probably sound, and are supported by work on *Chelophyes* as noted below.

4.1.2. Histology of the Stem: Discovery, Repudiation, and Rediscovery of Giant Axons

The siphonophore stem is a biepithelial structure. Both endoderm and ectoderm are composed of epitheliomuscular cells, but the ectoderm is a much thicker and more muscular layer. It provides the longitudinal musculature that shortens the stem in the escape response.

Nerves were first seen in a siphonophore by Claus (1878), who described a nerve plexus in a stem ectoderm of *Physophora*. Korotneff (1884) failed to find such a plexus in *Forskalia,* but he did discover two rows of enormous cells running along the whole stem on the dorsal side that he identified as neurons (Figure 8A). Schneider (1892) confirmed this identification in stem tissue dissociated by the osmium-acetic maceration technique, and he noted that the giant cells appear to be syncytially interconnected (Figure 8B) and to have lateral connections with a dispersed plexus similar to that seen by Claus in *Physophora*. Nuclei were visible in these dispersed elements. Schneider referred to the whole system as a *Riesensyncytium* (giant syncytium) and suggested that it functioned as the conduction pathway for the "lightening quick" contractions of the stem. He further suggested that the sluggish responses of *Apolemia* were due to the absence of a giant syncytium in this form. However, Korotneff's and Schneider's views were repudiated by Schaeppi (1898), who denied that the large dorsal cells were neurons. He regarded them as endodermal cells forming the wall of a secondary longitudinal canal connected to the main endoderm canal by transverse processes. Schaeppi's "dorsal canal" was supposed to provide for circulation and to serve as a sort of compensation sac for fluids expelled from the main canal during stem contractions. (This explained why it was present only in behaviorally active forms such as *Forskalia*.) Unfortunately, Schaeppi's interpretation was accepted and given prominence in an important contemporary review (Chun and Will, 1902) with the result that the giant fiber system in the siphonophore stem was for all practical purposes forgotten for 75 years.

In a new investigation on *Nanomia,* Mackie (1973) rediscovered

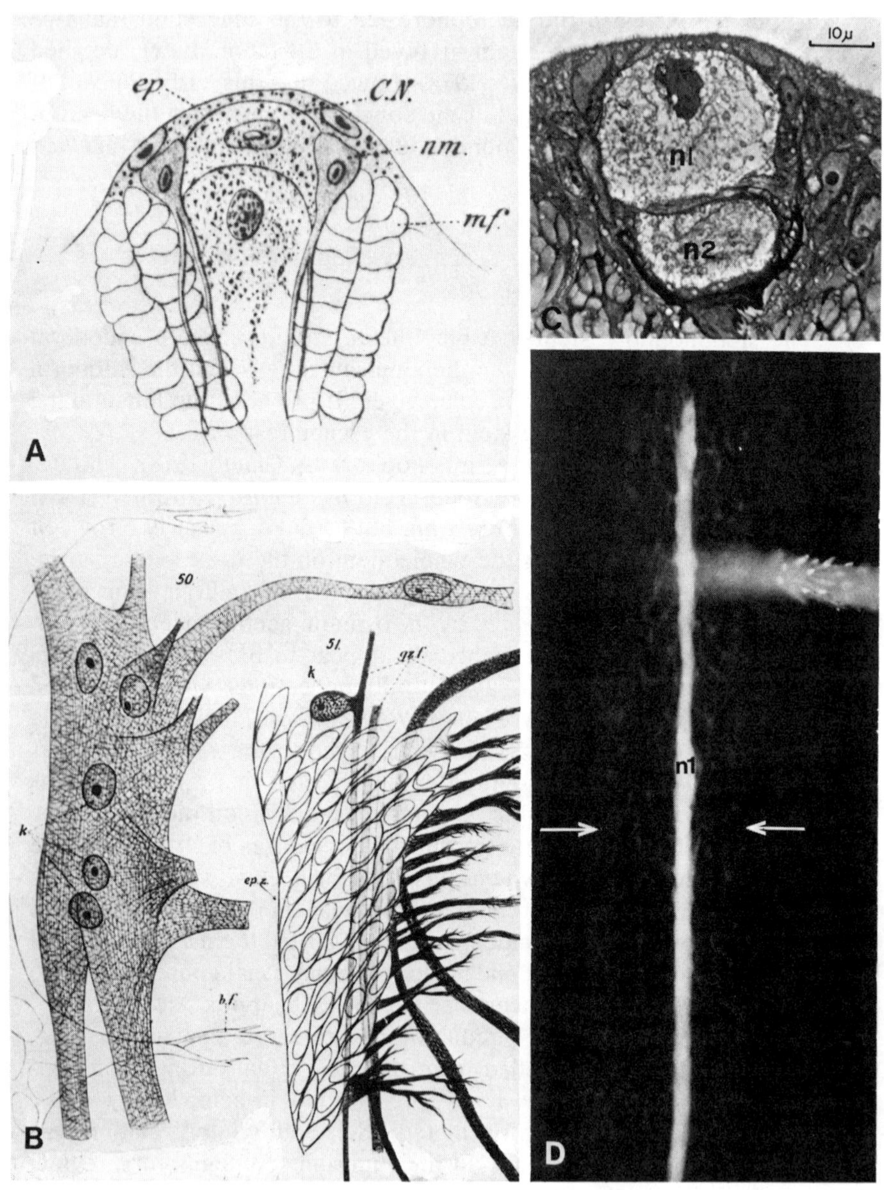
ep.
C.N.
nm.
mf.
A
50.
51.
k
gz.f.
k
ep.z.
b.f.
B
10μ
n1
n2
C
n1
D

Korotneff's giant axons and Schneider's dispersed plexus, and showed by electron microscopy that the nerves synapse with epitheliomyocytes in the stem ectoderm. There are consistently two giant fibers, one typically about 1.4 times the diameter of the other (Figure 8C). The larger (outermost) unit reaches 30 μm in *Nanomia* and 40 μm in *Forskalia*. The nuclei are elongated, about 10 μm in diameter, and cause no bulge in the axon. The apparent absence or rarity of transverse membrane barriers would indicate that the giant fiber is made up of one or perhaps a few syncytial units. If a single unit, it would be about 30 cm long in a large *Nanomia* and over a meter in a large *Forskalia*. The nature of the connections with the dispersed plexus elements is still uncertain, but Lucifer yellow injected into a giant axon readily penetrates the plexus (Figure 8D), suggesting that both components are part of a single giant syncytium as Schneider proposed, or that the plexus is coupled to the giants by permeable junctions.

It has not been possible to demonstrate conclusively that there are two dispersed plexuses, one for each giant axon, but the physiological evidence makes this appear extremely likely. Phase contrast microscopy shows many strands in the plexus to be double, as if two separate nets were present but tended to stick together during development. Further work with dye injection should clarify these relationships.

The giant axons are confined to the stem. Connections with the giant axons running across the nectophores are presumably made via dispersed network elements. Gastrozooids, palpons, and others have conventional nerve plexuses with no giant units.

The pattern of stem innervation described here for *Nanomia* probably represents the most advanced level of nervous organization in the group. Judging from Schneider's (1892) evidence, two giants are present in *Forskalia*, but they appear to be less well differentiated from the dispersed network elements than in *Nanomia*. There is good evidence for a double innervation in the stem of *Hippopodius* and in the tentacles of *Nanomia*, but there are no giant axons in these places. It would appear that double

Figure 8. Stem giant axons in *Forskalia* and *Nanomia*. (A) Korotneff's drawing of the two giant axons (c.n.) in a cross section of the stem of *Forskalia*. (B) Maceration preparations by Schneider showing parts of the giant nerve syncytium (left) and dissociated muscle tissue (right). The giant nerve contains numerous nuclei (k) and is connected syncytially to neurons running out into the dispersed plexus. (C) The two stem giants in *Nanomia* as seen in a 1-μm Epon section. (D) Whole amount of the stem of *Nanomia*. A giant axon (n1) has been injected with Lucifer Yellow. The dye has entered the dispersed nerve plexus. Arrows show edges of the preparation. A cactus spine holding the specimen projects at upper right (A from Korotneff 1884, B from Schneider 1892, C from Mackie 1973, and D by A. N. Spencer and G. O. Mackie, unpublished).

innervation is basic to the group, and that fast pathways evolved by "condensation" of network elements to form giant axons (Figure 9).

4.1.3. Physiological Analysis

Stimulation of the stem evokes propagated events in the two giant axons (*n1*, *n2*) and in a third, slow (*s*) conduction system identified as the stem endoderm, an excitable epithelium (Figure 10A). Velocities are much higher in the nerves (Table III). After measurement of conduction velocities, stems were sectioned and the giant axons measured to show the relationship between axon diameter and conduction velocity. Intracellular

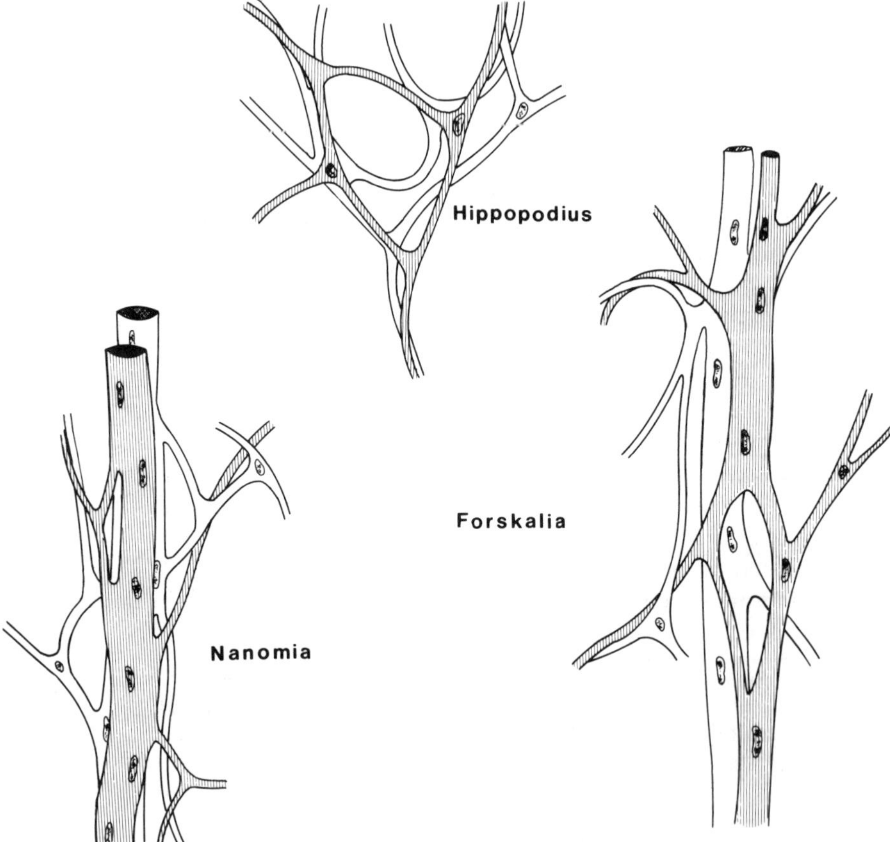

Figure 9. Double innervation patterns in the stems of three siphonophores, suggestive of a progressive concentration of the nerve nets to form giant axons.

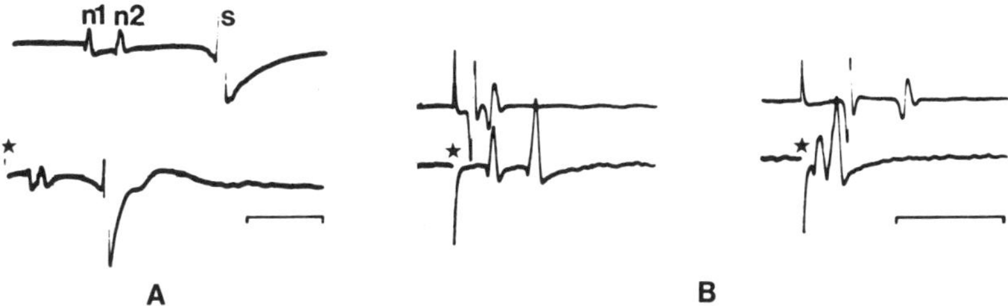

Figure 10. Three conduction systems in the stem in *Nanomia*. Mg^{2+} has been added to reduce muscle twitch potentials. (A) A shock to the stem evokes propagated events (n1, n2, s), which are recorded first at an electrode near the stimulus site (lower record) and then at a more distant electrode (upper). The intervals between the events increase with distance owing to their different conduction velocities. (B) Similar preparation to demonstrate two-way conduction of n1 and n2. The stimulating electrode was placed up stem from the two recording electrodes (left), then down stem (right). Scales: (A) 50 msec and (B) 50 msec (from Mackie, 1973).

recordings from all three conduction systems show conventional spikes, which propagate all-or-none along the whole length of the stem. Events in the two nervous subsystems are distinct, and there is no interaction between them or with events in the slow system. Events can travel in either direction (Figure 10B) and can pass going in opposite directions in *n1* and *n2*; a spike in the faster axon can overtake one initiated earlier in the slower. Events can propagate through regions in which all contractions are blocked by elevated Mg^{2+}.

Table III. Representative Conduction Velocities for Nerves and Epithelial Stem Systems in Various Siphonophores[a]

	Conduction velocity (m/sec)				
Species	Nerve 1	Nerve 2	Epithelial stem system	Temperature (°C)	Giant fibers
Suborder Calycophora					
Chelophyes	1.1	0.9	0.25	22	Present
Sulculeolaria	1.7	1.3	0.3	23	Assumed present
Hippopodius	0.5	0.3	0.2	22	Absent
Suborder Physonectae					
Forskalia	3.7	2.0	0.4	20	Present
Nanomia	2.7	1.5	0.3	14	Present

[a]From Mackie and Carré, 1983.

Evidence that activity in the giant axons spreads to the dispersed nerve nets comes from experiments in which both the giant axons were cut at one point. Following stimulation on the proximal side of the cut, recordings distal to it showed the two events associated with *n1* and *n2* continuing on in their normal relationships, and at velocities consistent with the conclusion that they had reinvaded the giants (Mackie, 1973). Together with the spread of Lucifer Yellow through the dispersed plexus from the giant, this evidence implies that functionally the plexus is just a peripheral extension of the giant (or alternatively, that the giant is simply a condensed region of the plexus). The fuction of the plexus is presumably to spread excitation over the muscle sheet. Coelenterates, unlike crabs or frogs, lack muscles with localized origins and insertions. The muscle sheet is a continuous layer covering the whole stem, and therefore requires a similar, essentially two-dimensional, sheet innervation along its whole length.

Intracellular recordings from the ectodermal muscle (Spencer, 1971; Mackie, 1976) failed to show propagated spikes, but showed graded depolarizations in response to input from all three conduction systems. Events in the two nervous subsystems cause similar, small (4–6 mV) excitatory junctional potentials (*n*EJPs) that sum and facilitate, producing twitch responses in the muscle. Twitches spread along the stem at a velocity determined by conduction velocities in the nervous system. The *s* system causes 10–12 mV *s*EJPs that sum but do not facilitate. (The long refractory period of the *s* system precludes a sufficiently high firing frequency.) Acting alone, the *s* system produces slow (? postural) contractions. Slow EJPs and *n*EJPs can sum, but there is no evidence of mutual facilitation. These general relationships are summarized in Figure 11.

A peculiar feature of the wiring is that the *s* system not only can excite the muscles, but also can itself be excited when the muscle is fired to a sufficient level of depolarization by *n* input. This is explicable on the assumption that the two tissues are coupled by nonrectifying electrical junctions. One effect of this is the production of a range of specious, high-conduction velocities in the *s* system. It is being fired by coupling potentials from the muscles that are being fired by nerves. The conduction velocity of the nerves sets the rate for spread of events in the muscles and hence in the *s* system; *s* events, once initiated in this way can be carried along "piggyback" for varying distances, but if the carrier potential in the muscle layer falls below a certain level, they "fall off" and continue to propagate at a slower rate (Mackie, 1976). They are then free to reexcite the muscle, prolonging its contraction. Stem contractions thus tend to be very complex events depending on the number of primary impulses initiated in the three conduction systems and their time rela-

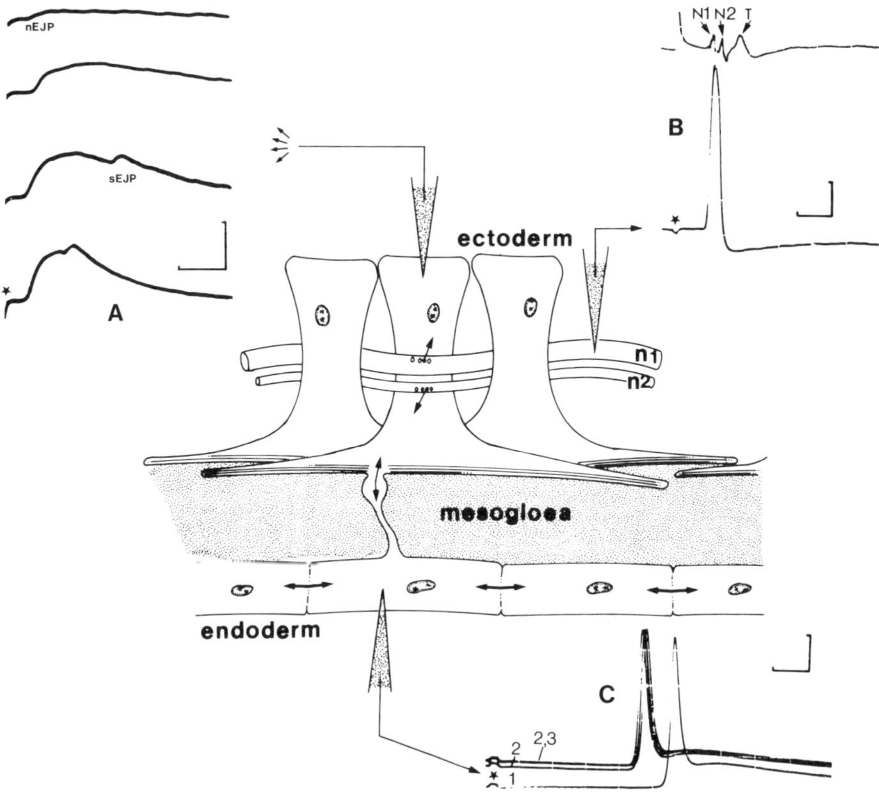

Figure 11. Stem contraction mechanism in *Nanomia* and *Forskalia*. The two nervous subsystems (n1, n2) synapse with ectodermal epitheliomuscular cells that are coupled to the endoderm, a conducting epithelium. (A) Intracellular records from the ectoderm showing graded depolarizations rising to twitch amplitude. These events are composed of summed excitatory junctional potentials representing input from either the nerves (nEJP) or the endoderm (sEJP). Top record shows a single nEJP. (B) Intracellular recording of n1 giant axon spike (below) with simultaneous extracellular recording from stem surface showing n1, n2, and muscle twitch potential (t). (C) Superimposed intracellular records of endodermal spikes (s events) evoked by four shocks at 250-msec intervals. Base line is elevated after the first response because of long lasting after depolarization characterizing these events. Scales: (A, C), 20 mV, 10 msec, (B) 20 mV, 20 msec (data from Mackie, 1973, 1978).

tionships and on the production and timing of secondary events in the *s* system, which may reexcite the muscle.

The question obviously arises: Why two nerve nets? There is no evidence that the two systems innervate different sectors of the muscle sheet. The picture is one of dual, overlapping excitatory input. Furthermore, all siphonophores so far studied, including forms such as *Hippopodius* which have no giant axons, have a double innervation in the stem, with one faster and one slower subsystem. The double innervation would therefore seem to be a fundamental feature of motor organization in the group as a whole, rather than having evolved in relation to escape behavior per se.

An explanation (still somewhat tentative) put forward to account for the double innervation (Mackie, 1976) is summarized diagrammatically in Figure 12. The stem muscle is depicted as a series of units innervated by

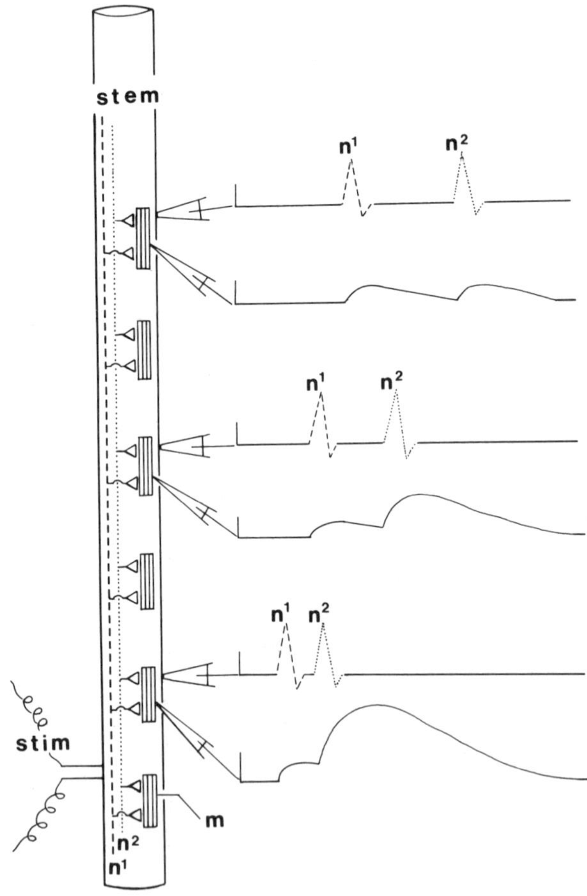

Figure 12. Suggested mechanism for gradation of muscle response intensity with distance from site of stimulus in *Nanomia* and *Forskalia* as explained in the text.

two nervous systems. The *s* system is omitted for simplicity. Impulses generated simultaneously in *n1* and *n2* at the same point will arrive at almost the same time at neuromuscular junctions close to this point, and will cause strongly facilitating muscle reponses. As the impulses travel along the stem, however, the interval between them will increase owing to the difference in conduction velocity between the two giant axons, so temporal summation and facilitation will decline progressively with distance from the stimulation point. The double innervation may thus be seen as a possible way of grading muscle contraction intensity with distance, using a nervous system whose elements are not synaptically interconnected and which cannot therefore conduct gradedly itself.

4.2. Chelophyes

Chelophyes (Figure 13) is one of the most streamlined and fast-moving of all plankton animals. Bone and Trueman (1982) have analyzed its locomotion in terms of biomechanics, comparing it with a slower species, *Abylopsis*. Stimulation of the stem in *Chelophyes* causes short bursts of escape swimming at frequencies up to 8 Hz in which velocities up to 30 cm/sec are achieved. A single contraction drives the animal forward 2.5 cm, and a typical short burst, 25 cm. Escape swimming is accompanied by stem contraction.

Though efficient in the sense that they rapidly remove the animal from potentially damaging contacts, these movements are costly in terms of the work requirement, ten times more so than in the case of *Abylopsis*, which swims at only 8 cm/sec. High values for muscle tension and muscle stress are recorded in *Chelophyes*. The maximum power developed probably exceeds that generated by frog twitch muscle fibers. As in other hydrozoans, the swimming muscles are striated and propagate action potentials (Chain *et al.*, 1981; Bone, 1981). Structures resembling a T system have been located by electron microscopy (Mackie and Carré, in 1983).

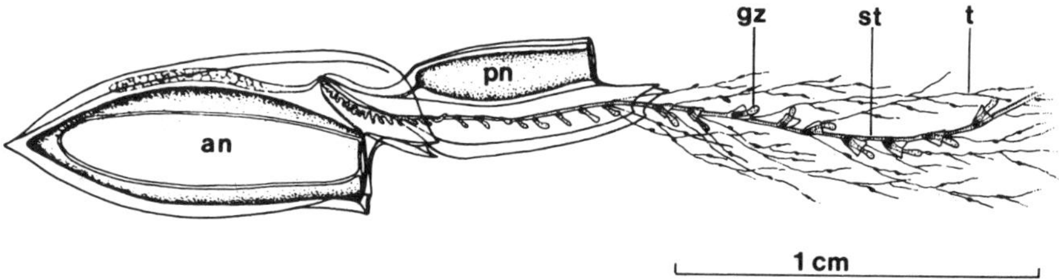

Figure 13. Chelophyes, with stem extended. Abbreviations: an, anterior nectophore; gz, gastrozooid; pn, posterior nectophore; st, stem; t, tentacle (from Mackie and Carré, 1983).

As in *Nanomia*, electrophysiological studies have shown there to be a double innervation and a slow conduction system in the stem (Figure 14), as well as two pathways between the stem and the locomotory centers in the nectophore margin, one pathway being nervous and the other epithelial. The nervous pathway consists of a giant axon and a number of small axons of unknown function. The giant probably functions as a fast pathway mediating escape swimming following stem stimulation. It is

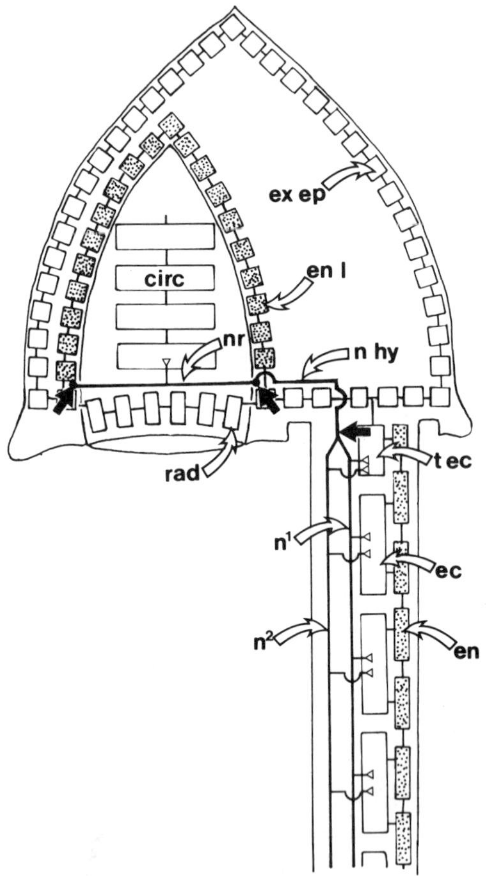

Figure 14. Action systems of nectophore and stem summarized in *Chelophyes*, showing nervous and nonnervous links between the two regions. Black arrows symbolize epithelioneural excitation, triangles neuromuscular excitation. Lines between excitable units represent coupling. Stimulation of the exumbrellar epithelium (ex ep) evokes impulses that propagate to the margin where they excite nerves in the nerve ring (nr) that excites the circular swimming muscle (circ). Epithelial impulses also propagate directly to the radial muscles (rad) of the velum, causing them to contract during swimming. On reaching the stem, epithelial impulses pass directly to ectoderm cells in a transitional zone (t ec) that has the property, when excited, of exciting the stem nervous systems (n1, n2), both of which contain a giant axon. Depolarizations of the transitional ectoderm also pass directly to the endoderm (en) where they propagate as s events. Throughout the length of the stem, activity in n1 and n2 causes twitch depolarizations in the ectodermal muscles (ec) but impulses do not propagate within this tissue. Two-way interactions occur between the ectoderm and the endoderm. Activity in the stem nerves leads to twitch depolarizations in the transitional ectoderm that, if sufficiently large, propagate to the exumbrellar epithelium as exumbrellar impulses. On reaching the nectophore margin, they cause swimming as before. The hydroecial nerve (n hy) is implicated as an alternate, but faster, pathway mediating escape swimming responses following stem stimulation (from Mackie and Carré, 1983).

absent in species that lack escape swimming behavior. The primary epithelial pathway is the exumbrellar ectoderm, but the endoderm is also excitable. Epithelial impulses can "jump" into the nervous system at the nectophore margin causing escape swimming and into *n1* and *n2* at the point where the stem joins the nectophore (Figure 14). Thus, stem contractions accompany escape swimming following stimulation of the nectophore. There appears to be a specialized region at the stem–nectophore junction ("transitional ectoderm") where (as just noted) epithelial impulses can excite the stem nerves and also where, going in the other direction, nerve-induced depolarizations of the ectoderm can propagate as spikes up through the ectoderm into the nectophore. Elsewhere in the stem, the ectoderm does not propagate. The *s* system does not participate in this two-way traffic.

Although *Chelophyes* can swim in only one direction, these findings nevertheless provide important clues as to the mechanism of reverse escape swimming in *Nanomia*. It is unlikely that there is a separate epithelial pathway for the reverse response in the stem as originally supposed. The two nervous subsystems and their giants are probably excited when the stem is touched giving rise to bursts of impulses that depolarize the transitional ectoderm at the nectophore attachment points that would then lead, as in *Chelophyes*, to propagation of events across the exumbrellar epithelium and thus to reverse swimming. Impulses coming up the stem in the other direction would not be so tightly grouped (owing to the velocity differential between *n1* and *n2*) and would fail to depolarize the transitional ectoderm to the requisite degree, but impulses would still cross to the nectophore margin in the exumbrellar nerve tract and lead to excitation of the swimming muscle causing forward swimming.

5. Conclusions

The hydrozoans described here show many resemblances to other animals having escape responses mediated by fast pathways. The responses themselves are simple and stereotypic in nature and involve near-simultaneous activation of muscles over a wide area. In *Aglantha* there is a "cascading" effect in the innervation: a single first-order giant fires eight second-order giants in a way reminiscent of the pattern in squids. The muscles, whether smooth or striated, give fast, twitch responses when activated in the escape mode, but the same fibers can give slower, weaker responses when fired through slow pathways.

A curious (but typically cnidarian) feature in *Nanomia* is that the slow responses of the stem muscle are mediated not by nerves but by a conducting epithelium coupled to the muscle sheet. By contrast, the slow pathway in *Aglantha* is thought to be a network of fine neurons that serves to spread graded responses across the muscle sheet. These are probably the same neurons that are coupled to the motor giants, and they would thus be activated during the escape response. This interpretation calls to mind the arrangement in the crayfish, where the FF cells are the exclusive outlet for all nongiant activity but are "commandeered" by the giant fiber system during (escape) tailflip responses (see Chapter 7, this volume).

The giant fibers in our hydrozoan examples are few in number and constitute a distinct size category, notably larger than those of the next size class down. Conduction velocity and axon diameter are linearly related much as in other invertebrates (Figure 15). The axons are syncytial and probably arise by fusion of smaller units. As for other animals, we would argue for *Nanomia* and *Aglantha* that the selective advantage of the short latency response outweighs the extra metabolic cost of maintaining outsize axons. Certainly, these forms live in an environment where rapid responses seem very likely to be beneficial in terms of survival.

What is perhaps most striking about the hydrozoans described here is that their well-developed giant axons have apparently evolved from the

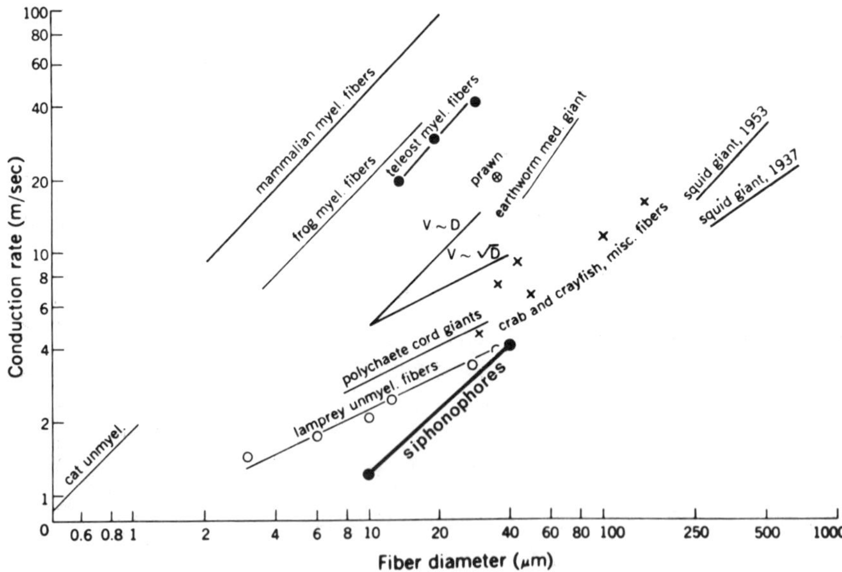

Figure 15. Conduction velocities plotted against fiber diameters for various animals from Bullock and Horridge (1965) with addition of data from *Nanomia* and *Forskalia*. Points for *Aglantha* would lie among the upper part of the same curve.

starting point of simple nerve nets spread out across the muscle fields they innervate. Current evidence from other hydromedusae (reviewed by Spencer and Schwab, 1982) suggests that the electrotonically coupled nerve net is basic to the class and that such nets have become oriented and condensed into large, and ultimately into giant units as need arose. No such process is in evidence in the other two classes of Cnidaria. Outsize neurons occur in some scyphomedusae, but they show no sign of being fusion products. Cnidarians have had a long evolutionary history in which to experiment with their conducting tissues and the result is that whereas some of their adaptations are highly peculiar, others parallel developments in higher animals to a degree that is only gradually becoming apparent as cellular-level studies advance.

6. References

Bassot, J.-M., Bilbaut, A., Mackie, G. O., Passano, L. M., and Pavans de Ceccatty, M., 1978, Bioluminescence and other responses spread by epithelial conduction in the siphonophore *Hippopodius, Biol. Bull.* **155**:473–498.

Batham, E. J., Pantin, C. F. A., and Robson, E. A., 1960, The nerve net of the sea anemone *Metridium senile:* the mesenteries and the column, *Q. J. Microscopical Sci.* **101**:487–510.

Bone, Q., 1981, The relation between the form of the action potential and contractions in the subumbrellar myoepithelium of *Chelophyes* (Coelenterata: Siphonophora), *J. Comp. Physiol.* **144**:555–558.

Bone, Q., and Trueman, E. R., 1982, Jet propulsion of the calycophoran siphonophores *Chelophyes* and *Abylopsis, J. Mar. Biol. Assoc. U.K.* **62**:263–276.

Bullock, T. H., and Horridge, G. A., 1965, *Structure and Function of the Nervous System of Invertebrates,* Freeman, San Francisco.

Chain, B. M., Bone, Q., and Anderson, P. A. V., 1981, Electrophysiology of a myoid epithelium in *Chelophyes* (Coelenterata: Siphonophora), *J. Comp. Physiol.* **143**:329–338.

Chun, C., and Will, L., 1902, Coelenterata, in: *Klassen and Ordnungen des Thier-Reichs,* Bd. 2, Abt. 2, (H. G. Bronn, ed.), C. F. Winter, Leipzig, pp. 327–371.

Claus, C., 1878, Uber *Halistemma tergestinum* n.sp. nebst Bemerkungen über den feinern Bau der Physophoriden, *Arb. Zool. Inst. Wien* **1**:1–56.

Donaldson, S., Mackie, G. O., and Roberts, A., 1980, Preliminary observations on escape swimming and giant neurons in *Aglantha digitale* (Hydromedusae: Trachylina), *Can. J. Zool.* **58**:549–552.

Guthrie, D. M., 1980, *Neuroethology: an Introduction,* Blackwell Scientific Publications, Oxford.

Hertwig, O., and Hertwig, R., 1878, *Das Nervensystem und die Sinnesorgane der Medusen,* Vogel, Leipzig.

Josephson, R. K., 1974, Cnidarian neurobiology, in: *Coelenterate Biology* (L. Muscatine and H. M. Lenhoff, ed.), Academic Press, New York, pp. 245–280.

Korotneff, A., 1884, Zur Histologie der Siphonophoren, *Mitt. Zool. Sta. Neapel* **5**:229–288.

Mackie, G. O., 1964, Analysis of locomotion in a siphonophore colony, *Proc. R. Soc. (London) Ser. B.* **159**:366–391.

Mackie, G. O., 1973, Report on giant nerve fibres in *Nanomia, Publ. Seto. Mar. Lab.* **20**:745–756.

Mackie, G. O., 1976, The control of fast and slow muscle contractions in the siphonophore stem, in: *Coelenterate Ecology and Behavior* (G. O. Mackie, ed.), Plenum Publishing Corp., New York, pp. 647–659.

Mackie, G. O., 1978, Coordination in physonectid siphonophores, *Mar. Behav. Physiol.* **5**:325–346.

Mackie, G. O., 1980, Slow swimming and cyclical "fishing" behavior in *Aglantha digitale* (Hydromedusae: Trachylina), *Can. J. Fish. Aquat. Sci.* **37**:1550–1556.

Mackie, G. O., and Carré, D., 1983, Coordination in a diphyid siphonophore, *Mar. Behav. Physiol.* **9**:139–170.

Mackie, G. O., and Mills, C. E., 1983, Use of the PISCES IV submersible for zooplankton studies in coastal waters of British Columbia, *Can. J. Fish. Aquat. Sci.* **40**:763–776.

McFarlane, I. D., 1973, Spontaneous contractions and nerve net activity in the sea anemone *Calliactis parasitica, Mar. Behav. Physiol.* **2**:97–112.

McFarlane, I. D., 1982, *Calliactis parasitica,* in: *Electrical Conduction and Behavior in "Simple" Invertebrates,* (G. A. B. Shelton, ed.), Clarendon Press, Oxford, pp. 243–265.

Pantin, C. F. A., 1935, The nerve net of the Actinozoa I: Facilitation, *J. Exp. Biol.* **12**:119–138.

Pantin, C. F. A., 1952, The elementary nervous system, *Proc. R. Soc. (London) Ser. B.* **140**:147–168.

Passano, L. M., 1982, Scyphozoa and Cubozoa, in: *Electrical Conduction and Behavior in "Simple" Invertebrates* (G. A. B. Shelton, ed.), Clarendon Press, Oxford, pp. 73–148.

Pickens, P. E., 1969, Rapid contractions and associated potentials in a sanddwelling anemone, *J. Exp. Biol.* **51**:513–528.

Roberts, A., and Mackie, G. O., 1980, the giant axon escape system of a hydrozoan medusa, *Aglantha digitale, J. Exp. Biol.* **84**:303–318.

Robson, E. A., 1961, A comparison of the nervous systems of two sea anemones, *Calliactis parasitica* and *Metridium senile, Q. J. Microscopical Sci.* **102**:319–326.

Robson, E. A., 1975, The nervous system in coelenterates, in: *"Simple" Nervous Systems* (P. N. R. Usherwood and D. R. Newth, eds.), Edward Arnold, London, pp. 169–209.

Robson, E. A., and Josephson, R. K., 1969, Neuromuscular properties of mesenteries from the sea anemone *Metridium, J. Exp. Biol.* **50**:151–168.

Ross, D. M., 1974, Behavior patterns in associations and interactions with other animals, in: *Coelenterate Biology* (L. Muscatine and H. M. Lenhoff, eds.), Academic Press, New York, pp. 281–312.

Schaeppi, T., 1898, Untersuchungen über das Nervensystem der Siphonophoren, Jena, *Zeit. Naturwiss.* **32**:483–550.

Schneider, K. C., 1892, Einige histologische Befunde an Coelenteraten, Jena, *Zeit. Naturwiss.* **27**:387–461.

Shelton, G. A. B., 1982, Anthozoa, in: *Electrical Conduction and Behaviour in "Simple" Invertebrates,* (G. A. B. Shelton, ed.), Clarendon Press, Oxford, pp. 203–242.

Singla, C. L., 1978, Locomotion and neuromuscular system of *Aglantha digitale, Cell Tissue Res.* **188**:317–327.

Spencer, A. N., 1971, Myoid conduction in the siphonophore *Nanomia bijuga, Nature* **223**:490–491.

Spencer, A. N., and Schwab, W. E., 1982, Hydrozoa, in: *Electrical Conduction and Behaviour in "Simple" Invertebrates* (G. A. B. Shelton, ed.), Clarendon Press, Oxford, pp. 73–148.

Weber, C., Singla, C. L., and Kerfoot, P. A. H., 1982, Microanatomy of the subumbrellar motor innervation in *Aglantha digitale* (Hydromedusae: Trachylina), *Cell Tissue Res.* **223**:305–312.

3

Escape Reflexes in Earthworms and Other Annelids

CHARLES D. DREWES

1. Introduction

Escape or startle reflexes are characteristically seen in many representatives of the Phylum Annelida (segmented worms), a group that comprises the earthworms, aquatic oligochaetes, leeches, and marine bristle or polychaete worms. From the standpoint of defense, the escape reflex represents one of the most important components of a worm's repertoire of locomotory behavior that, depending on the species, may also include undulating swimming or peristaltic creeping movements. The underlying control and coordination for all of these locomotory movements are carried out by the worm's central nervous system, which consists of a dorsal brain in anterior segments and a ventral nerve cord, the latter being composed of a chain of segmentally arranged ganglia joined by longitudinal connectives.

Although there is considerable diversity in the habitats and lifestyles of those annelid worms that possess escape reflexes, there appear to be a number of properties commonly shared in their escape reflex organization and function. These include (1) triggering of the reflex by sudden onset of threatening stimuli, (2) activation of giant nerve fibers within the ventral nerve cord, and (3) initiation of a short latency motor response, especially a synchronous, multisegmental shortening that effectively withdraws the animal from the source of the stimulus. Although one purpose

---

CHARLES D. DREWES • Zoology Department, Iowa State University, Ames, Iowa 50011.

of this chapter is to present a general overview of the organization of annelid escape reflexes, the main intent and therefore greatest emphasis will be placed on the design and function of rapid reflexes in oligochaete earthworms. Previous reviews that generally pertain to the earthworm or annelid escape reflexes include those by Laverack (1963), Bullock and Horridge (1965), Mill (1975, 1982), Gardner (1976), and Dorsett (1978, 1980).

1.1. Evidence That Giant Nerve Fibers Mediate Rapid Escape

Nicol (1948*c*) defined giant nerve fibers as "those fibres which have a much greater size than any other nerve-fibres of the animal" and "form an essential nervous component of a system which effects a rapid, widespread and synchronous response." But what is the evidence that giant nerve fibers are indeed essential and responsible for the initiation of rapid escape, in other words, that they function as so-called "command" neurons whose activity is both necessary and sufficient for the initiation of escape behavior (Kupfermann and Weiss, 1978)?

To demonstrate the criterion of necessity, the behavior should be abolished by removal of the giant fibers from the system. This has effectively been demonstrated by selectively severing giant fibers while sparing the rest of the ventral nerve cord in polychaetes (Nicol, 1948*a*), as well as in earthworms (Yolton, 1923; Stough, 1930; Ten Cate, 1938; Bullock, 1945*a*; Rushton, 1946). In all cases rapid escape movements initiated by tactile stimulation at one end of the animal progressed up to but not beyond the site of giant fiber lesion. Reversibility of this effect has also been demonstrated. That is, the recovery time required before rapid escape movements can again progress beyond the lesion corresponds to the time required for giant fiber regeneration, as determined by anatomical and physiological criteria (Bovard, 1918; Birse and Bittner, 1976, 1981; Pallas and Drewes, 1981).

Another study that pertains to the criterion of necessity is Bullock's (1948) systematic survey of 15 polychaete species in which correlations were made of giant nerve fiber morphology, electrophysiology, and associated escape behavior. In all ten species possessing giant fiber systems, as evidenced by histological and physiological observations, startle responses were significantly more pronounced than in the five species lacking such systems. Finally, it has been noted that establishment of giant nerve fiber pathways during embryonic development, as indicated by histological (Prosser, 1933) and electrophysiological observations (O'Gara *et al.*, 1982), correlates with the onset of rapid escape capabilities during the same stage of development.

Kupfermann and Weiss' (1978) criterion for sufficiency is satisfied if artificial stimulation of giant fiber spiking evokes a behavioral response indistinguishable from the response to an identical pattern of giant fiber spiking evoked by natural stimulation. This stringent criterion has not been dealt with in a rigorous fashion in any previous studies of rapid escape responses in annelids, although a number of studies indicate that some components of rapid escape, such as rapid tail flattening in the earthworm (Pallas and Drewes, 1981) and rapid longitudinal withdrawal in polychaetes and earthworms (Bullock, 1948; Horridge, 1959; Roberts, 1962*a,b,* 1966) can be evoked by just-threshold electrical stimulation of the giant nerve fibers. However, it has not yet been demonstrated whether such responses to artificial stimulation of giant fibers are indistinguishable from responses to natural stimulation of the fibers.

1.2. Experimental Utility of Annelid Escape Reflexes

By virtue of the metameric body plan of annelids, some species having hundreds of body segments, the neuronal networks underlying escape are typically arranged as sets of large, identifiable and serially homologous neurons. The large size and identifiability of neurons provide obvious advantages for studying neuronal connections in escape reflex circuitry using intracellular microelectrodes. The serially homologous arrangement of neurons presents opportunities for studying physiological and behavioral specializations related to this arrangement, such as (1) intersegmental integration of parallel sensory inputs, (2) conduction properties within intersegmental neuronal networks, and (3) segmental or regional variations in the patterns and coordination of motor responses.

An additional opportunity stems from the capability of noninvasively recording spiking activity from at least some of the neuronal units mediating rapid escape in annelids. In earthworms the capability of recording medial giant fiber spiking activity through the body wall of intact worms was first noted by Rushton and Barlow (1943). With further development of this approach, it has been demonstrated that spikes, not only from the medial giant fiber (MGF) but also from the lateral giant fibers (LGF) and associated giant motor neurons, can be resolved in surface recordings from intact unrestrained worms (Drewes *et al.*, 1978, 1980; Drewes and McFall, 1980). As shown in Figure 1, stereotyped patterns of all-or-none MGF and LGF spiking activity can be readily detected from the ventral surface of intact segments in response to tactile stimulation at either end of the animal. These stereotyped patterns of MGF and LGF activity are consistently seen not only in recordings from all specimens of the same species, but also in different species from a number of earthworm families

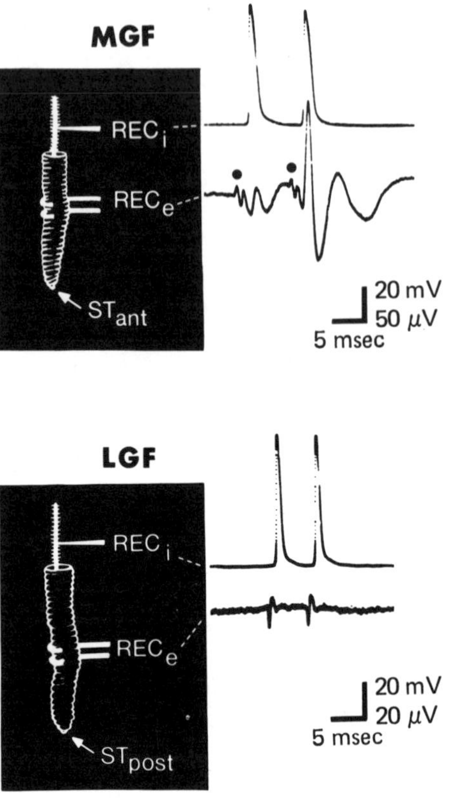

Figure 1. Simultaneous recordings of intracellular and surface electrical activity from the giant nerve fibers in two earthworm species. (Above) A pair of medial giant fiber (MGF) spikes evoked by light tactile stimulation (ST_{ant}) of head segments in *Eisenia foetida*. The two MGF spikes (•) were detected first by the pair of large hook electrodes (REC_e) in contact with the ventral surface of intact segments. The deflections following each extracellularly recorded MGF spike represent giant motor neuron and longitudinal muscle potentials (see Section 4.4). The same two MGF spikes were detected further posteriorly in intracellular microelectrode recordings (REC_i) from the exposed ventral nerve cord. (Below) A pair of lateral giant fiber (LGF) spikes evoked by tactile stimulation (ST_{post}) of posterior segments in *Lumbricus terrestris*. The two all-or-none LFG spikes were first detected extracellularly (REC_e) from the ventral surface of intact posterior segments, and then intracellularly (REC_i) from the LGF in the exposed ventral nerve cord in more anterior segments. For the intracellular recordings, the upward deflection is positive. For the MGF extracellular recording (lower trace), the upward deflection is negative whereas for the LGF extracellular recording (lower trace), the downward deflection is negative (courtesy of E. P. Vining, unpublished data).

(Figures 2 and 3). From an experimental standpoint, capabilities for noninvasive recording of giant fiber activity are useful because one can obtain a data base of direct correlations between nervous system activity *in vivo* and behavioral performance of the intact animal. Furthermore, repeated testing of the same animals is possible, thus permitting long-term studies of neural correlates of behavioral plasticity or escape reflex development.

2. Polychaete Escape Reflexes

A remarkable diversity of giant fiber number, size, and general design is found in the polychaetes (for reviews, see Nicol, 1948*b*; Bullock and Horridge, 1965). Nevertheless, all polychaete giant fiber systems seem to be classifiable into a few basic morphological patterns. According to Dorsett (1978, 1980), these include:

1. Unicellular fibers in which a neuron cell body gives rise to a large caliber axon that runs the length of the ventral nerve cord (e.g., giant fibers of *Protula*)
2. Multicellular fibers derived from serially homologous neurons whose enlarged axons either fuse longitudinally or form septate junctions with corresponding axons in adjacent segments (e.g., lateral giant fibers in *Nereis*)
3. Multicellular fibers derived from the lateral fusion of axons whose cell bodies originate in the same ganglion (e.g., medial giant fiber of *Nereis*)

In each case the design of the fiber appears well-suited for function as a single unit capable of rapid, uninterrupted conduction along the ventral nerve cord.

2.1. *Afferent Pathways*

Depending on the species, several stimulus qualities have been shown effective in evoking giant nerve fiber activity and associated escape behavior. These include tactile stimulation (e.g., light touch or body wall deformation), substrate vibration, noxious chemicals, and abrupt onset of light or shadow (Nicol, 1948*a*, 1950; Dorsett, 1964; Gwilliam, 1969; Seymour, 1972). However, measurements of the time from stimulus onset until initiation of giant fiber spiking have not been made in most species, a notable exception being *Branchiomma vesiculosum* in which the latency from touching until giant fiber spiking was 7–8 msec (Krasne, 1965). This time represents less than one half of the total reflex time in the animal.

Tactile sensory fields have been determined in a number of polychaete species (Bullock, 1945*b*, 1948, 1953; Horridge, 1959; Dorsett, 1964). Depending on the species and the specific giant fiber system, these sensory fields may be restricted to specific body regions or may involve virtually all segments of the animal. For example, in *Nereis,* the medial and paramedial giant fiber systems are selectively activated by touching anterior and posterior portions of the body, respectively, whereas the lateral giants are activated by touch anywhere along the body (Horridge, 1959). Thus it is apparent that significant overlapping can occur with respect to the sensory fields of different giant fiber systems in the same animal. However, it is not yet clear whether any given sensory unit in a segment can form divergent inputs onto two functionally distinct giant fiber systems. Also, it is not known whether inputs from tactile sensory units synapse

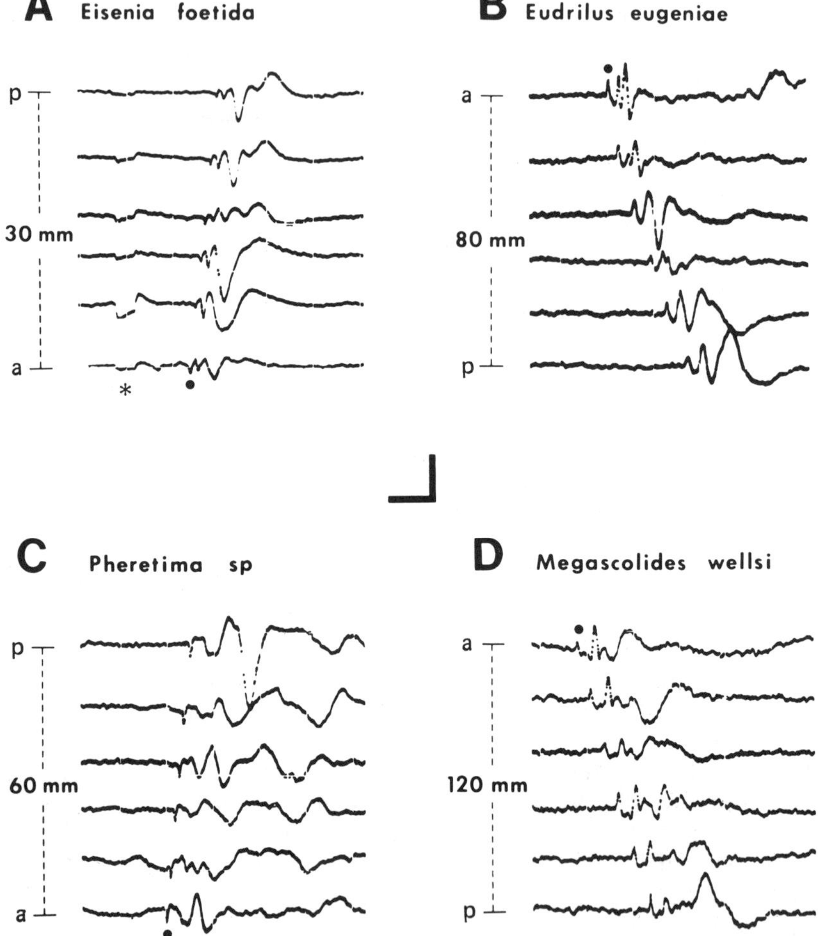
A Eisenia foetida
p
30 mm
a
*
B Eudrilus eugeniae
a
80 mm
p
C Pheretima sp
p
60 mm
a
D Megascolides wellsi
a
120 mm
p

directly onto giant fibers or whether inputs are preprocessed by nongiant interneurons interposed between sensory and giant fibers.

Whatever the specific arrangements for afferent to giant fiber transmission may be, it is clear that this transmission process is highly labile, thus providing a likely basis for the rapid habituation of escape responses to repeated stimuli. For example, in sabellid and nereid polychaetes, repetition of abrupt decreases in light intensity (Nicol, 1950; Gwilliam, 1969) or repeated tactile stimulation (Horridge, 1959; Krasne, 1965) caused rapid failure to evoke giant fiber spikes and escape withdrawal. Habituation of escape behavior and its possible adaptive significance have been discussed by Nicol (1950) and Dyal (1973). They suggested that the consequence of a worm's failure to eventually habituate to adequate but otherwise unimportant stimuli may be starvation or suffocation, especially in the case of certain tube-dwelling worms.

2.2. Central Conduction

An important factor contributing to the short latency with which escape responses can occur is rapid impulse conduction along the giant fibers. The highest values of conduction velocity for any polychaete giant fiber appear to be 20 m/sec in *Myxicola* (Nicol, 1948*b*). This fiber is 1000 μm in diameter, a relatively enormous size that has permitted extensive investigation into giant fiber membrane properties and excitability (for

Figure 2. Patterns of MGF spiking activity from four different species of intact, adult earthworms; family names are (A) Lumbricidae, (B) Eudrilidae, and (C, D) Megascoledidae. In all cases the recordings were obtained noninvasively from six recording sites along each animal by means of a printed circuit board consisting of pairs of differential recording electrodes in contact with the ventral surface of the animal (for methods, see O'Gara *et al.*, 1982). (A) A brief mechanical stimulus was applied to the anterior end of the worm using a fine glass rod attached to a small speaker. The speaker was driven by a 2-msec pulse from an electronic stimulator (*, stimulus artifact). The stimulus evoked a single, all-or-none MGF spike (•). In all records (A–D) the potentials that consistently followed a few milliseconds after each MGF spike probably represent giant motor neuron and longitudinal muscle potentials (see Section 4.4). (A) The MGF spike was progressively detected at each of the six recording sites along the animal as the spike conducted from anterior (a, bottom trace) to posterior (p, top trace). (B–D) A single MGF spike (•) was initiated in response to a very light tactile stimulus applied with a hand-held probe to the anterior end of the worm. The conduction velocites of the MGF spikes (A–D) were 11, 9, 25, and 15 m/sec, respectively. In this and later figures of this type, spike polarity depends on the direction of axonal conduction relative to the electrode grid: (A, C) The direction of propagation is from the bottom to the top trace; (B, D) from top to bottom. Thus these extracellular spikes, which are negative going, appear as (A, C) downward deflections, and (B, C) upward deflections. Voltage scale: (A) 50 μV (B, C) 35 μV, and (D) 20 μV. Time scale: 5 msec.

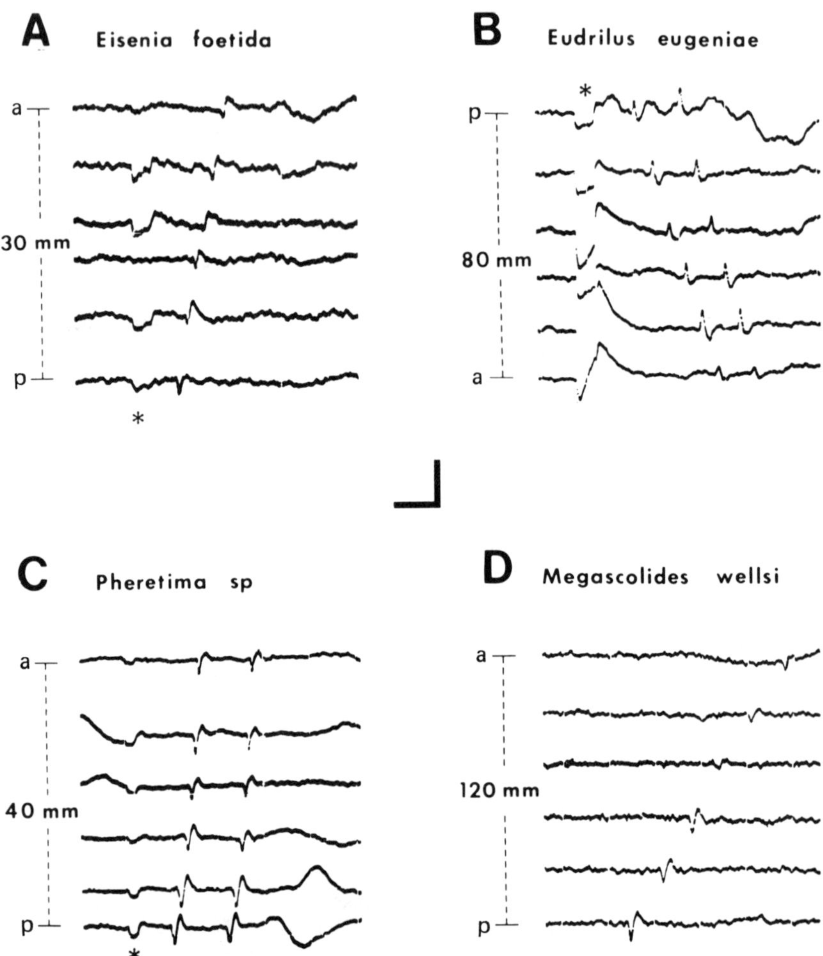

Figure 3. Patterns of LGF activity from four different species of intact, adult earthworms. (A–C) All-or-none LGF spikes were initiated by a brief mechanical stimulus applied to the posterior end of the animal with a fine glass rod attached to a speaker (*, stimulus artifact). For other details of stimulating and recording methods, see Figure 2. (D) The LGF spike was evoked by applying a light tactile stimulus to the posterior end with a hand-held probe. Conduction velocities: (A–D) 6, 9, 13, and 7 m/sec, respectively. Voltage scale: (A, D) 20 μV; (B, C) 35 μV. Time scale: 5 msec.

review, see Beleslin, 1982). In comparison to *Myxicola,* however, values of giant fiber velocity and diameter in most polychaetes are much lower, typically in the range of 2–10 m/sec and 15–200 μm (Bullock and Horridge, 1965). As noted by Nicol (1948*b*), ratios of conduction velocity to diameter in polychaete giant fibers tend to be substantially lower than in oligochaetes, probably due to the greater degree of sheath development in the latter group.

In some sedentary polychaetes, bilaterally synchronous escape withdrawal is mediated by activity in a single medial giant fiber (e.g., *Myxicola*), but in others such responses may be mediated by spikes in paired giant fiber systems (e.g., *Eudistylia, Protula, Branchiomma*). In these paired fibers, a high probability of bilateral synchrony of spiking is maintained by "cross talk" between the fibers. The structures mediating this interaction are decussations of giant fiber neurites and various anastomotic connections, especially in anterior segments close to the normal site of giant fiber spike initiation (Bullock, 1953; Hagiwara *et al.*, 1964; Krasne, 1965). Mellon *et al.* (1980) have studied the physiological properties of the segmentally arranged commissural junctions between paired giant fibers of *Sabella* and conclude that cross talk between the fibers results from direct electrical coupling (Figure 4). However, whether the commissural junctions involve cytoplasmic fusion or gap junctions is not clear.

2.3. Efferent Pathways and Behavioral Correlates

Because there have been so few studies of the properties of giant fiber efferent pathways in polychaetes, it is difficult to make generalizations about their function. In light microscopic studies of *Nereis,* Smith (1957) and Horridge (1959) described apparent junctions between the lateral giant fibers and the axons of segmentally arranged pairs of giant motor neurons which innervate the longitudinal musculature. Electrophysiological recordings by Horridge (1959) showed that there was consistent 1:1 spike transmission between the lateral giant and giant motor axons. Initially, the giant motor neuron spike triggered a large muscle potential and longitudinal muscle twitch, but with repeated stimulation the potentials became smaller and twitches rapidly failed. With further repetitive stimulation, the 1:1 transmission at the giant–motor junction eventually failed. Subsequent to this failure, recruitment of the medial giant fiber restored firing of the motor axon. This suggests possible summation of inputs from different giant fibers systems onto the same motor neuron.

As noted by Horridge (1968), the efficacy of lateral giant to giant motor transmission is affected not only by repetitive stimulation, but also by the ongoing behavior of the animal. For example, in the absence of

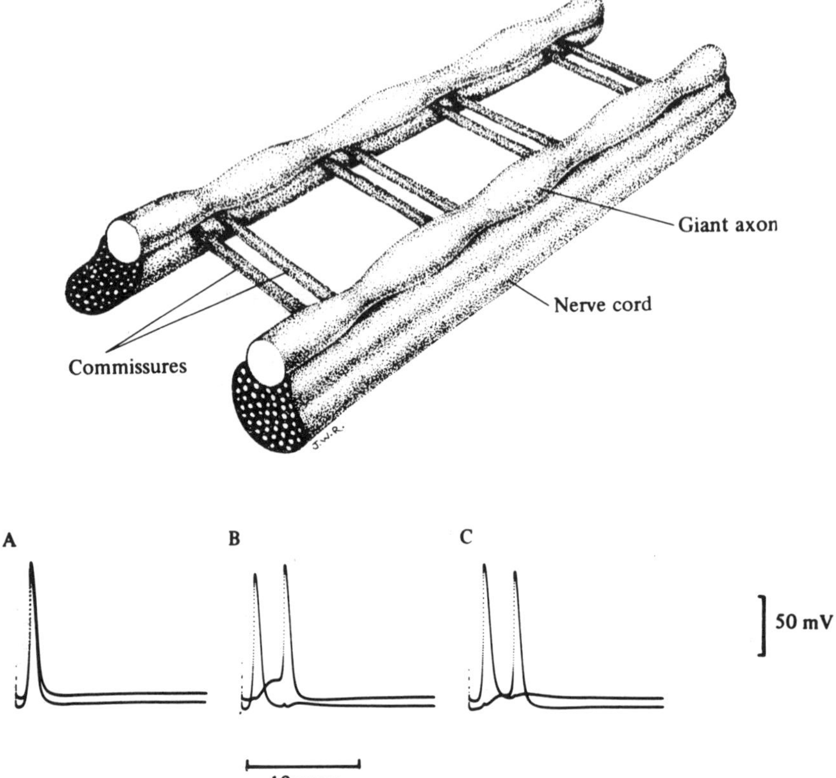

Figure 4. A diagram of the paired giant axons and lateral commissures in the nerve cord of the polychaete, *Sabella* (above). Superimposed intracellular electrical recordings from left and right giant axons of the same body segment are shown below. (A) Action potentials in the two giant axons occur synchronously in response to direct extracellular stimulation of both axons. (B) Direct stimulation of the right axon resulted in a delayed spike in the left axon. (C) Direct stimulation of the left axon resulted in a delayed spike in the right axon (from Mellon *et al.*, 1980).

crawling movements, spike transmission at the giant–motor junction was consistently 1:1, but during crawling movements lateral giant fiber spikes failed to excite the giant motor axon. These results suggest the possibility of a behavior-dependent depression of transmission at the giant–motor junction.

The timing of rapid escape movements in relation to giant fiber activity has not been extensively studied. Krasne (1965) noted that in *Branchiomma* escape withdrawal sometimes occurred without accompanying giant fiber activity, especially after repeated tactile stimulation. However, the latency of these nongiant-mediated reactions was always greater than 40 msec, or much longer than the shortest latency of giant fiber-mediated rapid responses. Such rapid withdrawal responses occurred 10–42 msec after the first spike in a train of synaptically evoked giant fiber spikes. The withdrawal response to a single spike was submaximal, whereas with repeated spiking, responses were graded in magnitude in relation to the number of giant fiber spikes. This gradedness of escape contractions has also been noted in *Nereis* (Dorsett, 1964), but is in contrast to the situation in *Myxicola* in which a single giant fiber spike evokes a rapid and complete withdrawal of the whole animal (Nicol, 1948*a*; Roberts, 1962*c*). The average latency from this single spike until onset of rapid withdrawal was 11 msec (Roberts, 1962*c*). This value, which is several milliseconds less than the minimal latency values for earthworm escape movements (Section 4.5), appears to represent the shortest efferent response time for any annelid escape reflex. One factor that undoubtedly contributes to minimization of this time in *Myxicola* is the anatomical arrangement of the efferent pathway in which innervation of the longitudinal muscle is by peripheral axonal branches that are in cytoplasmic continuity with the giant fiber system (Nicol, 1948*c*; Wells *et al.*, 1972, 1973).

In addition to longitudinal shortening, other rapid escape movements, especially involving parapodia, have been observed. For example, in nereids, a forward pointing of the parapodia accompanied giant fiber responses (probably MGF) to stimulation of anterior segments with touch (Horridge, 1959) or decreased light intensity (Gwilliam, 1969). In contrast, backward pointing of parapodia accompanied paramedial giant fiber responses to posterior tactile stimulation (Horridge, 1959). Unfortunately, the neural control of these movements has not been studied. In summary, there is a clear need for a better understanding of the precise timing and coordination of the various neural and behavioral events involved in polychaete rapid escape responses. At least some of these determinations could be made without employing surgical invasion, since many investigators have previously reported the possibility of detecting all-or-none giant fiber spikes from the body surface in a number of polychaete species (Bullock, 1945*b*, 1948; Nicol *et al.*, 1947; Mangum and Passano, 1964; Krasne, 1965).

3. Giant Fiber Reflexes in Leeches

Light tactile stimulation of anterior segments in a leech may evoke a variety of behavioral responses including a localized withdrawal involving only a few segments. Localized withdrawal would not necessarily require mediation by a rapid, through-conducting giant fiber system, and can be explained on the basis of the known monosynaptic connection from identified mechanoreceptive neurons (touch, pressure, and nociceptive) to segmentally arranged pairs of electrically coupled longitudinal motor neurons (Stuart, 1970; Nicholls and Purves, 1970; for review, see Muller, 1979). However, if an intense stimulus is applied to anterior segments, a rapid, symmetric shortening response involving the whole body may occur (Sawyer, 1981; Kramer, 1981; Kristan *et al.*, 1982). Such rapid responses, in which the body may shorten by as much as 70% (Sawyer, 1981), would appear to require involvement of some rapidly conducting, intersegmental pathway.

Leeches in fact do possess one relatively large fiber (Rohde's fiber = 3μm), or several such fibers, in the medial bundle (Faivre's nerve) of the ventral nerve cord (Coggeshall and Fawcett, 1964; Fernandez, 1978). In each ganglion the fiber arises from a single cell body (termed S cell; Frank *et al.*, 1975), with corresponding fibers from adjacent ganglia being electrically coupled to one another in the ganglionic connectives (Mistick, 1974; Bagnoli *et al.*, 1975*b*; Frank *et al.*, 1975). Electrical stimulation of the ventral nerve cord results in an all-or-none spike conducted in either direction along the S-cell network (or so-called "fast conducting system") at a velocity of 1.0–1.4 m/sec (Laverack, 1969; Bagnoli *et al.*, 1972, 1973). With afferent connections intact, S-cell spikes may be readily evoked by direct tactile stimulation anywhere along the animal, as well as by photic stimulation or low-amplitude water waves (Laverack, 1969; Bagnoli *et al.*, 1973, 1975*a*; Magni and Pellegrino, 1978*b*; Friesen, 1981).

Based on electrophysiological evidence, a direct excitatory connection between touch sensory neurons and S cells was proposed (Gardner-Medwin *et al.*, 1973; Bagnoli *et al.*, 1975*b*). Indeed, subsequent study by Muller and Scott (1981) confirmed that there is electrical coupling, albeit rectifying, between the touch neurons and S cells, but their morphological examination showed no areas of direct contact between processes of touch neurons and S cells (Figure 5). Rather, the cells appeared linked by a pair of small interneurons (Muller and Scott, 1981; Kramer and Goldman, 1981). The electrical junction between the touch neuron and small interneuron exhibited rectification, whereas the junction between the small interneuron and S cell was nonrectifying. The physiological significance of this arrangement is not yet clear, although one suggestion by Muller and Scott (1981) is that, depending on ongoing activity of the animal, the small connecting interneurons might function to appropriately modulate the efficacy of electrical transmission between touch neurons and S cells.

Friesen (1981) has shown that S-cell spikes are also readily evoked by stimulation of the body wall with low-amplitude water waves. Rather than activating touch neurons, these waves appear to activate a separate

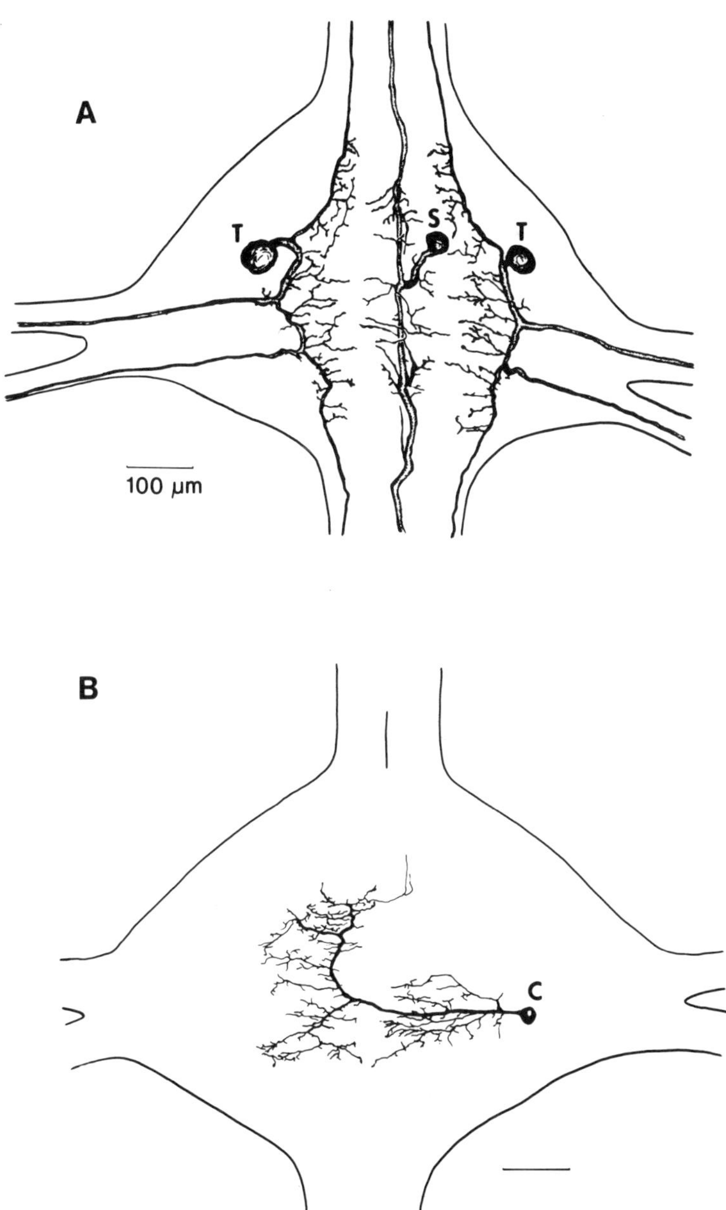

Figure 5. Morphology of touch cell, S cell, and a coupling interneuron in the ventral nerve cord of the leech, *Hirudo medicinalis*. (A) Two lateral touch cells (T) and the S cell (S) were injected with horseradish peroxidase (HRP). Note that the cells do not contact one another. (B) One of the two coupling interneurons (C) in the ganglion was injected with HRP. The processes of the interneuron span the gap between touch cell and S-cell processes (from Muller and Scott, 1981).

set of receptors, termed "sensillar movement receptors" (Friesen, 1981). Intracellular recordings of S-cell responses to wave stimulation indicate that spike initiation may require summation of movement receptor inputs. However, the details of the synaptic connections in this pathway and possible behavioral consequences of S-cell activation under these conditions have not been investigated.

The S-cell responses to photic stimulation are probably mediated by two different photosensory structures: five bilateral pairs of eyes on the dorsal surface of the head and seven bilateral pairs of sensilla on each of the 21 body segments (Kretz *et al.*, 1976). The rapid adaptation that characterizes S-cell responses to repeated photic stimulation (Laverack, 1969; Bagnoli *et al.*, 1973) is probably not attributable either to peripheral accomodation of the photoreceptors or to depressed excitability of the S cells (Kretz *et al.*, 1976). Since connections from photoreceptors to S cells are probably not monosynaptic (Bagnoli *et al.*, 1975*a*), failure of S-cell responses to repeated photic stimulation possibly could involve several sites along the afferent pathway.

There is some evidence for interaction between the converging mechanoreceptive and photoreceptive pathways. Bagnoli *et al.* (1973) found a suppression of S-cell responses to photic stimulation when this stimulation occurred within 120 msec after mechanical stimulation. However, the responses to photic stimulation were augmented if they occurred approximately 500–700 msec after mechanical stimulation. Chronic recordings from the ventral nerve cord of unrestrained leeches indicate that S-cell spiking does not occur during normal creeping, swimming, or respiratory movements, thus suggesting an absence of any reafferent excitation of S cells during these activities (Magni and Pellegrino, 1978*b*).

A recent study by Belardetti *et al.* (1982) has demonstrated another important sensory modality interaction in the S-cell system, namely between nociceptive and touch receptor inputs to the S cell. For example, if repeated mechanical stimulation is applied to the body wall, a rapidly developing and prolonged decrease in the number of S-cell spikes with each stimulus is observed. However, strong nociceptive stimulation to head segments can readily restore, or potentiate, the S-cell response to touch. Several kinds of experiments by Belardetti *et al.* (1982) demonstrate that serotonin and cyclic AMP (cAMP) may be involved in this potentiation effect. These include experiments showing that (1) serotonin application can potentiate previously decremented responses to either tactile stimulation of the body wall or electrical stimulation of segmental roots, (2) methysergide, an antagonist of serotonin, can block serotonin-induced potentiation of the S-cell response, (3) imidazole, a cAMP phosphodiesterase activator, can abolish the potentiation of S-cell firing by serotonin, and (4) application of db-cAMP, a cAMP analog, can potentiate the S-cell response to electrical stimulation of ganglionic roots.

The efferent connections of S cells include excitatory inputs onto motor neurons innervating annulus erector muscles and the so-called L-motor neurons innervating the longitudinal musculature (Gardner-Medwin *et al.*, 1973); however, only the S cell-to-L motor neuron pathway has been studied in any detail. Intracellular recordings have shown that the L cells receive excitatory input from the S cell (Magni and Pellegrino, 1978*a*; Kramer, 1981) and that the excitation probably involves an electrotonic junction (Gardner-Medwin *et al.*, 1973; Magni and Pellegrino, 1978*a*). Although Magni and Pellegrino (1978*a*) have shown that a single S-cell spike can frequently elicit a suprathreshold PSP (postsynaptic potential) in the L-motor neuron, detectable shortening does not usually accompany a single L-motor neuron spike (Stuart, 1970). The prerequisite for observable shortening appears to be a train of S-cell spikes. In dissected preparations of *Haementeria ghillianii,* intracellular stimulation of S cells at frequencies of 45 spikes/sec elicited shortening responses (Kramer, 1981). In contrast, in intact *Hirudo medicinalis*, a train of S-cell spikes at 60 spikes/sec (evoked by photic stimulation) was not accompanied by a clear-cut shortening response. However, higher frequency bursts of S-cell spiking in *Hirudo* (evoked by tactile stimulation) were accompanied by observable shortening (Magni and Pellegrino, 1978*a*). Taken together, these results seem to indicate that there is some critical number and frequency of both S-cell and L-motor neuron spikes required for evoking a just detectable shortening response; however, there may be important species differences in these critical values as well as alterations of these values due to dissection procedures. In any individual animal, gradations in the amount of shortening would occur as the number of frequency of S-cell and L-motor neuron spiking exceeded these critical values.

In addition to S-cell involvement in activating the body wall longitudinal musculature, Magni and Pellegrino (1975) have shown that repetitive S-cell firing is accompanied by a generalized contraction of longitudinal muscle fibers in the ventral nerve cord sheath. Subsequent study has shown that body wall and nerve cord shortening responses are mediated by the same set of segmentally arranged longitudinal motor neurons (Magni and Pellegrino, 1978*a*).

In summary, there is evidence that leeches possess a relatively large caliber, through-conducting pathway whose organization is reminiscent of the giant fiber systems of other annelids and whose functions include integration of inputs from a variety of stimulus modalities and subsequent mediation of a widespread reflexive shortening of the animal. Some particularly interesting topics for future study include further examination of the cellular basis of sensory modality interactions and plasticity of function in the shortening reflex, as well as study of the possible intrinsic modulation of reflex sensitivity in relation to ongoing behavior.

4. Rapid Escape Reflexes in Oligochaete Earthworms

Giant nerve fibers are found in the ventral nerve cords of many fresh water oligochaetes and nearly all earthworms (Stephenson, 1930; Prosser, 1934; Bullock and Horridge, 1965; Jamieson, 1981). However, it is only in the latter group that there have been any correlated anatomical, physiological, and behavioral studies relating to rapid escape reflexes. In this section some of the important anatomical specializations and physiological properties that underlie rapid escape reflexes of eathworms will be reviewed. Except when noted, discussion of the neuronal circuitry underlying earthworm escape reflexes will refer to the most extensively studied species, *Lumbricus terrestris* (Family Lumbricidae). Although detailed studies of escape reflex circuitry have not been made in most other earthworm families, we may infer from electrophysiological records, such as those in Figures 2 and 3, that there may be numerous common features of escape reflex design and function in most if not all earthworm families (Drewes *et al.*, 1983).

4.1. Anatomical Organization of Giant Fibers

Most earthworms possess three prominent giant nerve fibers: one medial fiber (MGF) and two lateral fibers (LGF). These are located in the dorsal portion of the ventral nerve cord (Figures 6 and 7). In *Lumbricus terrestris,* Günther (1971*b*) has also described a pair of smaller "ventral giant fibers" in the ventral portion of the cord, which appear to be organized as chains of segmentally arranged interneurons. The physiological role of the ventral giant fibers is unknown, although Günther noted that numerous collaterals of the ventral giants extend into the ventral longitudinal sensory bundles. In contrast to the dorsal giant fibers, there is no evidence that the ventral giants form a multisegmental through-conducting pathway.

In cross section, the MGF and LGF are easily identified by their large diameter and the presence of a well-developed myelinlike sheath that surrounds each fiber (for histological and ultrastructural descriptions of the sheath, see Taylor, 1940; De Robertis and Bennett, 1956; Hama, 1959; and Günther, 1976). In each segment Günther and Walther (1971)

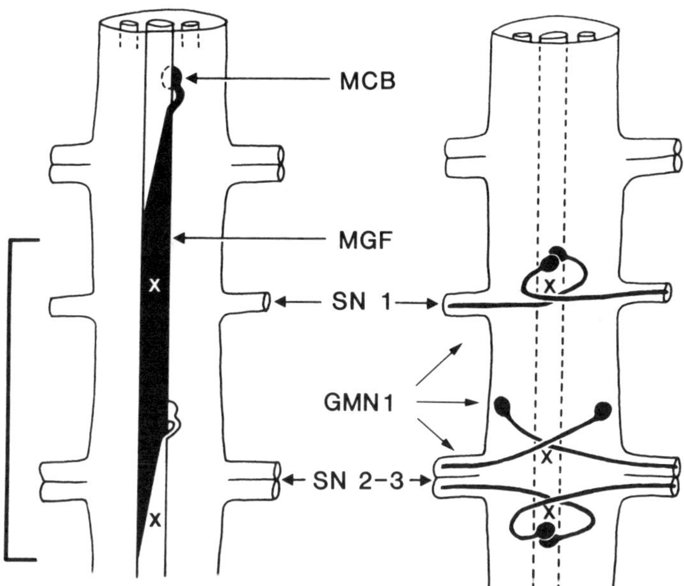

Figure 6. Diagram of general cellular organization of MGF and its associated giant motor neurons in the earthworm ventral nerve cord. (A) The medial giant cell body (MCB) and its neurite are shown displaced slightly to the right of the MGF proper. In each segment of the ventral nerve cord (bracket), three ventrally projecting MGF collaterals are found, their approximate positions being indicated by the two X's and by the neurite of the medial giant cell body. (B) Three pairs of giant motor neuron (GMN1) cell bodies are shown. The GMN1 axons decussate at the sites of the three MGF collaterals (X's), and pass out each of the segmental nerves (SN 1 and 2–3). See Section 4 for further details. (These diagrams are based on anatomical descriptions by Mulloney, 1970; Günther, 1971*a*, 1972; and Günther and Walther, 1971.)

noted that the sheath is interrupted ventrally by collateral projections arising from the MGF and LGF, the number and position of these collaterals being constant from segment to segment. Three ventral collaterals are associated with the MGF, two of which are relatively short and terminate in the central neuropil of the central nerve cord at the levels of the first and third segmental nerves. The remaining MGF collateral is longer and actually represents the neurite of the MGF cell body, which is located ventromedially in the nerve cord (Mulloney, 1970; Günther, 1971*a*; Günther and Walther, 1971; Schürmann and Günther, 1973).

Collectively, the MGF collaterals represent the sites for synaptic inputs and outputs (Günther and Walther, 1971; Günther and Schürmann, 1973). The main outputs of the MGF, to be discussed in Section 4.4,

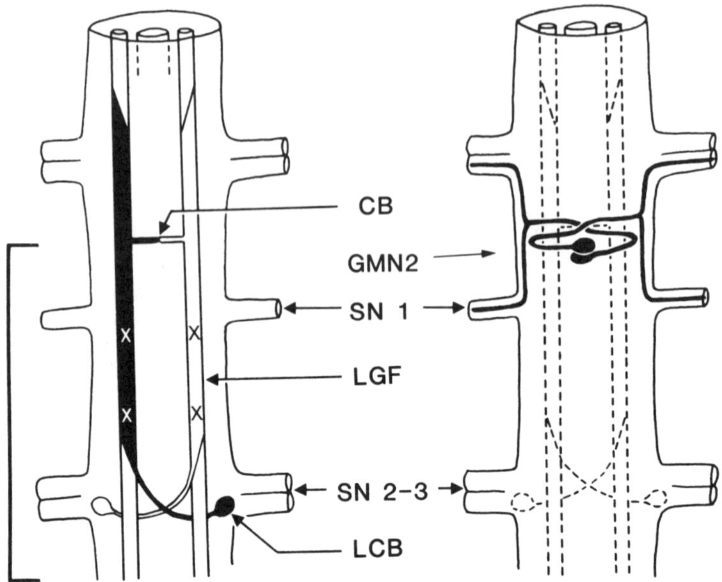

Figure 7. Diagram of the general cellular organization of the earthworm LGF and its associated giant motor neurons in the ventral nerve cord. (A) In each segment of the cord (bracket), the LGF system is derived from a pair of lateral giant cell bodies (LCB). The neurites of these cell bodies decussate and enlarge into the LGF proper. The four LGF collaterals in each segment include the lateral cross-bridge (CB), two ventral collaterals (X's), and the neurite of the lateral giant cell body. (B) The pair of giant motor neurons (GMN2) whose axons decussate and are closely associated with the LGF cross-bridge. Each GMN2 axon bifurcates and passes out segmental nerves 1 and 3 of adjacent segments. (These diagrams are based on anatomical descriptions by Mulloney, 1970; Günther, 1971*a*, 1972; and Günther and Walther, 1971.)

appear to be three pairs of giant motor neurons in each segment (Figure 6). With regard to possible afferent inputs, none of the three MGF collaterals appear to be directly associated with identifiable longitudinal sensory bundles in the ventral neuropil of the nerve cord. Rather, the association appears to be indirect, involving an interposed "giant interneuron," whose collaterals are associated with the ventral sensory neuropil and with the three MGF collaterals. The diameter of the paired, segmentally arranged giant interneurons, as well as the caliber and length of their collaterals, are greatest in anterior segments and steadily decrease in more posterior segments. Since these morphological gradients correlate with physiological evidence for anterior localization of the MGF sensory field, Günther and Walther (1971) have suggested that the giant interneurons function as afferent processing interneurons. Ultrastructural study of the junction between the giant interneuron and MGF suggests a possible electrical synapse (Günther and Schürmann, 1973).

Four identifiable collaterals project ventrally from each LGF in a segment. One of these collaterals, after branching, becomes the neurite of the LGF cell body, which is located ventrolaterally on the opposite side of the nerve cord (Günther, 1971*a*; Günther and Walther, 1971; Schürmann and Günther, 1973). Neurites arising from the paired LGF cell bodies in each segment usually come into close association with one another at a point of decussation ventral to the MGF (Mulloney, 1970; Günther, 1971*a*; Günther and Schürmann, 1973). Another of the four LGF collaterals projects somewhat medially where one of its branches forms a cross-bridge with the corresponding branch of the contralateral LGF. Thus there are two sites of close anatomical association between left and right LGFs in each segment and these involve branches from two different collaterals in each LGF. These collateral connections, in particular the cross-bridges (Mulloney, 1970), between left and right LGFs are undoubtedly the basis both for the observed electrotonic coupling between the two LGFs (Wilson, 1961) and the resultant bilateral synchronization of LGF spikes during propagation along the nerve cord (Rushton, 1945*a*).

Collectively, the LGF collaterals, as well as dendritic branches arising from the LGF cell bodies, represent the sites of synaptic inputs and outputs for the LGF (Günther and Walther, 1971; Günther and Schürmann, 1973). The outputs of the LGF system, to be discussed in a later section, include a pair of giant motor neurons in each segment (Figure 7). In regard to possible inputs, Günther and Walther (1971) noted that in posterior segments the LGF collaterals were relatively long and thick and projected into the ventral sensory neuropil. This suggests that afferent-to-LGF connections, at least in posterior segments, may be direct. However, apparent inputs onto the LGF from other sources, including the giant interneurons, were also noted (Günther and Walther, 1971; Günther and Schürmann, 1973). The mutual association of both LGF and MGF systems with the giant interneurons raises questions regarding possible MGF to LGF interactions, but there is yet no physiological evidence that spiking in one giant fiber system modifies in any way the properties of the other giant fiber system.

The anatomical associations between giant fiber cell membranes from adjacent segments typically consist of obliquely arranged partitions, or septa (Stough, 1926). There have been numerous ultrastructural studies describing these intersegmental membrane appositions and associated specializations in the MGF and LGF (Hama, 1959; Dewey and Barr, 1964; Coggeshall, 1965; Oesterle and Barth, 1973, 1981; Günther, 1975; Kensler *et al.*, 1979). In particular, the study by Kensler *et al.* (1979) using thin-section and freeze-fracture techniques has clearly demonstrated the presence of gap junctions between the LGFs of adjacent segments. Some demonstrated physiological properties that correspond with these ultrastructural observations include low transseptal resistance, nonrectification, and permeability of septa to small molecular weight dyes (Brink and Barr, 1977; Brink and Dewey, 1978, 1980; Kensler *et al.*, 1979). The nonrectifying properties of the septum account for the capability of bidirectional impulse conduction that was originally demonstrated in several classic studies of conduction in earthworm giant fibers (Eccles *et al.*, 1933; Bullock, 1945*a*; Rushton, 1945*a*).

4.2. Conduction Properties of Giant Fibers

In animals with highly elongated bodies, there can be a substantial saving in escape reflex time by maximization of giant fiber conduction velocity along the animal (for discussions, see Bullock, 1952; Kennedy, 1966; and Bennett, 1977). A number of expected, as well as unusual factors, contribute to the relatively high conduction velocities of earthworm giant fibers. These include: large diameter, influence of the sheath, formation of syncytia, and facilitation of conduction velocity.

Perhaps the most obvious factor contributing to rapid conduction is fiber diameter. In earthworm giant fibers the relationship between velocity and diameter appears to be linear over a wide range of fiber diameters (Adey, 1951; O'Gara *et al.*, 1982). The slope of the relationship is about 0.5; that is, a 10 m/sec increase in velocity for each 20 μm increase in diameter. This value is remarkably greater than for most other invertebrate giant fibers (cf. Table 3.4 in Bullock and Horridge, 1965). For example, in the giant axon of the squid stellar nerve, for which a linear velocity–diameter relationship was also found, the slope was only 0.065 (Hodes, 1953). The substantially higher value in the earthworm appears related, at least in part, to the presence of the myelinlike sheath that surrounds each giant fiber.

The sheaths of both MGF and LGF in earthworms are interrupted in each segment by ventrally projecting collaterals. However, in the MGF there are additional openings in the dorsal surface of the sheath (Günther, 1973*a*, 1976). These consist of two "nodes" (10–15 μm in diameter) in each segment. Günther's measurements of longitudinal currents along the surface of the MGF indicated the nodes act as current sources and sinks during action potential propagation. His records of electrically evoked spikes along the fiber indicated a step-wise increase in latency of 40–90 μsec at each node. Also, when a direct electrical stimulus was applied, the threshold current for an MGF spike at the internodal region was an order of magnitude greater than with stimulation at the nodal region. Based on these observations, Günther concluded that the dorsal nodes function as focal points for electrogenesis, and that there is a saltatory-type impulse conduction in the earthworm MGF.

Aside from giant fiber diameter and sheath influences, another factor that contributes to the minimization of conduction time is the formation of multineuronal syncytia by elimination of septal boundaries in the giant fibers (Günther, 1971*a*, 1975). Günther (1971*a*) found that about 60% of the septa were missing in the MGF and 20% in the LGF. One apparent consequence of septal elimination, as determined by Brink and Dewey (1980), is that conduction velocity across a septum is much slower (8 m/sec) than the velocity in the same fiber in the absence of a septum (25 m/sec). Assuming a conduction distance of 400 μm for both of these values and a uniform velocity of 25 m/sec along the fiber except across the septum, then the difference in conduction time (34 μsec) is an estimate of the actual delay time across the septum. This delay time, when multiplied by the number of missing septa in a worm (60% of MGF septa in a 150 segment worm), gives an overall time savings of more than 3 msec, which is a substantial percentage of the total MGF conduction time along the animal (usually about 5 msec).

Another factor that affects spike conduction velocity in earthworm giant fibers is antecedent spiking activity in the fibers. Bullock (1951) discovered that in the isolated ventral nerve cord of the earthworm, *Lumbricus terrestris* (as well as in nerve fibers of a polychaete, shrimp, and frog), giant fiber conduction velocity increased by as much as 10–20% for spikes occurring shortly after a "conditioning" spike. This phenomenon, referred to by Bullock as "facilitation of conduction velocity," was most pronounced about 6 msec after the conditioning spike, but was still evident at intervals greater than 100 msec. The same phenomenon, involving velocity increases of 20%, has been observed in recordings of naturally evoked MGF and LGF spikes in intact earthworms, *Lumbricus* and *Eisenia* (Drewes *et al.*, 1978; O'Gara *et al.*, 1982). One result of the phenomenon, as shown in Figure 8, is that the interspike interval between a given pair of spikes may gradually diminish as spikes are conducted along the animal. A physiological consequence of the reduced interspike interval, as discussed in a later section, is the increased probability for a pair of LGF spikes to trigger a giant motor neuron spike in anterior segments, where LGF interspike intervals are shortest. The biophysical mechanism that permits conduction at a supernormal velocity is unknown. However, as noted by Bullock (1951), the mechanism is probably unrelated to the presence of septa, since supernormal conduction velocity was also seen in a nonseptate polychaete giant fiber.

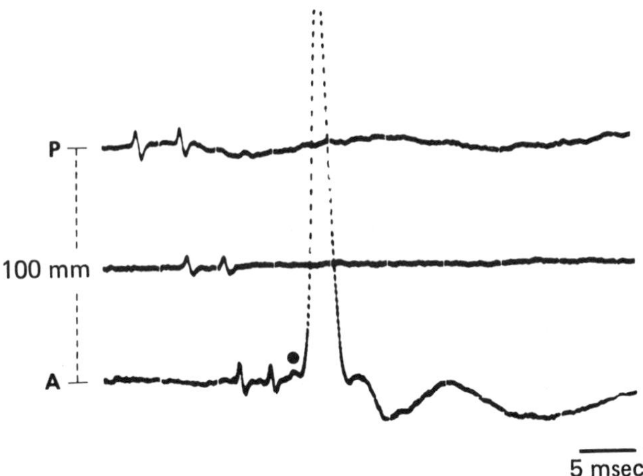

Figure 8. Reduction of interspike interval by facilitation of LGF conduction velocity in *Lumbricus terrestris*. A pair of LGF spikes, initiated by light tactile stimulation of the tail, was detected at three different recording sites along the intact animal: upper trace (segment 135), middle trace (segment 78), and lower trace (segment 33). The LGF interspike interval for the two spikes near the posterior end (P) was approximately 4 msec. The average conduction velocity of the first spike was about 11 m/sec and that of the second about 13 m/sec. Thus at the most anterior recording site (A), the interspike interval was reduced to less than 3 msec. At the anterior recording site the second LGF spike was followed by a GMN2 spike (•) and large longitudinal muscle potential (from Drewes and McFall, 1980).

Based on measurements from intact, unanesthetized worms (adult *Lumbricus terrestris*), typical values of normal, nonfacilitated conduction velocity are 30–38 m/sec for the MGF and 11–15 m/sec for the LGF (McFall *et al.*, 1977). Since these values are considerably higher than values from isolated cords (cf. Kao and Grundfest, 1957; Goldman, 1963, 1964; Lagerspetz and Talo, 1967), it appears that isolation procedures somehow adversely affect spike conduction properties. It is not yet clear to what degree such procedures might alter functioning of other escape reflex components such as synapses. Considering the marked lability of the reflex even in the intact animal, other substantial effects, either quantitative or qualitative in nature, are likely to accompany isolation procedures.

4.3. Afferent Pathways

In earthworms the MGF and LGF can be activated by a variety of abruptly delivered mechanical stimuli including body wall touch or pressure and substrate vibration (Bullock, 1945*a*; Adey, 1951; Morita and

Tateda, 1952; Günther, 1973*b*; Drewes *et al.*, 1978; Moore, 1979; Smith and Mittenthal, 1980). Unlike the giant fiber reflexes in polychaetes and leeches, there is not yet any experimental evidence for the adequacy of photic stimuli in evoking giant fiber spikes in earthworms, at least under laboratory conditions. However, a common field observation by those who nocturnally collect earthworms, such as *Lumbricus terrestris*, is that withdrawal responses of varying speeds are often evoked by abrupt onset of intense light. One question that remains unanswered is whether withdrawal responses to photic stimulation, under any circumstances, are giant fiber mediated. If so, are giant fibers excited directly via some photosensory to giant fiber pathway or, alternatively, could excitation involve an initial nongiant-mediated withdrawal followed by reafferent excitation of mechanosensory to giant fiber pathways, as suggested in some polychaetes (Gwilliam, 1969)? Failure of photic stimulation to evoke giant fiber spikes under any circumstances, of course, does not preclude the possibility that photic stimulation produces a subthreshold biasing of giant fibers toward spiking.

Functional distinctions between MGF and LGF afferent pathways are readily demonstrated by mapping of their respective sensory fields. Tactile stimulation of the most anterior segments invariably evokes MGF spikes, whereas stimulation of the most posterior segments invariably evokes LGF spikes (Figures 2 and 3). In *Lumbricus terrestris* the sensitivities of the MGF and LGF fields are especially high at the ends of the animals and diminish markedly toward middle segments (Günther, 1973*b*). Usually there is an area of substantial overlap between the two fields, this area being centered about one third of the way back along the animal (Günther, 1973*b*; Moore, 1979). The exact amount of overlap varies depending on factors such as interspecific differences, intraspecific variation, the nature of the stimulus, and the amount of previous stimulation.

At least some of the sensory neurons involved in giant fiber excitation have been identified and consist of serially homologous mechanosensory neurons with cell bodies in the ventral nerve cord and with axons projecting into segmental nerves. Based on anatomical criteria, these neurons have been classified into two groups (Günther, 1970, 1971*b*). One group, termed PN2, consists of five to seven bilaterally arranged pairs of neurons per segment. Axons from the PN2 neurons pass longitudinally through the third segmental nerve. The other group PN3, consists of two pairs of neurons, each neuron giving rise to two (or possibly three) axonal branches that exit ipsilaterally through the first and third segmental nerves of the same segment or through the first and third segmental nerves of adjacent segments.

Physiological differences have been demonstrated between the PN2 and PN3 groups (Günther, 1970; Smith and Mittenthal, 1980). Extracellular recordings from individual segmental nerves indicate that light tactile stimulation of the body wall (with a force greater than 3 *g*) evokes a phasic burst of spikes in which one or two large amplitude neural units can be readily distinguished. Simultaneous recordings from neighboring segmental nerves indicate that these spikes enter the cord through one segmental nerve and exit the cord via a neighboring ipsilateral segmental nerve (Figure 9). The large spike amplitudes, as well as the spatial and temporal distribution of these spikes in segmental nerves, indicate the spikes arise from the large-diameter, branched axons of the PN3 sensory neurons. Based on their responsiveness to relatively light mechanical stimulation, the PN3 neurons have been physiologically classified as touch-sensitive neurons (Günther, 1970; Smith and Mittenthal, 1980).

Physiological criteria for identification of PN2 spiking are also consistent with anatomical features (Günther, 1970; Smith and Mittenthal,

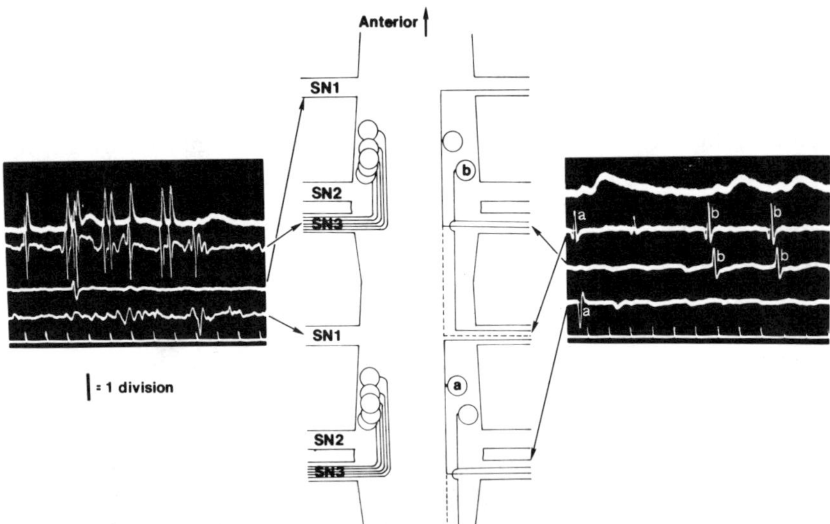

Figure 9. Afferent spiking and associated postsynaptic potential (PSP) activity in the LGF of *Lumbricus terrestris*. Two segments of ventral nerve cord, with the cell bodies and processes of touch neurons shown in the right half of the cord and those of pressure neurons shown in the left half. (Right) Recordings show typical responses of touch neurons from SN1 (intact) and two adjacent SN3s (cut). The top trace shows the associated PSPs in the LGF. Spikes labeled "a" derive from processes of cell "a" in the diagram; spikes labeled "b" derive from processes of cell "b." (Left) Recordings show typical responses of pressure neurons from SN3 (intact) and subsequent PSPs in the LGF (top trace). Time mark = 10 msec/div. One division = 2 mV for intracellular traces and 500 μV for extracellular records from segmental nerves (from Smith and Mittenthal, 1980).

1980). Extracellular electrical records from the third segmental nerve indicate that PN2 spikes are evoked by tactile stimulation of the body wall and consist of relatively large amplitude spikes from several different units (Figure 9). Unlike the responsiveness of PN3 neurons, PN2 spiking activity is tonic and requires relatively strong tactile stimulation (forces from 10–20 *g*). Based on these response patterns, the PN2 neurons have been classified as pressure-sensitive neurons.

The synaptic inputs from touch- and pressure-sensitive neurons onto MGF and LGF have been studied by Smith and Mittenthal (1980). In both MGF and LGF each touch neuron spike evokes a small depolarizing PSP that peaks after the touch neuron spike enters the ventral nerve cord. In some cases this short latency PSP is accompanied by a later and a slower PSP of more variable waveform. As shown in Figure 10, the amplitude of the early PSP in the MGF is greatest in anterior segments and absent in posterior segments. Conversely, in the LGF, the PSP is largest in tail segments and barely detectable in anterior segments. In middle segments both MGF and LGF receive inputs from touch neurons.

Synaptic inputs from pressure neurons onto MGF and LGF consist of unitary depolarizing PSPs, each being initiated approximately 3 msec after a pressure neuron spike. The constancy of PSP amplitude and the

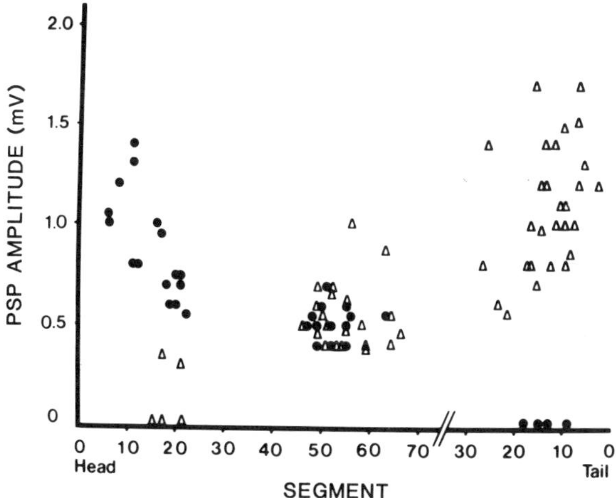

Figure 10. Amplitudes of early phase of giant fiber PSPs generated by touch neurons in *Lumbricus terrestris*. The ordinate indicates the PSP amplitude in mV (see Figure 9). Numbers before the break in the abscissa indicate the number of segments away from the head; numbers after the break indicate the number of segments away from the tail. Solid circles show values for MGF PSPs and triangles for LGF PSPs (from Smith and Mittenthal, 1980).

absence of PSP failures during high frequency bursts of pressure neuron spikes are suggestive of an electrical synapse between afferent and giant nerve fibers. Longitudinal gradients in PSP amplitude for pressure neuron inputs closely parallel those seen in Figure 10 for touch neuron inputs. The monotonic longitudinal gradients in the strength of the synaptic connections would appear to provide a physiological basis for the spatial arrangements and sensitivity gradients of the MGF and LGF sensory fields as determined by Günther (1973*b*). However, there is the possibility that other, as yet unidentified, mechanoreceptors may excite the giant fibers (Smith and Mittenthal, 1980), and perhaps these contribute to the observed gradients in giant fiber sensory fields. In addition, possibilities for segmental variations in the locus or threshold of giant fiber spike initiation have not been eliminated.

One important consideration in Smith and Mittenthal's study of the afferent to giant fiber pathway is the relatively small amplitude of giant fiber PSPs evoked by touch and pressure neuron spiking activity and the usual failure of high-frequency trains of PSPs to evoke giant fiber spikes in their dissected preparations. These observations are not easily reconciled with results from intact animals that show that giant fiber spikes are readily initiated by focal mechanical stimulation, with latencies from stimulus onset to giant fiber spikes being as short as 5 msec (Figure 3) (see also Günther, 1970; Moore, 1979). This amount of time is sufficient to allow generation of no more than one or perhaps two spikes in any given afferent neuron. Possible reasons for failure of PSP trains to evoke giant fiber spikes in dissected preparations might be a general depression of afferent to giant transmission or reduction in excitability of the giant fibers due to surgical trauma.

In view of the segmental repetition of afferent to giant connections, future studies of the intersegmental integration of afferent inputs would appear especially apropos. Intersegmental integration of inputs may be particularly important during stimulus conditions such as substrate vibration, which could result in nearly synchronous activation and spatial summation of afferent inputs from many different segments. Considering the relatively long space constant of the giant fibers (approximately 3–6 mm) (Goldman, 1964; Dierolf and Brink, 1973), spatial summation of inputs from points as far as 5–10 segments apart would be expected. Also, one would expect giant fibers to be particularly responsive to multisegmental inputs and less responsive to unisegmental input. This is because synaptic input currents occurring at a single locus would be more diffusely shunted throughout the giant fiber network. As a consequence of this shunting (loading), PSP amplitude would be reduced due to an effectively low input resistance and short time constant (Berry and Pentreath, 1977; Dorsett, 1980). In contrast, with synchronous inputs at several loci, there would be less loading of synaptic input currents because little or no voltage difference would exist at the points of input along the giant fiber. Consequently, the effective input resistance and time constant would be relatively great, making PSPs larger and longer.

4.4. Efferent Pathways

As shown in Figure 6, the MGF system is closely associated with three pairs of serially homologous giant motor neurons, termed GMN1 (Günther and Walther, 1971; Günther, 1972). In the middle of the nerve cord, each GMN1 axon makes contact with an MGF collateral and with the corresponding contralateral GMN1 axon at the point of axonal decussation. These connections undoubtedly help to assure bilateral synchrony in the excitation of GMN1 axons. Diameters of the GMN1 axons are 15 μm in the first and third segmental nerves and 9 μm in the second nerve. Corresponding to these relatively large diameters, extracellular spikes in GMN1 axons are distinguishable by their large amplitudes (1.5–3.0 mV) and rapid conduction velocities (1.25–1.85 m/sec) (Günther, 1972).

Electrophysiological studies of dissected as well as intact animals have shown that each MGF spike is coupled in 1:1 fashion to a GMN1 spike (Günther, 1972; Drewes *et al.*, 1980). The short time lag from the MGF spike in the cord until the GMN1 spike in segmental nerves (minimum = 0.6 msec), as well as the consistent 1:1 coupling with repetitive stimulation, suggest that an electrical junction is involved (Günther, 1972). If the MGF is electrically coupled to all three GMN1 pairs in each segment, then the sequential excitation of GMN1 axons during spike propagation along the animal may impose a substantial shunting or loading of MGF spike currents. The large diameter of the MGF and the resultant large spike currents may help to offset this tendency (Dorsett, 1980).

Descriptions of the innervation patterns of GMN1 axons along the animal are only partially complete. In anterior segments at least some GMN1 axons appear to innervate the longitudinal musculature, as indicated by the negative-going longitudinal muscle potentials that accompany GMN1 spikes (Günther, 1972); however, the possibility of some GMN1 axons innervating other muscle units, such as septal muscles, has not been eliminated. In tail segments of *Lumbricus terrestris,* Pallas and Drewes (1981) have shown that each MFG spike is coupled in 1:1 fashion with a large amplitude spike in a motor axon that appears to innervate the septal musculature. It is not yet clear whether these motor axons arise from a uniquely derived set of motor neurons or from motor neurons that are serial homologues of the GMN1 neurons innervating longitudinal muscle in other segments.

With repetitive stimulation in dissected preparations, GMN1-mediated muscle potentials show marked facilitation in amplitude (Günther, 1972). Likewise, in noninvasive recordings from the ventral surface of intact animals, facilitation of these potentials is clearly evident (Drewes *et al.*, 1980). Such recordings also show there is a pronounced longitudinal gradient in the amount of facilitation along the animal, the greatest amount occurring in the most anterior 60 segments and gradually less in more posterior segments (Figure 11). One hypothesis that would explain these observations is the occurrence of a longitudinal gradient in the tendency for facilitated transmitter release at GMN1 terminals along the animal.

As shown in Figure 7, a separate set of serially homologous giant motor neurons (termed GMN2) is associated with the LGF system. Each GMN2 axon makes contact both with the LGF system, at its collateral cross-bridge, and with the corresponding contralateral GMN2 axon, at the point of axonal decussation. The GMN2 axon then bifurcates, a rostral branch exiting the nerve cord via the third segmental nerve and a caudal branch exiting via the first segmental nerve of the next posterior segment. The diameter of GMN2 axons in the segmental nerves is about 9 μm. Corresponding to this relatively large diameter, spikes in GMN2 axons are identifiable in extracellular records by their large amplitude (about 1.5 mV) and rapid conduction velocity (1.25 m/sec) (Günther, 1972).

The transmission between LGF and GMN2 appears less reliable than between MGF and GMN1. In dissected preparations, the first LGF spike in a train does not trigger a GMN2 spike, but each LGF spike thereafter successfully triggers a GMN2 spike. The time lag between LGF and GMN2 spikes is somewhat variable, ranging from a minimum of 0.9 msec to more than 2.0 msec (Günther, 1972). The exact nature of the transmission process between LGF and GMN2 is unknown, but the close anatomical association of LGF and GMN2 is consistent with some type of monosynaptic transmission (Günther and Walther, 1971).

The peripheral connections of GMN2 axons appear to involve the longitudinal musculature in all segments. In dissected animals each GMN2 spike evokes a relatively large, negative-going longitudinal muscle potential that decreases in amplitude, rather than facilitates, during repeated stimulation (Günther, 1972). The initially large amplitude and subsequent antifacilitation of GMN2-mediated muscle potentials correspond to properties of the so-called "fast-axon"-mediated longitudinal muscle potentials evoked by direct segmental nerve stimulation, as described by Drewes and Pax (1974). The initially rapid antifacilitation of these potentials may

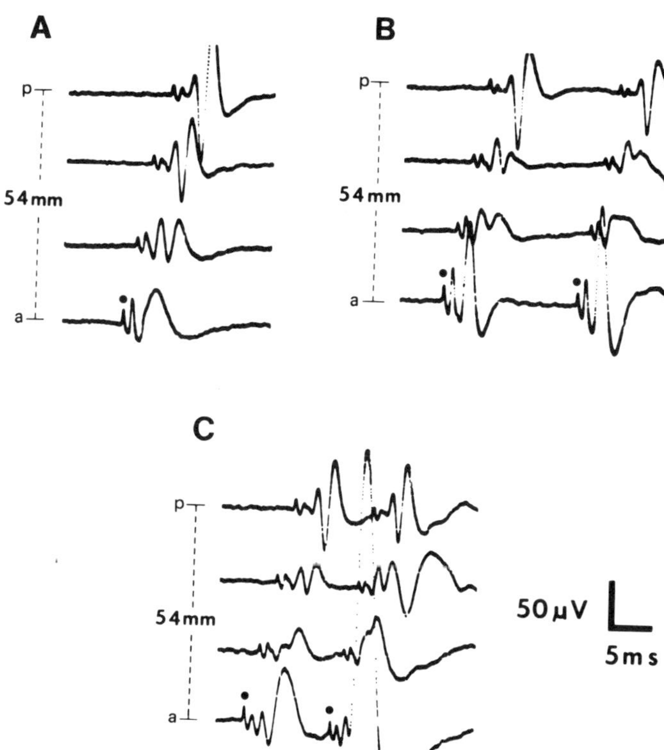

Figure 11. MGF activity in an adult *Eisenia foetida* recorded simultaneously at four sites along the intact animal. The most anterior recording site (a) (approximately segment 20) was 54 mm from the most posterior site (p) (approximately segment 85). The distance between adjacent recording sites was 18 mm. (A) The single MGF spike (•) was evoked by very light tactile stimulation of the head. The spike was conducted posteriorly at a velocity of 10 m/sec. At each recording site the MGF spike was followed after approximately 1.5 msec by an apparent motor neuron spike and longitudinal muscle potential. (B) The muscle potentials accompanying two widely spaced (interspike interval, 16 msec) MGF spikes (•) showed little or no facilitation in amplitude. (C) The muscle potentials accompanying two closely spaced (interspike interval 10 msec) MGF spikes (•) showed marked facilitation, especially in the most anterior record (from O'Gara *et al.*, 1982).

be attributed to some type of synaptic depression at the GMN2 neuromuscular junctions.

The potentials generated by LGF and GMN2 spiking are sufficiently large to permit detection with electrodes in contact with the ventral surface of intact animals (Figure 8). Such *in vivo* recordings have shown that the effectiveness of LGF to GMN2 transmission is not uniform along the animal. The transmission is most effective in anterior segments where a single LGF spike occasionally triggers a GMN2 spike and a pair of closely spaced LGF spikes (e.g., 4–6 msec interspike interval) is nearly always effective in triggering a GMN2 spike. The efficacy of transmission diminishes in a monotonic anterior-posterior gradient along the animal, except in extremely posterior segments where there is an increase in efficacy (Drewes and McFall, 1980).

Besides the longitudinal gradient, another factor that influences the likelihood of GMN2 activation in a segment is the interspike interval between a pair of LGF spikes. For example, a pair of LGF spikes with an interspike interval of 2–4 msec effectively triggers a GMN2 spike in more than 80% of the tests in segments 31–60. However, if interspike intervals are 12–14 msec, GMN2 spikes are triggered in only about 5% of the tests from the same region (Drewes and McFall, 1980). Because of the previously mentioned effects of facilitation of giant fiber conduction velocity, the interspike interval between pairs of closely spaced LGF spikes is not constant as spikes propagate along the animal (Figure 8). For spikes initiated in posterior segments, the reduction in interspike interval may be as much as 1–2 msec in anterior segments. This effect tends to further accentuate the high probability of LGF to GMN2 transmission due to longitudinal gradient effects.

4.5. Rapid Escape Movements

The rapid response of an earthworm to abrupt tactile stimulation of anterior segments consists of several components, the most obvious being a rapid and vigorous longitudinal shortening away from the stimulus. In some species the shortening is accompanied by a characteristic dorsoventral flattening and a forward pointing of setae in tail segments, which undoubtedly function to anchor the animal's tail (Bovard, 1918; Studnitz, 1937; Rushton, 1945*b*). Neural correlates and timing relationships have been examined in regard to the shortening and flattening components in some species, but there has been no study of the neural control and coordination of setal movements during escape.

In intact earthworms (*Lumbricus terrestris* and *Eisenia foetida*) tested under laboratory conditions, the firing of a single MGF spike evoked by light tactile stimulation is always accompanied by the sequential firing of GMN1 spikes and small longitudinal muscle potentials in each segment along the body, but this amount of activity is insufficient to produce rapid shortening of the body (Drewes *et al.*, 1980; O'Gara *et al.*, 1982). Likewise, GMN1-mediated muscle potentials remain small, and no shortening is observed in response to two widely spaced MGF spikes (e.g., 20 msec interspike interval). However, when two more closely spaced MGF spikes occur (8–12 msec interspike interval), GMN1-mediated muscle potentials facilitate in anterior segments and a corresponding shortening of these segments is seen. Thus a prerequisite for MGF-mediated shortening appears to be a longitudinal muscle potential of some critical amplitude, and due to the anterior-posterior gradient in muscle potential facilitation, this shortening tends to be localized within anterior segments. Rapid shortening can occur in middle segments, but usually requires a train of three or more rather than two, closely spaced MGF spikes. Furthermore, responses in middle segments are consistently weaker and typically occur 5–10 msec later than in anterior segments. The focusing of MGF-mediated shortening into anterior segments seems reasonable because these segments are most vulnerable to attack as the animal's head is extended from its burrow.

The tail flattening that accompanies responses to anterior stimulation is also mediated by the MGF, the effectors for flattening being the septal muscles (Pallas and Drewes, 1981). As in the case with anterior shortening, tail flattening in intact animals requires at least two antecedent MGF spikes. However, for detectable flattening to occur, the spikes need not be as closely spaced. For example, some flattening may occur if the interval between a pair of MGF spikes is as long as 20 msec. A comparison of the timing of tail flattening in relation to anterior shortening has been made by Pallas and Drewes (1981). Their results from intact animals indicated that the onset of tail-flattening responses to anterior tactile stimulation occurred 1–16 msec before the onset anterior shortening, in spite of the fact that the antecedent MGF spikes were initiated in head segments. This difference in timing may enable the animal to escape more effectively because tail flattening would help anchor the animal's tail in the burrow before longitudinal shortening of anterior segments occurs. The physiological mechanisms that permit flattening responses to precede shortening are not known, but could involve differences in the electrical properties of the septal and longitudinal muscle groups or differences in the excitation–contraction coupling properties of these muscle groups.

One similarity between the shortening and tail-flattening components of escape is that the amplitude and onset latency of both components are graded in relation to the number and frequency of antecedent MGF spiking (Roberts, 1962*a*, 1966; Drewes *et al.*, 1980; Pallas and Drewes, 1981). The grading of rapid escape movements would seem advantageous for a burrowing animal that by its own locomotory movements may inadvertently encounter stimuli that are suprathreshold for giant fiber activation, but are otherwise not sufficiently threatening to require a vigorous withdrawal of the entire animal.

Based on data from studies of the MGF afferent and efferent pathways, a total reflex time can be calculated for the earliest behavioral component of escape (i.e., tail flattening). The minimal time from the delivery of a mechanical stimulus to onset of an MGF spike in the most anterior segments is approximately 5 msec. Assuming an average MGF velocity of 30 m/sec and an animal length of 150 mm, the total conduction time along the animal is 5 msec. Finally, the minimal time from the occurrence of the first MGF spike in tail segments until onset of tail flattening is approximately 15 msec (Pallas and Drewes, 1981). Thus a minimal estimate of total reflex time for rapid tail flattening in response to anterior stimulation is 25 msec, with longitudinal shortening occurring a few milliseconds thereafter.

Under laboratory conditions, the most obvious response to strong tactile stimulation of the earthworm's tail is a rapid and vigorous end-to-end shortening, which is especially pronounced in posterior segments (Studnitz, 1937; Rushton, 1945*b*; Drewes and McFall, 1980). The underlying neural basis of this shortening appears to be a high-frequency train of LGF spikes accompanied by repetitive firing of GMN2 spikes and longitudinal muscle potentials along the entire animal (Drewes and McFall, 1980). Although other movements, such as setal movements, may accompany high frequency LGF firing, these movements have not been carefully described.

The laboratory observation that vigorous withdrawal to tail stimulation is focused into posterior segments is difficult to reconcile with field observations that suggest that an anterior withdrawal occurs in response to tail stimulation. For example, Rushton (1945*b*), in describing nocturnal escape responses of worms extended from their burrows, noted that attempts to seize the worm near its posterior end (presumably within the LGF sensory field) resulted in an anterior withdrawal resembling that seen in response to touching the head. In the absence of any neurophysiological data, these observations are not easily explained, but several possibilities exist. One suggestion, by Rushton, is that posterior stimulation activates the MGF rather than LGF system due to some switchover in the escape reflex circuitry. Possibly, some type of long-lasting excitation or bias of the MGF in posterior segments could enhance or unmask MGF responsiveness to afferent inputs in these segments (Günther, 1973*b*; Smith and Mittenthal, 1980).

An alternative is that only the LGF system is activated by posterior touch, but LGF responses in the field are different than under laboratory conditions. For example, following LGF activation in the field, GMN2-mediated muscle potentials in anterior segments could be larger and more effective in triggering contraction than potentials occurring under laboratory conditions. Since LGF–GMN2 coupling is known to be highly effective in anterior segments (Drewes and McFall, 1980), LGF spiking might then trigger a pronounced anterior shortening, resembling that evoked by anterior stimulation. A similar anterior focusing of LGF- and GMN2-mediated shortening may provide the basis for possible LGF-mediated escape responses to substrate vibration as the animal is extended from its burrow.

Finally, there is a possibility that posterior seizing of the animal could initiate responses mediated by both MGF and LGF systems, either simultaneously or sequentially. Sequential activation could involve some type of reafference, whereas simultaneous initiation of MGF and LGF responses could occur following substrate vibration of the whole animal. Simultaneous initiation of MGF and LGF spikes at opposite ends of the animal, in response to substrate vibration, has been often observed under laboratory conditions (C. D. Drewes, unpublished data).

4.6. Habituation

The giant-fiber-mediated reflexes in earthworms, as in other annelids, readily habituate during repeated stimulation (for review, see Dyal, 1973). Roberts (1960, 1962*b*) concluded that the main sites of MGF reflex lability were the sensory–giant and giant–motor junctions. However, Roberts' studies were done without the benefit of more recently obtained information regarding physiological properties of identifiable afferent and efferent neurons associated with the earthworm giant fiber systems (Günther, 1970, 1972; Günther and Walther, 1971). For example, coupling between MGF and GMN1 spikes remains 1:1 during trains of MGF spikes and fails only after many cycles of repetitive stimulation (Günther, 1972; Drewes *et al.*, 1980). Thus, assuming there is no intrinsic modulation of the efficacy of this coupling, the MGF–GMN1 junction does not appear to be a likely site for MGF reflex lability. On the other hand, if the MGF afferent pathway involves a preprocessing giant interneuron, as suggested by Günther and Walther (1971), then several possibilities exist for labile transmission on the afferent side of the reflex. An additional, and perhaps no less critical, site of MGF reflex lability is the GMN1–muscle junction. As shown by Günther (1972), there is an initial facilitation of GMN1-mediated muscle potentials with the first few cycles of repetitive stimulation, but this is soon followed by a pronounced and long-lasting fatigue of the muscle electrical responses. Reflex lability has not been studied in any detail in the LGF system, but could involve one or more sites in the afferent or efferent pathway, including LGF–GMN2 transmission.

Clearly, there is a need for further study of escape behavior modification in earthworms including the possibility of phenomena such as dishabituation. Future investigators in this area may look forward, either happily or with regret, to a variety of complicating factors.

1. There are two different giant fiber systems, both highly sensitive to some of the same as well as different stimulus modalities.
2. The reflexes involve at least three probable sites of lability including sensory–giant, giant–motor, and motor–muscle synapses.
3. There are important segmental variations in the functional properties of the afferent and efferent pathways of each giant fiber reflex; thus qualitative and quantitative differences in reflex lability are likely to occur in different segments of the animal.
4. Reflex properties may be modulated in relation to ongoing locomotory activity (e.g., forward or reverse peristalsis) or possibly during copulatory activity (Andrews, 1895).
5. Physiological properties of the reflex may be significantly altered by dissection procedures.
6. Intraspecific, interspecific, age-dependent, daily, or seasonal variations in reflex properties are likely to occur.

Despite this plethora of potentially complicating factors, the demonstrated capability of obtaining repeated and long-term electrophysiological assessments of giant fiber reflex function from intact animals offers some promise for systematic study of the influence of virtually all of these factors on escape reflex function and behavioral plasticity.

5. Growth and Development of Earthworm Giant Fiber Systems

Although there have been relatively few studies of giant fiber reflex development in earthworms, there appear to be several experimental advantages offered for such studies. First, anatomical studies of neuronal differentiation and maturation are favored due to the relatively large size and identifiability of the giant nerve fibers throughout development (Ogawa, 1939). Second, recordings of giant fiber spiking activity can be readily obtained from intact animals throughout postembryonic development and even during embryonic stages when giant fiber diameters are only a few microns (O'Gara *et al.*, 1982). Third, the metameric body plan offers advantages for studying longitudinal gradients in the structure and function of serially homologous neurons involved in the reflex. Fourth, earthworms possess a significant capacity for rapid regeneration of giant fibers following either specific nervous system lesions or loss of entire body segments.

5.1. Embryonic and Postembryonic Development of Escape Reflexes

The development of earthworm giant fibers has been most extensively studied in *Eisenia foetida*. Prosser (1933) noted that giant fiber pathways in *Eisenia* become histologically identifiable in anterior segments during the next to last embryonic stage, termed weak-crawling stage. During the last embryonic stage, termed normal-crawling stage, the giant fibers extend the entire length of the cord and end-to-end contractions become evident in response to tactile stimulation. These results correlate well with electrophysiological studies by O'Gara *et al*. (1982), who concluded that MGF and LGF reflex pathways are functional in normal-crawling but not weak-crawling embryonic stages. These conclusions were based on the observations that in normal-crawling embryos, MGF and LGF spikes are triggered by tactile stimulation, are conducted along the entire animal, and are accompanied by stereotyped motor activity patterns and rapid contractions (Figure 12).

Velocities and diameters of *Eisenia* giant fibers during the normal-crawling stages are MGF, 1.1–1.6 m/sec (6–7 μm) and LGF, 0.7–1.1 m/sec (2.0–3.5 μm) (O'Gara *et al*., 1982). Although these values are considerably less than those in adult worms, numerous other features of the escape

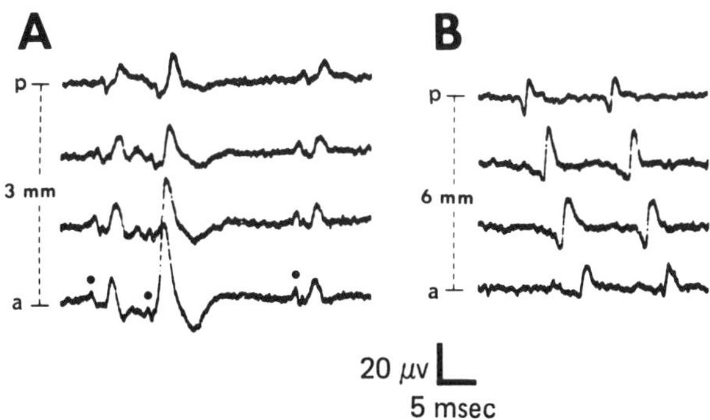

Figure 12. MGF and LGF activity recorded from the normal-crawling stage of intact embryos of *Eisenia foetida*. (A) A light tactile stimulus to the head evoked three posteriorly conducted MGF spikes (•). Each spike was followed by a larger, slower potential, presumably a GMN1 mediated longitudinal muscle potential. The interspike interval (about 7 msec) between the first two MGF spikes was sufficiently short that facilitation of the second muscle potential was seen, especially at the most anterior recording sites. (B) Light tactile stimulation to the tail evoked two anteriorly conducted, all-or-none LGF spikes. The conduction velocity of the LGF spike (about 0.9 m/sec) was slightly less than the MGF velocity (about 1.1 m/sec) (from O'Gara *et al*., 1982).

reflex activity in embryonic worms appear adultlike. These include (1) the localization of MGF and LGF receptive fields into anterior and posterior regions, (2) a longitudinal gradient in MGF conduction velocity, (3) 1:1 coupling of MGF spikes to giant motor neuron spikes and muscle potentials, (4) an anterior-posterior gradient in facilitation of longitudinal muscle potentials during repetitive MGF spiking, and (5) occurrence of rapid shortening only in response to repeated giant fiber spiking and its accompanying large amplitude muscle potentials. The observation that each MGF spike is invariably coupled to motor neuron activity, even at about the time that MGF spikes are initially through-conducted, suggests that functional connections between MGF and its segmentally arranged giant motor neurons are completed at about the same time or before the MGF system becomes through-conducting.

The most obvious changes in the functional properties of the giant fibers during postembryonic development are age-dependent increases in MGF and LGF conduction velocity (Figure 13). Velocity increases are

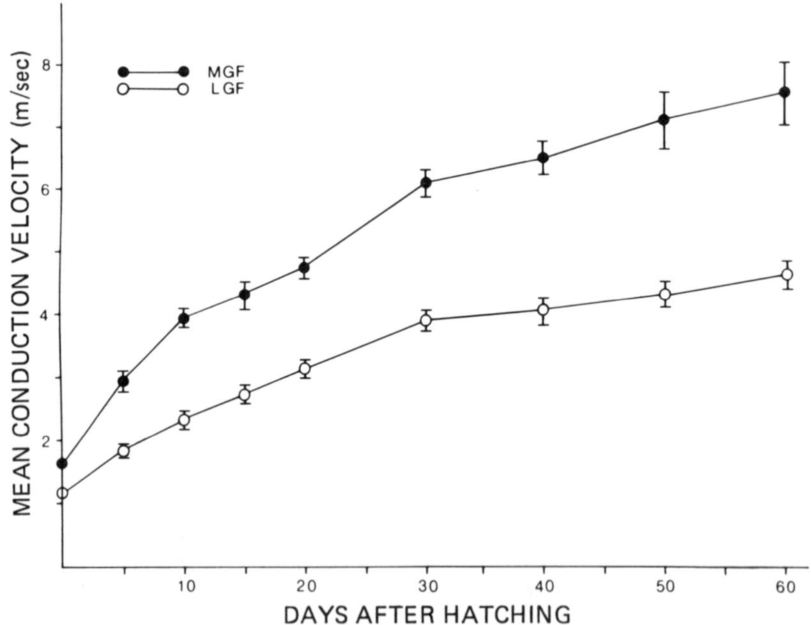

Figure 13. Relationship between giant fiber conduction velocity and age (days after hatching). Each point represents the mean velocity (± SEM) of 12 animals (*Eisenia foetida*). The same worms were tested repeatedly for the 60 days after hatching (from O'Gara *et al.*, 1982).

most rapid during the first 10 days after hatching, and then more gradual for the next 50 days or more. These velocity increases correlate with increases in fiber diameter, such that the velocity–diameter relationship appears linear for fibers of different ages. Although diameter appears to be the main factor contributing to velocity increases, it may not be the only maturational change involved. For example, age-dependent increases in sheath thickness may contribute to the observed increase in giant fiber conduction velocity. In addition, Günther (1971*a*, 1975) suggested that there may be an age-dependent dissolution of giant fiber septa during postembryonic maturation. If septal dissolution along the whole worm occurred within a relatively short time span during postembryonic development, this could account for the slight upward inflection in the age-dependent increases in MGF velocity seen between 15 and 30 days after hatching (Figure 13). On the other hand, if elimination of septa occurred gradually throughout development, then velocity increases due to septal elimination would not be easily resolved in the plots of velocity vs. age.

Aside from age-dependent changes in giant fiber function, a number of other important questions pertaining to escape reflex development remain unanswered. For example, it is not known whether there are age-dependent changes in the structure or function of the sensory neurons involved in escape reflex circuitry. Such changes might be expected in view of the scanning electron microscopic studies by Moment and Johnson (1979) that showed increases in the number and size of epithelial sensory buds during development from newly hatched to adult stages in *Eisenia foetida*. Although Moment and Johnson (1979) have suggested a possible mechanoreceptive function for these buds, there have been no studies that would implicate these structures in giant fiber escape reflex function.

5.2. Food Deprivation Effects on Giant Fiber Growth

Studies of the influence of food deprivation on growth and maturation of earthworm giant fibers are facilitated because worms are easily reared under a variety of growth conditions and because repeated assessments of giant fiber conduction parameters can be made from the same animals throughout postembryonic development. In well-fed animals, MGF and LGF velocities and diameters increase by a factor of five during postembryonic growth. By comparison, body weight increases by more than 100 times during the same period (Vining *et al.*, 1982).

As one might expect, growth of giant fibers (indicated by conduction velocity increases) as well as body weight are significantly reduced, or stunted, by food deprivation. Figure 14 shows that progressively greater

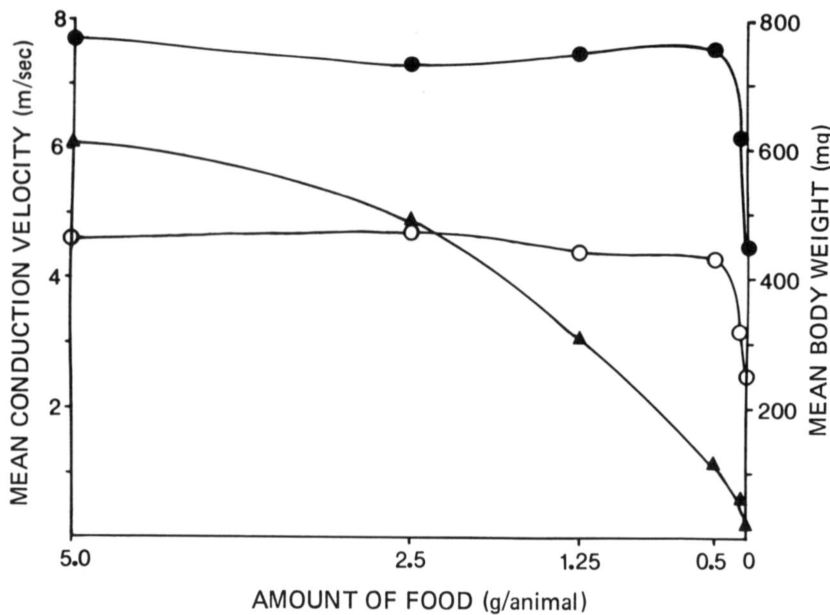

Figure 14. Effects of food deprivation on giant fiber conduction velocity and somatic growth in *Eisenia foetida*. The results were obtained from six different groups of animals (six animals/group). Beginning at hatching, each group received a different amount of food, ranging from 5 g food/worm to 0.1 g food/worm. The group means for conduction velocity and body weight were then obtained for all worms after 42 days of rearing under the various levels of food deprivation. MGF (●) and LGF (○) conduction velocities were essentially unaffected at reduced food levels ranging from 5.0 to 0.5 g food/worm, but marked decreases in velocity occurred at levels below 0.5 g. The SEM values for velocities in all groups were within the range of 0.2–0.3 m/sec for MGF and 0.1–0.2 m/sec for LGF. Body weights (▲) were markedly reduced at each food level below 5 g. Standard deviations of the body weights for the six groups, from highest to lowest food levels, were 79, 128, 96, 27, 18, and 9 mg, respectively) (from Vining *et al.*, 1982).

stunting of body weight occurred as food deprivation increased from moderate to severe levels. In contrast, stunting of giant fiber conduction velocity occurred only during the most extreme levels of food deprivation. These results suggest that there is a priority of giant fiber growth, in particular radial growth of the fibers, during periods of low food availability. The mechanisms that permit the giant fibers to be spared the growth-stunting effects of food deprivation are not known; nor is it known whether this sparing involves the entire nervous system in general or only the giant nerve fibers.

Based on experiments in which food is replenished after long periods of deprivation, it is evident that neither stunting of giant fiber conduction velocity nor stunting of body weight are irreversible (Vining *et al.*, 1982). For example, when newly hatched worms were severely food deprived for more than 60 days following hatching, marked stunting of both giant fiber conduction velocity and body weight occurred. When food was then suddenly made available, rapid somatic and giant fiber growth ensued, with growth rates resembling those seen in newly hatched, well-fed worms. In contrast, when newly hatched worms were only moderately food deprived, body weight but not giant fiber conduction velocity appeared stunted. Subsequently, replenishment of food was accompanied by rapid somatic growth, but there was no concomitant increase in giant fiber conduction velocity. This provides further evidence for the independent control of somatic growth and giant fiber growth.

Despite the stunting of giant fiber conduction velocity during extreme food deprivation, the capability of giant-fiber-mediated reflex responses to tactile stimulation persisted for many weeks in starving worms (Vining *et al.*, 1982). This maintenance of escape reflex function during food deprivation might be expected in view of the wide variation of food availability that is likely to occur in nature and the importance of the rapid escape reflex to the worm's survival.

5.3. Regeneration of Giant Fibers

One factor contributing to the rapidity of giant nerve fiber regeneration in annelids is the lack of significant giant fiber degeneration in segments anterior and posterior to a ventral nerve cord lesion, a situation undoubtedly related to the segmental origin of the interneurons forming the giant fiber systems. This lack of degeneration has been demonstrated in the giant fibers of several species of earthworms (Bovard, 1918; Hall, 1921; Birse and Bittner, 1976, 1981), in polychaete giant fibers (Nicol, 1948c; Hulsebosch and Bittner, 1981), and in the S cell of leeches (Frank *et al.*, 1975; Carbonetto and Muller, 1977). In the case of leeches, the failure to degenerate includes not only the S cells in segments adjacent to the lesion, but also the distal stump of the specific S-cell axon that is severed (Frank *et al.*, 1975; Carbonetto and Muller, 1977).

The survival of giant fibers in segments anterior and posterior to a lesion creates a favorable situation for successful regeneration because the distances over which reconnecting fibers must grow are relatively short, often less than 1 mm. It should be noted, however, that failure of giant fiber degeneration in segments adjacent to a lesion does not necessarily ensure successful regeneration, as shown in several species of polychaetes (Nicol, 1948*c*; Hulsebosch and Bittner, 1981).

When regeneration does occur, it generally involves outgrowth of multiple processes from one end or in some cases from both ends, of the severed fiber (Frank *et al.*, 1975; Carbonetto and Muller, 1977; Birse and Bittner, 1976, 1981; Hulsebosch and Bittner, 1981). It is these small-caliber processes that reunite the severed ends of the fibers and form the anatomical substrate for re-establishment of spike conduction across the lesion. As recovery progresses, these processes may increase in diameter and decrease in number (Birse and Bittner, 1981).

In earthworms, the time required for establishment of successful spike transmission across the lesion is variable and depends on individual variations in the regeneration process and on testing procedures employed. Use of noninvasive electrophysiological recording methods has shown that recovery of spike through-conduction occurs very rapidly, especially in the MGF. In *Lumbricus terrestris,* MGF through-conduction is re-established as early as 5 days and usually within 7 days after severing the giant fibers (Balter *et al.*, 1980; Pallas and Drewes, 1981). Even more rapid recovery is seen in *Eisenia foetida,* in which through-conduction of MGF spikes is re-established as early as 20 hr and usually within 48 hr after severing the ventral nerve cord (Vining and Drewes, 1982). During the initial period of functional recovery, properties of spike transmission across the lesion in *Lumbricus terrestris* include: (1) intermittently successful spike transmission across the lesion, (2) conduction delays of several milliseconds within the lesioned area, (3) occurrence of "rebound" spikes (Figure 15), and (4) similarity of conduction properties in both directions across the lesion (Balter *et al.*, 1980). Interestingly, all of these properties are highly reminiscent of the transmission properties associated with the small-diameter transverse junctions that normally connect pairs of nonregenerating giant fibers in polychaetes (Bullock and Turner, 1950; Bullock, 1953; Mellon *et al.*, 1980). After several weeks of recovery, spike transmission across lesioned earthworm giant fibers appears more normal, as indicated by an increased tendency for 1:1 spike transmission across the lesion, a reduction in the conduction delay within the lesion, and a decreased tendency for rebound spiking (Balter *et al.*, 1980).

The *in vivo* conduction properties of regenerating earthworm giant fibers, as determined by Balter *et al.* (1980), do not coincide exactly with *in vitro* properties of regenerated giant fibers, as determined in isolated ventral nerve cord preparations (Birse and Bittner, 1976, 1981). In the latter studies, somewhat longer times were required for successful through-conduction, and spike transmission was often not bidirectional. These differences between *in vivo* and *in vitro* performance of regenerated giant

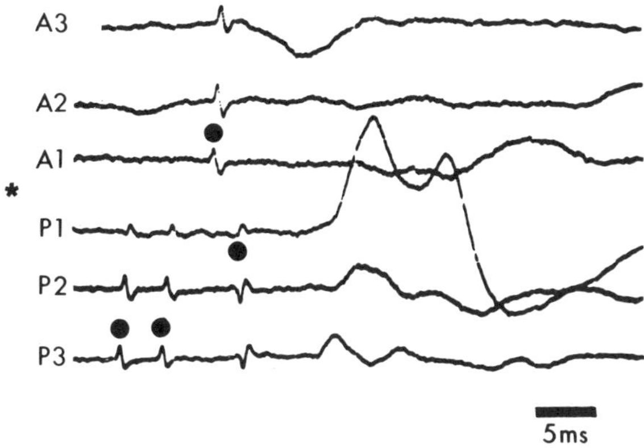

Figure 15. Spiking activity in the regenerating LGF of *Lumbricus terrestris* 14 days after ventral nerve cord transection. The LGF spiking was initiated by tactile stimulation of the tail and recorded at six sites along the ventral surface of the intact animal. The three recording sites A1–A3 were just anterior to the regeneration site (*), whereas the other three P1–P3, were just posterior to the regeneration site. The distance between adjacent recording sites was 4 mm. The response to touch consisted of an initial pair of LGF spikes (dots in P3) that were conducted toward the regeneration site. The first spike failed to conduct across the site of regeneration, whereas the second spike was successfully conducted across after a delay of approximately 3 msec (dot in A1). An LGF rebound spike (dot in P1) was initiated approximately 2 msec after the spike at A1. The waveform of the rebound spike was predictably inverted since it was conducted posteriorly. Intermittent spike conduction, conduction delays, and rebound spiking were also characteristic of the regenerating MGF (from Balter *et al.*, 1980).

fibers are likely an indication that the trauma associated with nerve cord isolation reduces the efficacy of spike transmission across the newly formed and relatively labile connections joining the severed giant fibers. This idea is supported by noninvasive testing of giant fiber through-conduction in *Lumbricus terrestris* at various times following ventral nerve cord transection (E. P. Vining, unpublished data). Such recordings have shown that even after through-conduction of giant fiber spikes is re-established across the regenerated portion of the nerve cord, a blocking of through-conduction at the transection site can be readily induced by slight trauma elsewhere in the animal, such as a dorsal puncture or cut in the worm's body wall.

Despite these effects, there seems to be uniform agreement from *in vivo* and *in vitro* studies that there is a high degree of precision and specificity with regard to the re-establishment of connections between the severed fibers; that is, anterior and posterior portions of a severed MGF always reconnect to one another, as is the case for the LGF. Based on electrophysiological recordings from regenerating giant fibers, there is also general agreement that transmission between regenerated earthworm giant fibers probably involves some type of electrical junction (Balter *et al.*, 1980; Birse and Bittner, 1976, 1981). Biophysical and ultrastructural studies of giant fibers during early stages of regeneration would help to confirm and further resolve this issue.

In addition to the earthworm's capabilities for regeneration after a simple transection of the ventral nerve cord, there is evidence that giant fiber regeneration can occur after ablation of one to three segments of ventral nerve cord (Bovard, 1918; Birse and Bittner, 1981). Except for the longer time required for recovery, the regeneration process appears similar to that seen after transection. That is, highly specific reconnection of giant fibers occurs via outgrowth of multiple processes arising from the severed ends of the fibers. Birse and Bittner (1981) found no evidence for reformation of the ablated cell bodies in earthworm giant fibers. This also appears to be the situation in the polychaete, *Nereis,* although reformation of giant fiber cell bodies apparently does occur in another polychaete, *Clymenella torquata* (Hulsebosch and Bittner, 1981).

The compensatory regeneration of whole body segments at the site of removal of head or tail segments has been well documented in many classical studies involving earthworms and polychaetes (for older reviews see, Stephenson, 1930 and Hyman, 1940). From the standpoint of developmental neurobiology, the capability for new segment formation raises interesting questions in regard to the interfacing and interaction between the newly formed and original portions of the nervous system, as well as influences of axial gradients on nervous system and segment differentiation. Recently, some of these questions have been addressed in a preliminary study of the electrophysiological properties of giant fibers in regenerating tail segments of *Eisenia foetida* (Drewes and Marty, 1982). By nine days after removal of 36–44 tail segments, a delineation of new segments has occurred in the tail bud and by 14 days all-or-none LGF spikes can be noninvasively detected in bud segments in response to tactile stimulation of either bud segments or older adjacent segments. The LGF conduction velocity in bud segments at this time is nearly identical to that of embryonic worms at the normal crawling stage. Subsequent bud growth involves increase in bud segment size but no increase in segment number (Moment, 1979). During this growth the rates of increase in LGF velocity are nearly identical to those seen during normal development, suggesting that the processes underlying giant fiber growth in regenerating tail segments recapitulate those occurring during normal embryonic and postembryonic development.

Interestingly, tail regeneration is not necessarily restricted to conditions of posterior segment removal, but can occasionally occur following removal of anterior segments (Morgan, 1899; Gates, 1950). In such cases these misplaced, or so-called "heteromorphic" tails appear to be exclusively subserved by the LGF sensory field (Drewes and Marty, 1982). This indicates that formation of LGF sensory field in regenerated segments can occur independently of the location of regeneration along the body axis.

6. Conclusions

Rapid escape reflexes in polychaetes, leeches, and oligochaetes are mediated by one or several relatively large-caliber and through-conducting giant fiber networks. The ensemble of neurons involved in generating and processing afferent inputs to the giant fibers, as well as the motor neurons controlling escape movements, consist of serially repeated, identifiable neurons. Appropriate variations in stimulus sensitivity and escape reflex movements along the length of the animal are achieved by segmental gradients or transitions in the properties of both peripheral and central components in the reflex. Although some progress has been made in characterizing the reflex circuitry and segmental variations in reflex function in a few annelids (e.g., leeches and earthworms), we have a very limited understanding of escape reflex organization in the vast majority of annelid species.

Several features pertaining to escape reflex development and regeneration have been revealed in studies of oligochaete earthworms. For example, noninvasive electrophysiological recordings indicate that many adultlike features of earthworm escape reflex function are established in late embryonic stages and are maintained throughout postembryonic maturation. In under-nourished worms postembryonic growth of giant fibers, and possibly growth of other neuronal elements, are given priority in relation to somatic growth. Finally, very rapid regeneration of earthworm giant fibers and escape reflex function occurs following either specific nervous system lesions or loss of entire body segments.

ACKNOWLEDGMENTS. I thank Dr. S. Shen and E. Vining for their comments on this manuscript and Drs. E. Neuhauser and C. Callahan for supplying the animals. I also thank M. Nims for typing the manuscript. This research was supported in part by the Hazardous Materials Assessment Program of the United States Environmental Protection Agency (EPA Cooperative Agreement CR810006-01).

7. References

Adey, W. R., 1951, The nervous system of the earthworm *Megascolex, J. Comp. Neurol.* **94**:57–103.

Andrews, E. A., 1895, Conjugation of the brandling, *Am. Nat.* **29**:1021–1027.

Bagnoli, P., Brunelli, M., and Magni, F., 1972, A fast conducting pathway in the central nervous system of the leech *Hirudo medicinalis, Arch. Ital. Biol.* **110**:35–51.

Bagnoli, P., Brunelli, M., and Magni, F., 1973, Afferent connections to the fast conduction pathway in the central nervous system of the leech *Hirudo medicinalis, Arch. Ital. Biol.* **111**:58–75.

Bagnoli, P., Brunelli, M., Magni, F., and Pellegrino, M., 1975*a*, Suprasegmental inputs to the fast conducting system in the central nervous system of *Hirudo medicinalis, Arch. Ital. Biol.* **112**:307–329.

Bagnoli, P., Brunelli, M., and Magni, F., 1975*b*, The neuron of fast conducting system in *Hirudo medicinalis:* Identification and synaptic connections with primary afferent neurons, *Arch. Ital. Biol.* **113**:21–43.

Balter, R. J., Drewes, C. D., and McFall, J. L., 1980, *In vivo* conduction properties of regenerating giant nerve fibers in earthworms, *J. Exp. Zool.* **211**:395–405.

Belardetti, F., Biondi, C., Colombaioni, L., Brunelli, M., and Trevisani, A., 1982, Role of serotonin and cyclic AMP on facilitation of the fast conducting system activity in the leech *Hirudo medicinalis, Brain Res.* **246**:89–103.

Beleslin, B. B., 1982, Membrane physiology of excitable cells in annelids, in: *Membrane Physiology of Invertebrates* (R. B. Podesta, ed.), Marcel Dekker, New York, pp. 199–260.

Bennett, M. V. L., 1977, Electrical transmission: A functional analysis and comparison to chemical transmission, in: *Handbook of Physiology,* Sect. 1, Vol. 1, Part 1 (E. R. Kandel, ed.), American Physiological Society, Bethesda, pp. 357–416.

Berry, M. S., and Pentreath, V. W., 1977, The integrative properties of electrotonic synapses, *Comp. Biochem. Physiol.* **57A**:289–295.

Birse, S. C., and Bittner, G. D., 1976, Regeneration of giant axons in earthworms, *Brain Res.* **113**:575–581.

Birse, S. C., and Bittner, G. D., 1981, Regeneration of earthworm giant axons following transection or ablation, *J. Neurophysiol.* **45**:724–742.

Bovard, J. F., 1918, The function of the giant fibers in earthworms, *Univ. Calif. Publ. Zool.* **18**:135–144

Brink, P., and Barr, L., 1977, The resistance of the septum of the median giant axon of the earthworm, *J. Gen. Physiol.* **69**:517–536.

Brink, P. R., and Dewey, M. M., 1978, Nexal membrane permeability to anions, *J. Gen. Physiol.* **72**:67–86.

Brink, P. R., and Dewey, M. M., 1980, Evidence for fixed charge in the nexus, *Nature* **285**:101–102.

Bullock, T. H., 1945*a*, Functional organization of the giant fiber system of *Lumbricus, J. Neurophysiol.* **8**:55–71.

Bullock, T. H., 1945*b*, Organization of the giant nerve fiber system in *Neanthes virens, Biol. Bull. (Woods Hole)* **89**:185–186.

Bullock, T. H., 1948, Physiological mapping of giant nerve fiber systems in polychaete annelids, *Physiol. Comp. Oecol.* **1**:1–14.

Bullock, T. H., 1951, Facilitation of conduction rate in nerve fibers, *J. Physiol. (London)* **114**:89–97.

Bullock, T. H., 1952, The invertebrate neuron junction, *Cold Spring Harbor Symp. Quant. Biol.* **17**:267–273.

Bullock, T. H., 1953, Properties of some natural and quasi-artificial synapses in polychaetes, *J. Comp. Neurol.* **98**:37–68.

Bullock, T. H., and Horridge, G. A., 1965, *Structure and Function in the Nervous Systems of Invertebrates,* Vol. 1, W. H. Freeman, San Francisco.

Bullock, T. H., and Turner, R. S., 1950, Events associated with conduction failure in nerve fibers, *J. Cell. Comp. Physiol.* **36**:59–81.

Carbonetto, S., and Muller, K. J., 1977, A regenerating neurone in the leech can form an electrical synapse on its severed axon segment, *Nature* **267**:450–452.

Coggeshall, R. E., 1965, A fine structural analysis of the ventral nerve cord and associated sheath of *Lumbricus terrestris, J. Comp. Neurol.* **125**:393–438.

Coggeshall, R. E., and Fawcett, D. W., 1964, The fine structure of the central nervous system of the leech *Hirudo medicinalis, J. Neurophysiol.* **27**:229–289.

De Robertis, E. D. P., and Bennett, H. S., 1956, Some observations on the fine structure of the giant nerve fibres of the earthworm, in: *Proceedings of the Third International Conference on Electron Microscopy* (R. Ross, ed.), Royal Microscopical Society, *London* pp. 431–436.

Dewey, M., and Barr, L., 1964, A study of the structure and distribution of the nexus, *J. Cell Biol.* **23**:553–585.

Dierolf, B. M., and Brink, P. R., 1973, Effects of temperature acclimation on cable constants of the earthworm median giant axon, *Comp. Biochem. Physiol.* **44**:401–406.

Dorsett, D. A., 1964, The sensory and motor innervation of *Nereis, Proc. R. Soc. (London) Ser. B* **159**:652–667.

Dorsett, D. A., 1978, Organization of the nerve cord, in: *Physiology of Annelids* (P. J. Mill, ed.), Academic Press, New York, pp. 115–160.

Dorsett, D. A., 1980, Design and function of giant fibre systems, *Trends Neurosci.* **3**:205–208.

Drewes, C. D., and Marty, B. L., 1982, Giant nerve fiber function in regenerating tail segments of earthworms, *Soc. Neurosci. Abstr.* **8**:869.

Drewes, C. D., and McFall, J. L., 1980, Longitudinal variations in the efficacy of lateral giant fiber to giant motor neuron transmission in intact earthworms, *Comp. Biochem. Physiol.* **66A**:315–321.

Drewes, C. D., and Pax, R. A., 1974, Neuromuscular physiology of the longitudinal muscle of the earthworm, *Lumbricus terrestris* II. Patterns of innervation, *J. Exp. Biol.* **60**:453–467.

Drewes, C. D., Landa, K. B., and McFall, J. L., 1978, Giant nerve fibre activity in intact, freely moving earthworms, *J. Exp. Biol.* **72**:217–227.

Drewes, C. D., McFall, J. L., Vining, E. P., and Pallas, S. L., 1980, Longitudinal variations in MGF-mediated giant motor neuron activity and rapid escape shortening in intact earthworms, *Comp. Biochem. Physiol.* **67A**:659–665.

Drewes, C. D., Callahan, C. A., Fender, W. M., 1983, Species specificity of giant nerve fiber conduction velocity in oligochaetes, *Can. J. Zool.* **61**:2688–2694.

Dyal, J. A., 1973, Behavior modification in annelids, in: *Invertebrate Learning,* Vol. 1 (W. C. Corning, J. A. Dyal, and A. O. D. Willows, eds.), Plenum Press,New York, pp. 225–290.

Eccles, J. C., Granit, R., and Young, J. Z., 1933, Impulses in the giant nerve fibers of earthworms, *J. Physiol. (London)* **77**:23–24.

Fernandez, J., 1978, Structure of the leech nerve cord: Distribution of neurons and organization of fiber pathways, *J. Comp. Neurol.* **180**:165–191.

Frank, E., Jansen, J. K. S., and Rinvik, E., 1975, A multisomatic axon in the central nervous system of the leech, *J. Comp. Neurol.* **159**:1–13.

Friesen, W. O., 1981, Physiology of water motion detection in the medicinal leech, *J. Exp. Biol.* **92**:255–275.

Gardner, C. R., 1976, The neuronal control of locomotion in the earthworm, *Biol. Rev. Camb. Philos. Soc.* **51**:25–52.
Gardner-Medwin, A. R., Jansen, J. K. S., and Taxt, T., 1973, The "giant axon" of the leech, *Acta Physiol. Scand.* **87**:30A–31A.
Gates, G. E., 1950, Regeneration in an earthworm, *Eisenia foetida* (Savigny) 1826. II. Posterior regeneration, *Biol. Bull. (Woods Hole)* **98**:36–45.
Goldman, L., 1963, The effects of stretch on impulse propagation in the median giant fiber of *Lumbricus, J. Cell. Comp. Physiol.* **62**:105–112.
Goldman, L., 1964, Effects of stretch on cable and spike parameters of single nerve fibres; some implications for the theory of impulse propagation, *J. Physiol. (London)* **173**:424–444.
Günther, J., 1970, Zur Organisation der exteroceptiven Afferenzen in den Korpersegmenten des Regenwurms, *Verh. Dtsch. Zool. Ges.* **64**:261–265.
Günther, J., 1971*a*, Der cytologische Aufbau der dorsalen Riesenfasern von *Lumbricus terrestris* L., *Z. Wiss. Zool.* **183**:51–70.
Günther, J., 1971*b*, Mikroanatomie des Bauchmarks von *Lumbricus terrestris* L. (Annelida, Oligochaeta), *Z. Morphol. Tiere* **70**:141–182.
Günther, J., 1972, Giant motor neurons in the earthworm, *Comp. Biochem. Physiol.* **42**:967–974.
Günther, J., 1973*a*, A new type of "node" in the myelin sheath of an invertebrate nerve fiber, *Experientia* **29**: 1263–1265.
Günther, J., 1973*b*, Overlapping sensory fields of the giant fiber systems in the earthworm, *Naturwissenschaften* **11**:521–522.
Günther, J., 1975, Neuronal syncytia in giant fibres of earthworms, *J. Neurocytol.* **4**:55–62.
Günther, J., 1976, Impulse conduction in the myelinated giant fibers of the earthworm. Structure and function of the dorsal nodes in the median giant fiber, *J. Comp. Neurol.* **168**:505–531.
Günther, J., and Schürmann, F. W., 1973, Zur Feinstruktur des dorsalen Riesenfasersystems im Bauchmark des Regenwurms II. Synaptische Beziehung der proximalen Riesenfaserkollateralen, *Z. Zellforsch, Mikrosk. Anat.* **139**:369–396.
Günther, J., and Walther, J. B., 1971, Funktionelle Anatomie der dorsalen Riesenfaser-Systeme von *Lumbricus terrestris* L., *Z. Morphol. Tiere* **70**:253–280.
Gwilliam, G. F., 1969, Electrical response to photic stimulation in the eyes and nervous system of nereid polychaetes, *Biol. Bull. (Woods Hole)* **136**:385–397.
Hagiwara, S., Morita, H., and Naka, K., 1964, Transmission through distributed synapses between two giant axons of a sabellid worm, *Comp. Biochem. Physiol.* **13**:453–460.
Hall, A. R., 1921, Regeneration in the annelid nerve cord, *J. Comp. Neurol.* **33**:163–177.
Hama, K., 1959, Some observations on the fine structure of the giant nerve fibers of the earthworm, *Eisenia foetida, J. Biophys. Biochem. Cytol.* **6**:61–66.
Hodes, R., 1953, Linear relationship between fiber diameter and velocity in giant axon of squid, *J. Neurophysiol.* **16**:145–154.
Horridge, G. A., 1959, Analysis of the rapid responses of *Nereis* and *Harmothoe* (Annelida), *Proc. R. Soc. (London) Ser. B* **150**:245–262.
Horridge, G. A., 1968, *Interneurons,* W. H. Freeman, San Francisco.
Hulsebosch, C. E., and Bittner, G. D., 1981, Regeneration of axons and nerve cell bodies in the CNS of annelids, *J. Comp. Neurol.* **198**:77–88.
Hyman, L. H., 1940, Aspects of regeneration in annelids, *Am. Nat.* **74**:513–527.
Jamieson, B. G. M., 1981, *The Ultrastructure of the Oligochaeta,* Academic Press, New York.
Kao, C. Y., and Grundfest, H. J., 1957, Postsynaptic electrogenesis in septate giant axons, *J. Neurophysiol.* **20**:553–573.
Kennedy, D., 1966, The comparative physiology of invertebrate central neurons, in: *Ad-*

vances in Comparative Physiology and Biochemistry (O. Lowenstein, ed.), Academic Press, New York, pp. 117–184.

Kensler, R. W., Brink, P. R., and Dewey, M. M., 1979, The septum of the lateral axon of the earthworm: A thin section and freeze-fracture study, *J. Neurocytol.* **8**:565–590.

Kramer, A. P., 1981, The nervous system of the glossiphoniid leech *Haementeria ghillianii* II. Synaptic pathways controlling body wall shortening, *J. Comp. Physiol.* **144**:449–457.

Kramer, A. P., and Goldman, J. R., 1981, The nervous system of the glossiphoniid leech *Haementeria ghilianii* I. Identification of neurons, *J. Comp. Physiol.* **144**:435–448.

Krasne, F. B., 1965, Escape from recurring tactile stimulation in *Branchiomma vesiculosum*, *J. Exp. Biol.* **42**:307–322.

Kretz, J. R., Stent, G. S., and Kristan, W. B., 1976, Photosensory input pathways in the medicinal leech, *J. Comp. Physiol.* **106**:1–37.

Kristan, W. B., McGirr, S. J., and Simpson, G. V., 1982, Behavioral and mechanosensory neurone responses to skin stimulation in leeches, *J. Exp. Biol.* **96**:143–160.

Kupfermann, I., and Weiss, K. R., 1978, The command neuron concept, *Behav. Brain Sci.* **1**:3–39.

Lagerspetz, K. Y. H., and Talo, A., 1967, Temperature acclimation of the functional parameters of the giant nerve fibers in *Lumbricus terrestris*, *J. Exp. Biol.* **47**:471–480.

Laverack, M. S., 1963, *The Physiology of Earthworms*, McMillan, New York.

Laverack, M. S., 1969, Mechanoreceptors, photoreceptors, and rapid conduction pathways in the leech, *Hirudo medicinalis*, *J. Exp. Biol.* **50**:129–140

Magni, F., and Pellegrino, M., 1975, Nerve cord shortening induced by activation of the fast conducting system in the leech, *Brain Res.* **90**:169–174.

Magni, F., and Pellegrino, M., 1978*a*, Neural mechanisms underlying the segmental and generalized cord shortening reflexes in the leech, *J. Comp. Physiol.* **124**:339–351.

Magni, F., and Pellegrino, M., 1978*b*, Patterns of activity and the effects of activation of the fast conducting system on the behavior of unrestrained leeches, *J. Exp. Biol.* **76**:123–135.

McFall, J. L., Landa, K. B., and Drewes, C. D., 1977, Parameters of giant fiber conduction in intact, freely moving earthworms, *Am. Zool.* **17**:105.

Mangum, C. P., and Passano, L. M., 1964, Giant fibers in madanid polychaetes, *Nature* **201**:210–211.

Mellon, D., Treherne, J. E., Lane, N. J., Harrison, J. B., and Langley, C. K., 1980, Electrical interactions between the giant axons of a polychaete worm (*Sabella penicillus* L.), *J. Exp. Biol.* **84**:119–136.

Mill, P. J., 1975, The organization of the nervous system in annelids, in: *"Simple" Nervous Systems* (P. N. R. Usherwood and D. R. Newth, eds.), Edward Arnold, London, pp. 211–264.

Mill, P. J., 1982, Recent developments in earthworm neurobiology, *Comp. Biochem. Physiol.* **73A**:641–661.

Mistick, D., 1974, Rohde's fibre: A septate axon in the leech, *Brain Res.* **74**:342–348.

Moment, G. B., 1979, Growth, posterior regeneration and segment number in *Eisenia foetida*, *Megadrilogica* **3**:167–175.

Moment, G. B., and Johnson, J. E. Jr., 1979, The structure and function of external sense organs in newly hatched and mature earthworms, *J. Morphol.* **159**:1–16.

Moore, M. J., 1979, The rapid escape response of the earthworm *Lumbricus terrestris* L.: Overlapping sensory fields of the median and lateral giant fibers, *J. Exp. Biol.* **83**:231–238.

Morgan, T. H., 1899, A confirmation of Spallanzani's discovery of an earthworm regenerating a tail in place of a head, *Anat. Anz.* **15**:407–410.

Morita, H., and Tateda, H., 1952, Giant nerve fiber and its functional organization in the earthworm, *Mem. Fac. Sci. Kyushu Univ. Ser. E. Biol.* **1**:89–100.
Muller, K. J., 1979, Synapses between neurones in the central nervous system of the leech, *Biol. Rev. Camb. Philos. Soc.* **54**:99–134.
Muller, K. J., and Scott, S. A., 1981, Transmission at a "direct" electrical connexion mediated by an interneurone in the leech, *J. Physiol. (London)* **311**:565–583.
Mulloney, B., 1970, Structure of the giant fibers of earthworms, *Science* **168**:994–996.
Nicholls, J. G., and Purves, D., 1970, Monosynaptic chemical and electrical connexions between sensory and motor cells in the central nervous system of the leech, *J. Physiol (London)* **209**:647–667.
Nicol, J. A. C., 1948*a*, The function of the giant axon of *Myxicola infundibulum, Can. J. Res.* **26**:212–222.
Nicol, J. A. C., 1948*b*, The giant axons of annelids, *Q. Rev. Biol.* **23**:291–319.
Nicol, J. A. C., 1948*c*, The giant nerve-fibres in the central nervous system of *Myxicola* (Polychaeta, Sabellidae), *Q. J. Microsc. Sci.* **89**:1–45.
Nicol, J. A. C., 1950, Responses of *Branchiomma vesiculosum* (Montagu) to photic stimulation, *J. Mar. Biol. Assoc. U. K.* **29**:303–320.
Nicol, J. A. C., Smyth, C. N., and Whitteridge, D., 1947, Conduction velocity in relation to axon diameter in *Myxicola infundibulum,* in: *XVII International Physiology Congress. Abstracts of Communications, Oxford,* pp. 243–244.
Oesterle, D., and Barth, F. G., 1973, Zur Feinstruktur einer elektrischen Synapse. Die septum der dorsalen Riesenfasern von Regenwurmern *(Lumbricus terrestris, Eisenia foetida), Z. Zellforsch, Mikrosk. Anat.* **136**:139–152.
Oesterle, D., and Barth, F. G., 1981, Dorsal giant fiber septum of earthworm: Fine structural details and further evidence for gap junctions, *Tissue Cell* **13**:9–18.
O'Gara, B., Vining, E. P., and Drewes, C. D., 1982, Electrophysiological correlates of rapid escape reflexes in intact earthworms, *Eisenia foetida*. I. Functional development of giant nerve fibers during embryonic and postembryonic periods, *J. Neurobiol.* **13**:337–353.
Ogawa, F., 1939, The nervous system of earthworm *(Pheretima communissima)* in different ages, *Sci. Rep. Tohoku Univ. Fourth Ser. (Biol.)* **13**:395–488.
Pallas, S. L., and Drewes, C. D., 1981, The rapid tail flattening component of MGF-mediated escape behavior in the earthworm, *Lumbricus terrestris, Comp. Biochem. Physiol.* **70A**:57–64.
Prosser, C. L., 1933, Correlation between development of behavior and neuromuscular differentiation in embryos of *Eisenia foetida,* Sav., *J. Comp. Neurol.* **58**:603–641.
Prosser, C. L., 1934, The nervous system of the earthworm, *Q. Rev. Biol.* **9**:181–200.
Roberts, M. B. V., 1960, Giant fibre reflex of the earthworm, *Nature* **186**:167.
Roberts, M. B. V., 1962*a*, The giant fibre reflex of the earthworm *Lumbricus terrestris L.* I. The rapid response, *J. Exp. Biol.* **39**:219–227.
Roberts, M. B. V., 1962*b*, The giant fibre reflex of the earthworm *Lumbricus terrestris* L. II. Fatigue, *J. Exp. Biol.* **39**:229–237.
Roberts, M. B. V., 1962c, The rapid response of *Myxicola infundibulum* (Grube), *J. Mar. Biol. Assoc. U.K.* **42**:527–535.
Roberts, M. B. V., 1966, Facilitation in the rapid response of earthworms *Lumbricus terrestris* L., *J. Exp. Biol.* **45**:141–150.
Rushton, W. A. H., 1945*a*, Action potentials from the isolated nerve cord of the earthworm, *Proc. R. Soc. (London) Ser. B* **132**:423–437.
Rushton, W. A. H., 1945*b*, Motor responses from giant fibers in earthworms, *Nature* **156**:109–110.

Rushton, W. A. H., 1946, Reflex conduction in the giant fibers of the earthworm, *Proc. R. Soc. (London) Ser. B.* **133**:109–120.

Rushton, W. A. H., and Barlow, H. B., 1943, Single fibre response from an intact animal, *Nature* **152**:597–598.

Sawyer, R. T., 1981, Leech biology and behavior, in: *Neurobiology of the Leech* (K. J. Muller, J. G. Nicholls, and G. S. Stent, eds.), Cold Spring Harbor Laboratory, Cold Spring Harbor, New York, pp. 7–26.

Schürmann, F. W., and Günther, J., 1973, Zur Feinstruktur des dorsalen Riesenfasersystems in Bauchmark des Regenwurms. I. Die Somata der Riesenfasern, *Z. Zellforsch. Mikrosk. Anat.* **139**:351–368.

Seymour, M. K., 1972, The giant nerve fibres of *Arenicola marina* (L.), *Comp. Biochem. Physiol.* **41**:457–464.

Smith, J. E., 1957, The nervous anatomy of the body segments of nereid polychaetes, *Philos. Trans. R. Soc. London Ser. B* **240**:135–196.

Smith, P. H., and Mittenthal, J. E., 1980, Intersegmental variation of afferent pathways to giant interneurons of the earthworm, *Lumbricus terrestris* L., *J. Comp. Physiol. A*, **140**:351–353.

Stephenson, J., 1930, *The Oligochaeta*, Clarendon Press, Oxford.

Stough, H. B., 1926, Giant nerve fibres of the earthworm, *J. Comp. Neurol.* **40**:409–463.

Stough, H. B., 1930, Polarization of the giant fibres of the earthworm, *J. Comp. Neurol.* **50**:217–229.

Stuart, A. E., 1970, Physiological and morphological properties of motoneurones in the central nervous system of the leech, *J. Physiol. (London)* **209**:627–646.

Studnitz, G. von, 1937, Der Zuckreflex der Regenwurmer, *Zool. Jahrb. Abt. Allg. Zool. Physiol. Tiere* **38**:127–158.

Taylor, G. W., 1940, The optical properties of the earthworm giant fiber sheath as related to fiber size, *J. Cell. Comp. Physiol.* **15**:363–371.

Ten Cate, J., 1938, Sur la fonction des neurochordes de la chaine ventrale du ver de terre *(Lumbricus terrestris)*, *Arch. Neerl. Physiol.* **23**:136–140.

Vining, E. P., and Drewes, C. D., 1982, Changing conduction properties of earthworm giant fibers during regeneration, *Soc. Neurosci. Abstr.* **8**:868.

Vining, E. P., O'Gara, B., and Drewes, C. D., 1982, Electrophysiological correlates of rapid escape reflexes in intact earthworms, *Eisenia foetida* II. Effects of food deprivation on the functional development of giant nerve fibers, *J. Neurobiol.* **13**:355–367.

Wells, J., Besso, J. A., Boldosser, W. G. and Parsons, R. L., 1972, The fine structure of the nerve cord of *Myxicola infundibulum* (Annelida, Polychaeta), *Z. Zellforsch.* **131**:141–148.

Wells, J., Besso, J. A., Boldosser, W. G., and Parsons, R. L., 1973, Morphology of the escape system in *Myxicola infundibulum*, *Anat. Rec.* **175**:468.

Wilson, D. M., 1961, The connections between giant fibers of earthworms, *Comp. Biochem. Physiol.* **3**:274–284.

Yolton, L. W., 1923, The effects of cutting the giant fibers in the earthworm, *Eisenia foetida* (Sav.), *Proc. Natl. Acad. Sci. U.S.A.* **9**:383–385.

4

The Cockroach Escape Response

ROY E. RITZMANN

1. Introduction

In the American cockroach *Periplaneta americana*, remarkably low velocity wind puffs evoke a running escape response oriented away form the source of the wind puff. This behavior and its underlying neural components provide considerable advantages for neurobiological studies. These include a quantifiably predictable behavior and numerous neural elements that are identifiable from animal to animal. The behavior requires that the cockroach not only must detect a wind stimulus, but also it must be able to determine the direction from which the wind originated. The information on wind direction is then used to establish a behavioral decision; that is, where and how to move. Far from being a simple movement, even the initial turn involves the coordination of six legs each of which possesses a complex musculature controlling five separate leg parts.

Several of the neurons and associated structures involved in this response have distinct advantages for experimental analysis. The sensory structures are hairs located in distinct columns on two abdominal appendages, the cerci (Nicklaus, 1965). Many of the motor neurons that control leg movements are identifiable both by intracellular and less technically demanding extracellular techniques (Ritzmann and Camhi, 1978). Finally, the information received by the sensory hairs is transported to the motor centers via two distinct populations of interneurons, the dorsal and ventral giant interneurons, and an as yet undetermined group of nongiant interneurons. The relatively large axonal diameter of the giant interneurons

ROY E. RITZMANN • Department of Biology, Case Western Reserve University, Cleveland, Ohio 44106.

(GI) makes them particularly accessible to intracellular analysis. Moreover, in the abdominal ganglia, each GI assumes a reproducible position relative to the other GIs. Thus, using intracellular dye injection techniques, each GI can be unambiguously identified in any preparation.

1.1. Historical Background

Roeder (1948, 1967), in the first quantitative study of the escape system of the cockroach, calculated that the response from onset of wind stimulation to the initial leg movement takes 28–90 msec with a mean of 54 msec. He suggested that the GIs described by Pumphrey and Rawdon-Smith (1936) probably carried the wind-activated stimulus signal to the leg motor neurons of the thoracic ganglia. Clearly, the GIs are excited by wind, and their large axonal diameter results in a rapid conduction velocity. Rapid conduction was believed to be an important factor for an escape system (Roeder, 1967).

The simple model in which sensory neurons from cercal hairs excite GIs that in turn excite leg motor neurons was accepted until Dagan and Parnas (1970) presented data that questioned the basic assumption that GIs excite leg motor neurons. They found that stimulation of the abdominal nerve cord at a current amplitude just suprathreshold for GI activation failed to evoke action potentials in metathoracic nerve 5 (N_5), which contains both the fast and slow depressor motor neurons of the coxa (D_f and D_s, respectively). Stimulus pulses of higher amplitude are required to elicit action potentials in these motor neurons. Moreover, in preparations missing GIs, action potentials in N_5 could still be evoked by stimulating the ventral nerve cord. These preparations were generated by severing the A5–A6 connective. Because the somata of GIs are located in the terminal ganglion (A6), this operation causes the axons of GIs to degenerate anterior to the cut. Finally, nicotine applied to the abdominal nerve cord reduced the effectiveness of cord stimulation in evoking a motor response. Nicotine blocks synaptic transmission in insects. Thus, this experiment suggests that cord stimulation evokes motor activity via a chain of neurons rather than by continuous interneurons such as the GIs.

Dagan and Parnas (1970) concluded from these data that GIs do not synapse directly with leg motor neurons. They suggested that a population of smaller interneurons excite the leg motor neurons. The GIs might prepare the animal for escape by evoking movements such as lowering of antennae and inhibiting all other on-going activities.

More recent data, which I will discuss in detail, demonstrate that several of the GIs can, in fact, excite leg motor neurons as suggested by

Roeder. However, as the observations of Dagan and Parnas (1970) indicate, the excitation does not occur over monosynaptic connections. Moreover, the GIs are not a homogeneous population of neurons. Rather, different GIs are excited by different wind directions and play distinct roles in evoking motor outputs. As a result, consistent with the observations of Dagan and Parnas (1970), a single action potential in a single GI is totally ineffective in evoking a motor response. However, gentle wind puffs actually cause high frequency trains of action potentials in multiple GIs and such activity in individual GIs can indeed evoke motor responses.

2. The Escape Behavior

Prior to discussing the neurophysiological data on escape, it is essential to understand, in detail, the animal's behavioral response to wind stimuli. A considerable amount of quantitative data is now available describing the wind-mediated escape response.

2.1. Directionality of Escape

Camhi and Tom (1978) employed high-speed cinematography and single frame analysis to quantify several aspects of wind-mediated escape. Free-ranging cockroaches were placed in an arena where they could be stimulated by a machine designed to deliver reproducible wind puffs from anywhere in the arena. Under these conditions, the initial response to a wind puff is highly directional (Figure 1). Cockroaches tend to turn away from the wind stimulus. This assures that the cockroach is at least initially directed away from an approaching predator. Thus, a wind originating from behind the cockroach causes a forward escape, and wind from the front causes a rearward escape. Likewise, wind from the right makes the animal turn left, and wind from the left elicits a right turn. After the initial turn, the animal's movements are considerably more variable.

2.2. Minimum Stimulus

The minimum velocity of wind needed to evoke escape movements is remarkably low. A velocity of 0.012 m/sec has been found sufficient to evoke a turn (Camhi and Nolen, 1981). With such a low minimum velocity, one might expect cockroaches to be constantly escaping. Certainly breezes in excess of 0.012 m/sec are quite often encountered in the animal's natural environment. In fact, they do not escape from these winds because acceleration is also a critical parameter for the wind stimulus.

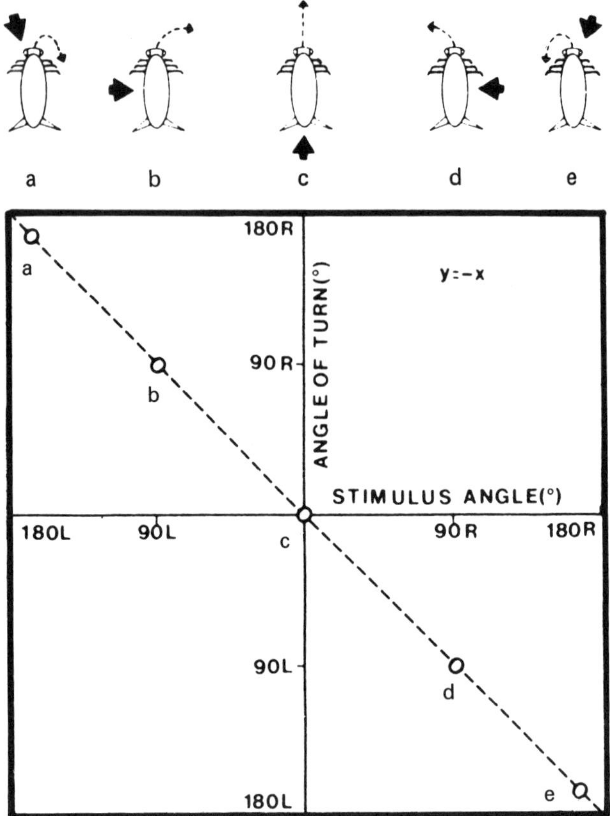

Figure 1. (Left) Conventions for plotting angle of turn vs. stimulus angle. For illustrative purposes, the cockroach is assumed to turn exactly away from the wind source. (a–e) Thick arrows represent stimulus angles, 170°L, 90L°, 0°, 90° R, and 170° R, respectively; thin arrows represent angle of turn, 170° R, 90° R, 0°, 90° L, and 170° L, respectively. The regression line for these idealized data points has the formula $y = x$. (Right) Turning

Plummer and Camhi (1981) developed a preparation for testing the importance of acceleration. The cockroach was held to a lightly greased plate by four insect pins inserted through the animal's cuticle and into a piece of wax. In this preparation the cockroach appears to walk in a very normal fashion. The plate was placed inside a movable box mounted on two tracks. Moving the box created a wind on the cockroach. By varying the acceleration of the box, the acceleration of the wind could be varied.

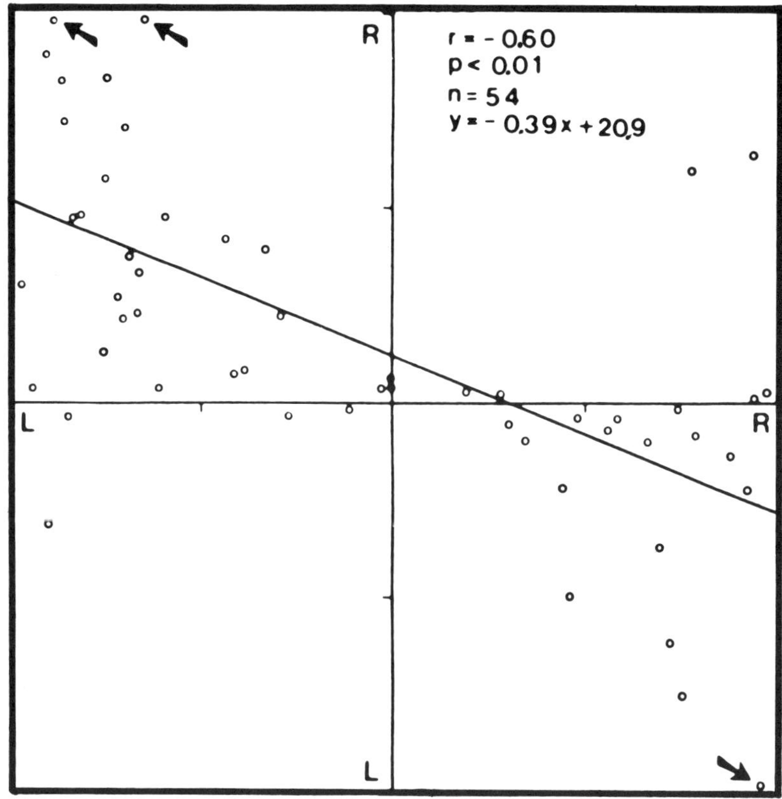

responses of ten free-ranging cockroaches. Angle of turn vs. stimulus angle plotted by conventions shown in top figure. Each point represents one turning response. Points indicated with an arrow represent turns larger than 180° R or 180° L (all 180°–245°). The regression line has a correlation coefficient (r), that is highly significant (from Camhi and Tom, 1978).

Acceleration of the wind stimulus is in fact positively correlated with success in generating escape responses. Stimuli accelerating at 0.6 m/sec^2 usually evoke running responses. Those accelerating between 0.3 and 0.6 m/sec^2 usually evoke a pause but no escape.

The cockroach's behavior at the time of wind stimulation has important consequences for the escape response (Camhi and Nolen, 1981). If the cockroach is walking slowly (4 steps/sec), the mean latency from stimulus onset to the initial escape movements is 14 msec compared with 54–58 msec for standing cockroaches. However, the latency increases as the speed of walking increases. The threshold wind velocity is also reduced during slow walking (Camhi and Nolen, 1981).

2.3. Sensory Structures That Evoke Escape

Is wind and indeed wind impinging on the cerci the critical stimulus for initiating escape? The wind machine used in the experiments described above is quite massive and would provide visual and auditory cues that could initiate or direct the escape rather than the actual wind it produces. To control for this possibility, Camhi and Tom (1978) placed a smaller wind tube in the arena along with the wind machine. The shutter valve of the wind machine was activated in the same way as in previous experiments. However, the fans were turned off so that no wind was released. In all cases the escape was directed away from the wind emanating from the smaller tube rather than from the windless wind machine. This established that wind is the critical stimulus.

Many structures in addition to the cercal hairs are wind sensitive (e.g., antennae). To establish that the cerci provide the principle sensory cue for escape, the cercal hairs were covered with glue. This prevented initiation of the escape response. The same amount of glue applied to the dorsal surface of the cerci, which is devoid of hairs, had no effect on escape.

Another experiment established that inputs from cerci not only initiate the escape but also direct the initial turn (Camhi and Tom, 1978). The cerci were rotated in their sockets 60–70° to the left. The turning angles in these animals was shifted by approximately 65° left. In addition, in manipulated animals, wind from 60° right to 180° would be interpreted as wind from the animal's right side by any structure other than the cerci and therefore would evoke a left turn. However, the receptors on the rotated cerci would interpret these stimuli as coming from the left side and evoke right turns. In fact the mean angle for turns to stimuli in this range was 28° right, thus demonstrating that the important structures for directing turns are the cercal hairs.

2.4. Predator–Prey Encounters

Does the wind-mediated escape response really increase the cockroach's ability to survive predator–prey encounters? To answer this question, Camhi *et al.* (1978) studied encounters between *P. americana* and a natural predator, the toad *Bufo marinus*. Intact cockroaches successfully escape attacks from toads 50% of the time; that is, about half of the strikes by toads fail to catch a cockroach. Moreover, 48% of the cockroaches tested managed to escape at least once. The success rate decreased significantly in cockroaches that had their cerci covered with wax. With these cockroaches, 92% of the toad strikes were successful and only 5% of the tested cockroaches escaped at least once. A control group had an equivalent amount of wax applied to abdominal sternites. This had no significant effect on the success of toad strikes.

Cinematic analysis revealed that the escape is initiated as the toad lunges prior to any movement of its tongue. This part of the toad's attack generates a wind of approximately 0.02 m/sec at the cockroach, well above the threshold for initiation of escape.

Thus, the wind-mediated escape system does impart a significant benefit to the cockroach. Moreover, the response threshold appears to be tuned to the cues generated by the attacker.

3. *Neural Elements of the Escape System*

With the knowledge of the precise behavioral repertoire of escape in the cockroach, we can begin to determine what neural elements control this response and how that control is effected.

3.1. *Sensory Structures*

Two types of hairs are located on the ventral surfaces of the cerci, thread hairs and bristle hairs (see Gnatzy, 1976). The approximately 200 thread hairs are freely deflected in response to wind (Nicklaus, 1965; Gnatzy, 1976). Deflection excites the single bipolar neuron associated with each hair. The thread hairs and their associated sockets are located in 14 distinct columns (Figure 2A) (Nicklaus, 1965). Their sockets are asymmetric, causing the hair to bend preferentially in one plane. All hairs in any single column have the same preferential planes of movement (Nicklaus, 1965).

The asymmetry of movement in the hairs confers a directionality in the response properties of their associated sensory neurons. Extracellular records from single hairs show a best response in one direction of the preferred plane and an inhibition in the opposite direction (Nicklaus, 1965). Intracellular recordings from sensory axons in the cercal nerve confirm the monopolar response characteristics of the hairs (Westin, 1979). Each axon is excited by wind from one set of directions and inhibited by wind in the opposite directions. Winds from directions not in the preferred

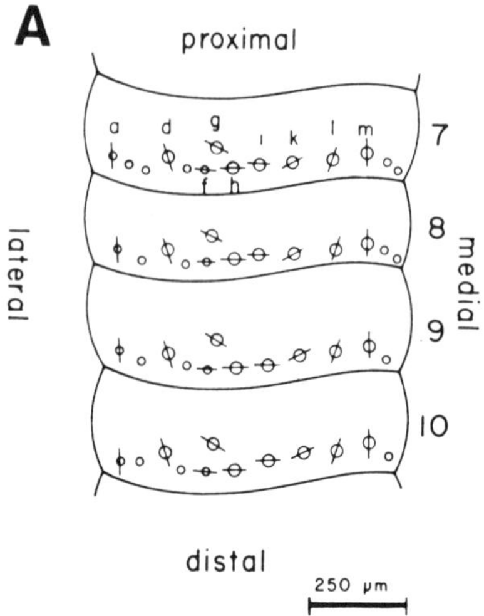

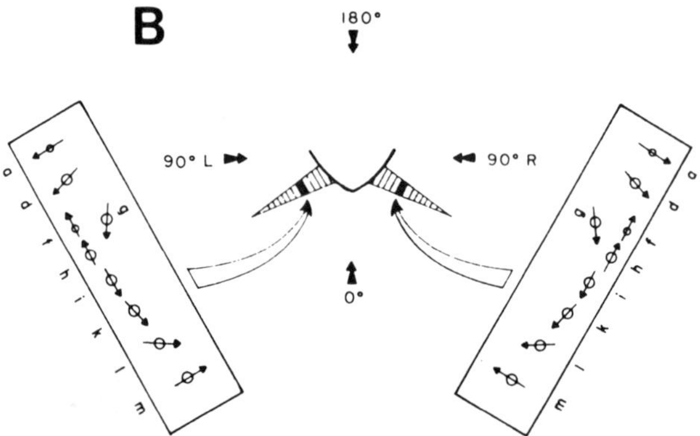

Figure 2. (A) Distribution of filiform hairs on the ventral segments 7–10 in an adult *Periplaneta americana*. Each circle shows the position of one hair. Larger circles indicate larger hairs. A line through a circle indicates the two opposite wind directions that are most effective in deflecting the hair. Letter designations are given for those columns of hairs studied by Nicklaus (1965) and Dagan and Camhi (1979). Modified from Nicklaus (1965) as shown in Dagan and Camhi (1979). (B) Best excitatory wind directions (thin arrows) for hairs of the nine most prominent columns on the left and right cercus as determined by cercogram recordings. Schematic representation of one cercal segment on each side, dorsal view, drawn at normal orientation to the body. Larger circles represent larger hairs. Letter designation for each hair corresponds to that in (A) (from Dagan and Camhi, 1979).

planes did have an effect, but in these cases the hair still moved in its preferred plane (Nicklaus, 1965). Polar plots of the responses to wind of individual sensory axons revealed seven to nine distinct groups of curves having different mean best angles (Figure 3) (Westin, 1979).

Dagan and Camhi (1979) determined the response for 9 of the 14 columns of hairs by recording extracellularly from the cercal nerve (cercogram) after immobilizing all but one column. When stimulated by air movements generated by a speaker, each column responds best to movement in the preferred plane and not at all to wind orthogonal to that plane. In the preferred plane, one direction causes an initial depolarization and the other direction causes an initial hyperpolarization. The absolute nature of the cercogram records suggests that the directional response properties of all hairs in any column are in fact the same.

As in the intracellular study, the cercograms revealed nine different preferred wind directions, one for each column studied. The best direction for each of the nine columns is consistent from animal to animal (Figure 2B).

Taken together, the best directions for all nine columns cover all four quadrants surrounding the animal (Figure 4A). Some quadrants are represented by only one or two columns. However, because each hair responds to a broad range of angles (Westin, 1979), all angles of wind should generate activity in the cercal nerve. Indeed, cercograms from an uncoated left cercus reveal a continuous curve, with smaller responses from approximately 30° right to 120° right (Figure 4B). A minimum response occurs at 90° right.

As a result of the directional properties of the thread hairs located on the cerci, the information on wind stimuli entering the CNS encodes the directional origin of wind and not just the time of onset. This information is then available to the CNS for directing the escape movements away from a potential predator.

3.2. Sensory Structures of First Instar Nymphs

In contrast to adults that possess approximately 200 thread hairs on each cercus, first instar nymphs have only four hairs, two on each cercus (Dagan and Volman, 1982). Nevertheless, these nymphs turn away from wind as well as adults do. In test cases, 84.0% of the first instar nymphs make correct turns. This compares with 83.7% for adults. Moreover, the regression line depicting angle of turn vs. angle of wind presentation for first instar nymphs is not significantly different from that calculated for adults.

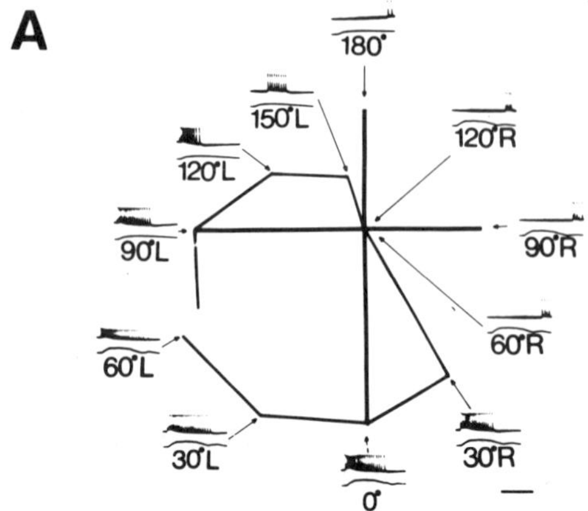

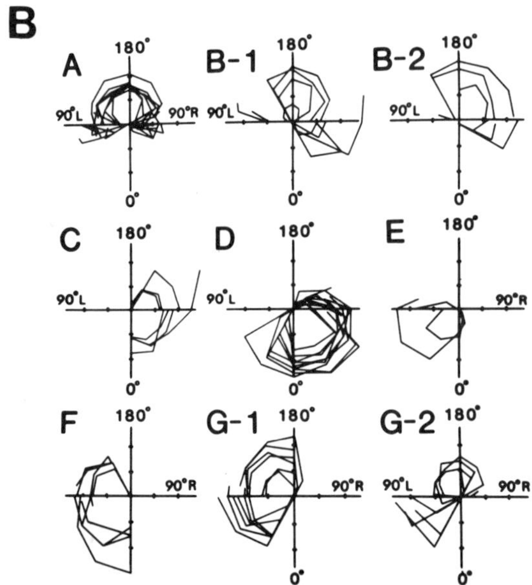

Figure 3. (A) Responses from a single sensory cell to different angles of wind. Polar plot of number of action potentials vs. wind angle. 0° represents wind from directly behind animal, 90°L represents wind flowing from animal's left to right, 180° represents wind from in front, and 90°R represents wind from right to left. Bars mark units of ten action potentials. Each point represents mean of three trails. A sample record for most wind positions is shown. Top trace shows action potentials recorded intracellularly from sensory axon. Bottom trace shows wind speed. For 60°L, 30°L, 0°L, 30°R, 60°R, 90°R, and 120°R peak speed (at the cerci) = 2.6 m/sec, acceleration = 130 m/sec^2 (outward puffs); and for 90°L, 120°L, 150°L, and 180°L peak speed (at the cerci) = 0.6 m/sec, acceleration = 25 m/sec^2 (inward puffs). Time bar = 50 msec. (B) Sensory response types. Polar plots of numbers of action potentials vs. wind angle, constructed as in part (A). Responses from several different animals are plotted on each set of axes (from Westin, 1979).

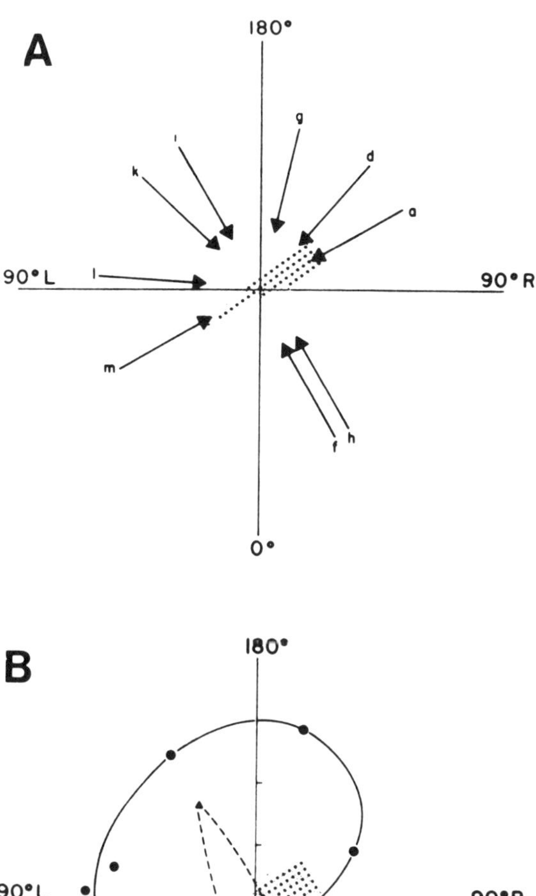

Figure 4. (A) Best excitatory directions relative to the body's axes of all columns of filiform hairs on the left cercus only. For wind stimulus from 60°L–0°–120°R, one finds the best excitatory direction of only two columns (f and h). Stippling represents orientation of left cercus. (B) Polar plot of filtered, summated cercograms obtained from an uncoated, intact left cersus (— and ●) and repeated after covering all columns except f, g, h, i, and as well as all the most proximal and distal cercal segments (--- and ▲). Numbers on abscissa indicate relative size of peak positive response. The smallest response of the uncoated cercus occurs for wind from the range of directions represented by the fewest columns of hairs. Stippling represents orientation of left cercus (from Dagan and Camhi, 1979).

As in adults, each hair is depolarized in one direction and hyperpolarized in the opposite direction (Dagan and Volman, 1982). On both cerci the more lateral hair responds primarily to wind from the ipsilateral front and the medial hair responds primarily to wind from the ipsilateral rear. As a result, although there is some overlap, the four hairs do cover all four quadrants surrounding the animal. The morphology of the ventral giant interneurons in first instar nymphs (Blagburn and Beadle, 1982) is similar to that found in adults (Daley *et al.*, 1981).

The success of this system raises the as yet unanswered question of why an adult cockroach requires so many hairs to accomplish the same task that a first instar nymph does with four hairs.

3.3. The Giant Interneurons

Information on wind stimuli is conducted from the terminal ganglion (A_6) anteriorly to motor centers in the thoracic ganglia via 14 giant interneurons (seven bilateral pairs) (Figure 5A). These are found in two morphologically distinct groups, the ventral (vGIs) and dorsal (dGIs) giant interneurons (Figure 5B). A considerable amount of data (to be presented later) indicate that these are functionally as well as morphologically distinct groups.

Each GI has a single cell body that is located in A_6 (Figure 6). In all cases a single neurite exits the soma, crosses the midline, then turns anteriorly (Daley *et al.*, 1981). The axon exits A_6 in the connective contralateral*. The vGIs generally have most of their dendritic branches located at the point where the neurite turns anteriorly. The side branches of the dGIs tend to be more evenly distributed along the fiber leading out of A_6 as well as on the side of the ganglion in which the soma is located. The soma position and the general distribution of dendritic branches in A_6 is sufficient for positive identification of each GI (Daley *et al.*, 1981).

Positive identification can also be made in cross sections of abdominal ganglia (Figure 5B). (Camhi, 1976; Westin *et al.*, 1977). In abdominal connectives, the axons of GIs often change position. However, on entering an abdominal ganglion, each GI consistently assumes a typical position (Camhi, 1976). In this array, GIs 1–4 form a close-fitting ventral group and GIs 5–7 form a more dorsal triangular-shaped group. Occasionally, a fourth axon is seen between GIs 5 and 7 ventral to GI-6. However, no physiological characteristics have been ascribed to this neuron.

*The convention in this system is that the words "ipsilateral" and "contralateral" are used with reference to the position of the axon of the giant fiber, not the soma. Thus, a stimulus to the contralateral cercus would be delivered to the side of the animal opposite the axon.

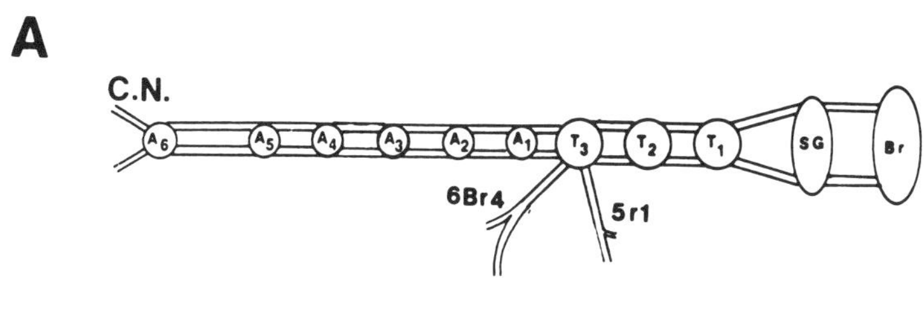

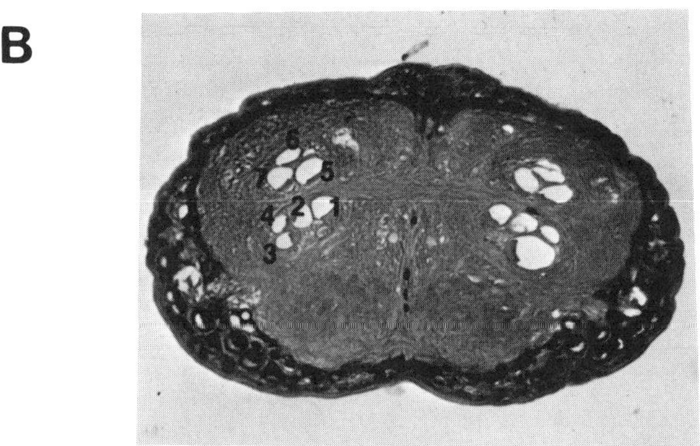

Figure 5. (A) Diagramatic representative of the cockroach CNS. Br, brain; SG, subesophageal ganglion; T_1–T_3, thoracic ganglia 1–3; A_1–A_6, abdominal ganglia; CN, cercal nerves; 6Br4, nerve in metathoracic leg containing levator axons, and 5rl, nerve in metathoracic leg containing depressor axons. (B) Cross section of an abdominal ganglion stained with hematoxylin and eosin. Numbers that have been assigned to each giant interneuron are indicated on the left side. The same numbers apply to the mirror image GIs on the right (from Ritzmann and Camhi, 1978).

The axons of vGIs, with the exception of GI-4, are larger in diameter than those of dGIs. In the abdominal connectives, GIs 1–3 have diameters of approximately 60 μm. Axons of dGIs are 25–30 μm in diameter (Spira *et al.*, 1969*a*). Giant interneuron 4 is approximately the same size or slightly smaller than the dGIs. Conduction velocities are predictably faster for the large vGIs (6–7 m/sec) than for the dGIs (4–5 m/sec). All of the GIs taper to a narrower diameter in ganglia and are smaller in thoracic segments than in abdominal connectives (Parnas and Dagan, 1971). The vGIs have been followed uninterrupted to the supraesophageal ganglion (Farley and Milburn, 1969; Spira *et al.*, 1969*b*). However, dGIs have not been detected anterior to T_2 (Farley and Milburn, 1969).

A considerable amount of information on the biophysics of GIs has

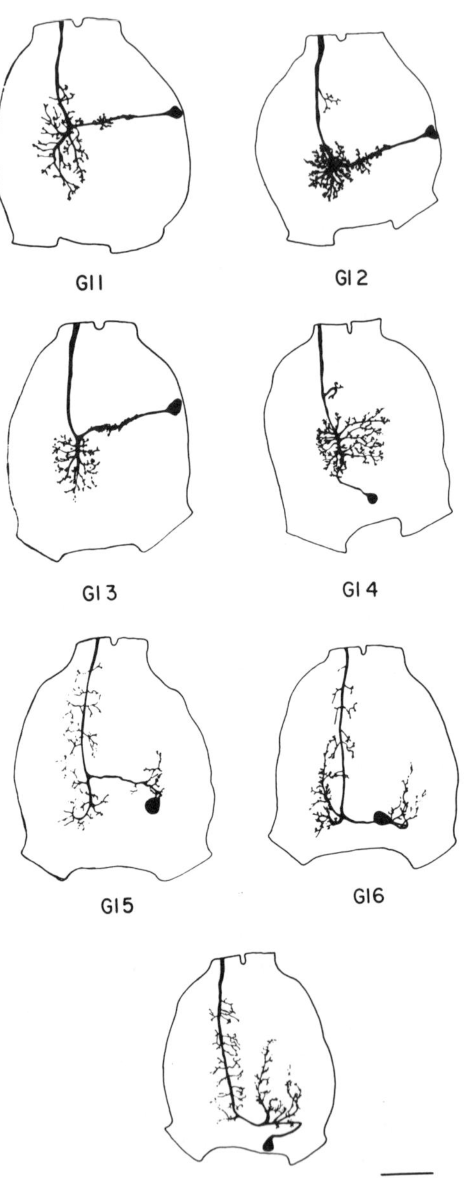

Figure 6. Morphology of each GI in the terminal ganglion (A_6) from camera lucida drawings of interneurons filled intracellularly with cobalt chloride or nitrate (from Daley *et al.*, 1981).

been documented. However, these data are beyond the scope of this review. The reader is referred to the works of Parnas, Spira, and their colleagues for this information (Parnas and Dagan, 1971; Dagan and Parnas, 1974; Spira *et al.*, 1976; Yarom and Spira, 1982).

3.3.1. Inputs to GIs

On entering A_6, the sensory neurons of the cerci make connections with the 14 GIs. In some cases the connections between sensory neurons and GIs has been shown to be monosynaptic (Callec *et al.*, 1971). The GIs tested in these experiments were not positively identified but were probably vGIs. Wind-mediated activity in dGIs tends to have a longer latency than that of vGIs (Westin *et al.*, 1977). Because only part of this discrepancy could be due to differences in conduction velocity, the sensory-to-dGI pathway may be polysynaptic.

The pharmacology of the monosynaptic connection between cercal afferents and vGIs has been studied extensively. The data indicate that these are cholinergic synapses (for a review, see Callec, 1974). The GIs are also cholinergic. This has been demonstrated by the finding that choline acetyltransferase (CAT), which is the rate-limiting enzyme in the synthesis of ACh, is significantly reduced following specific degeneration of GIs (Dagan and Sarne, 1978).

3.3.2. GI Response to Wind

All GIs respond positively to gentle wind puffs, but they do not all respond to wind from the same directions. Each GI has a specific set of wind directions to which it will respond (Westin *et al.*, 1977). The directional nature of each GI was determined by impaling individual GIs with microelectrodes filled with Procion yellow M4RS and recording the number of action potentials in response to wind stimuli from various directions. Other parameters including burst duration, average frequency, and reciprocal latency had directional properties similar to number of action potentials.

The stimulus was delivered by a device capable of repeatedly generating wind puffs of the same amplitude, duration, and acceleration. The delivery tube was mounted on a track that could be rotated around the animal. Thus, the wind stimulus could be presented from almost every angle around the animal. The presence of micromanipulators prevented the tube from rotating into the front quadrants. To stimulate from the front of the animal, wind was drawn into the delivery tube. This resulted in a similar, but lower velocity, wind stimulus. With this device it was possible to map the response to wind from all directions for a single GI.

Procion yellow could then be injected into the GI for subsequent identification in cross sections of abdominal ganglia. Response curves for all 14 GIs were generated, and they were consistent for each GI from animal to animal (Westin *et al.*, 1977).

As shown in Figure 7, only GIs 2 and 4 respond equally well to wind from any direction. Giant interneuron 1 responds to wind from all directions, but wind ipsilateral to its axon evokes a stronger response than wind from the contralateral side. Giant interneuron 7 also responds best to ipsilateral wind, but in addition it responds to wind from some contralateral front angles. It is insensitive to wind only from the contralateral rear. Giant interneurons 3, 5, and 6 are the most directional, and GIs 3

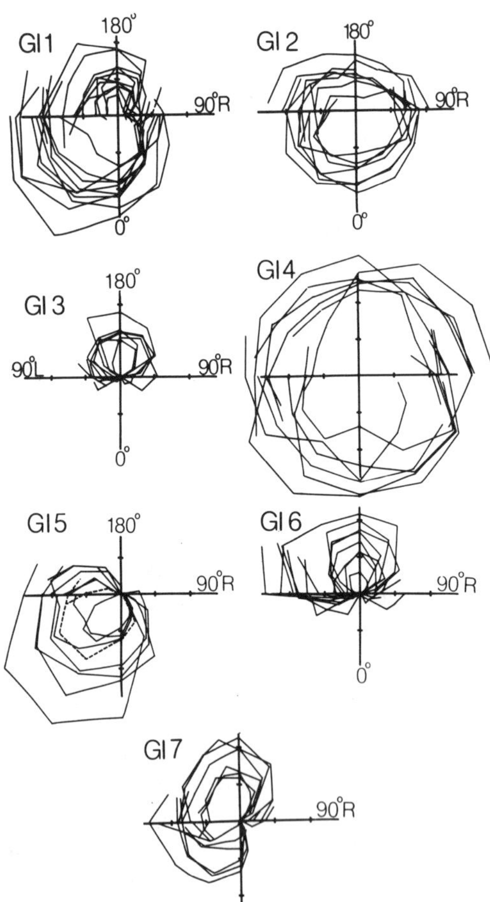

Figure 7. Directional selectivity curves for all seven histologically identified GIs. Polar plots of number of action potentials vs. wind angle constructed as in Figure 3. For each GI, curves from several different animals are plotted as if recorded from the left connective. Each point represents the mean of five trials, all σ's < 1.5 (from Westin *et al.*, 1977).

and 6 respond to wind only from the front. Each of the two GI-5s respond primarily to winds from their contralateral rear quadrant.

The mean frequency of action potentials in response to winds of 2.6 m/sec ranged from 207 ± 61/sec for GI-6 to 354 ± 79/sec for GI-2. The first two action potentials of GI-2 reach an instantaneous frequency of 900/sec. All of the GIs respond to winds of 0.1 m/sec (the lowest wind velocity tested in these experiments). With GI-1 and all of the dGIs, the number of action potentials in a response increases as the wind velocity is increased between 0.1 m/sec and 2.6 m/sec. However, GIs 2, 3, and 4 (all vGIs) reach a plateau at 0.5 m/sec. Camhi and Nolen (1981) have subsequently demonstrated that the threshold wind velocity for the GIs is much lower than 0.1 m/sec. Indeed, puffs of 0.012 m/sec (the behavioral threshold for escape) are suprathreshold for all of the vGIs and for GI-5 of the dGIs.

The frequency of action potentials for the dGIs remains relatively constant throughout the duration of wind stimulation. In contrast, the instantaneous frequency for the vGIs declines throughout the response burst. With GIs 1 and 4 the decline is gradual. However, GIs 2 and 3 are quite phasic.

To demonstrate that the responses recorded in GIs were totally due to cercal inputs, experiments similar to those described in the behavioral section were performed (Westin *et al.*, 1977). Covering the two cerci with petroleum jelly completely abolished the wind response in the GIs. Covering the animal's entire cuticle except the cerci had no influence on the response.

If only one cercus was covered or removed, the response of each GI was reduced. Indeed, the responses of GIs 1, 2, 3, and 6 ipsilateral to the ablated cercus were virtually eliminated, whereas the responses in these GIs contralateral to the ablated cercus were not appreciably affected. These GIs must receive little if any input from the contralateral cercus. In contrast, GIs 4, 5, and 7 are still excited, albeit much less, in the absence of their ipsilateral cercus. Responses for these GIs are also reduced following removal of the contralateral cercus. Therefore, they must receive inputs from both cerci.

The manner in which the sensory neurons confer directionality on the GIs is not totally understood. Certainly some differential connection of various columns of hairs must exist. However, the difference between the wind response curves of GIs 1 and 2 appear to result from differential strength of sensory inputs (Daley, 1982). Both GIs 1 and 2 receive inputs from all nine major columns of hairs. The ipsilateral bias of GI-1 arises directly from the wind input represented in cercograms of one intact cercus (cf. Figures 4B and 7). In contrast, hairs in columns *a* and *h* have proportionally stronger inputs to GI-2 than to GI-1. This increases the responses of GI-2 to wind from the contralateral front and rear quadrants, resulting in a symmetrically omnidirectional wind response curve.

The information on wind direction encoded in the relative activity of the various GIs is rapidly conducted to the thoracic ganglia. By comparing the activity of all members of either the dGI or the vGI populations, the animal should be capable of determining the direction from which the wind originated. This can then be used to direct leg motor neurons in the ultimate behavioral response.

3.3.3. Cricket GI System

Very similar and probably homologous GI systems exist in crickets. As in *P. americana,* the cricket *Acheta domesticus* has two populations of GIs (Mendenhall and Murphey, 1974). In cross sections of connectives, only three vGIs are found, one of which is small (as in GI-4 in *P. americana*). The two large vGIs are referred to as the medial and lateral giant interneurons (MGIs and LGIs, respectively) and are probably homologous to two of the large vGIs in *P. americana*. There are four dorsal interneurons Although most cross sections of *P. americana* connectives show three obvious dGIs, occasionally a fourth one is also seen. The morphology of the cricket GIs in the terminal ganglion (Mendenhall and Murphey, 1974) is also similar to that of cockroach GIs. A single soma is located contralateral to the axon, and in the vGIs the major dendritic branches are also contralateral to the soma.

There are only two sets of hairs in the cricket. L hairs bend longitudinally with respect to the long axis of the cercus. T hairs bend transversely across the long axis (Palka *et al.*, 1977). As in the cockroach, these respond to low frequency sound (Palka *et al.*, 1977) and to wind (Tobias and Murphey, 1979). The sensory neurons and GIs have been studied in more detail in cricket (Palka and Olberg, 1977; Matsumoto and Murphey, 1977*a*) than in cockroach, and this system has been used extensively in developmental experiments (Palka and Edwards, 1974; Matsumoto and Murphey, 1977*b*; Murphey and Levine, 1980; Levine and Murphey, 1980). However, less is known about the motor outputs and the behavioral role of these systems in the cricket than in the cockroach.

3.3.4. Motor Outputs from GI Activation

The primary problem with models that have implicated GIs in escape has been the difficulty in eliciting motor responses by stimulating GIs. Dagan and Parnas (1971) demonstrated that in a dissected preparation a single stimulus delivered extracellularly to the abdominal cord at threshold for activating GIs fails to evoke a motor response. This was later confirmed by Iles (1972), who further could not detect synaptic potentials in D_f (the fast depressor of the coxa) associated with GI action potentials. Intracellular stimulation of individually identified GIs also consistently failed to evoke motor responses when single stimulus pulses were used (Ritzmann and Camhi, 1978). However, the responses of GIs to wind stimulation, as determined by Westin *et al.*, (1977), indicate that a single action potential in a single GI is in fact a very weak stimulus. Wind puffs of 2.8 m/sec generate trains of action potentials with frequencies as high as 300/sec. Moreover, this occurs in 8–12 GIs depending on the direction of the wind stimulus. When high-frequency stimulus trains were tested even in individual GIs, several of them consistently evoked motor responses (Ritzmann and Camhi, 1978). Moreover, the motor neurons that were excited by each GI would produce movements consistent with the behavioral outputs observed in response to wind directions to which that GI responded.

The basic experimental paradigm for the motor response experiments of Ritzmann and Camhi (1978) was as follows. The animal was pinned to a cork dorsal-side-up, and a window was opened in the dorsal cuticle. After removing the gut, fat, and extraneous muscle tissue, the ventral nerve cord was raised onto a wax-covered stainless steel platform. This supported the cord so that a GI could be impaled with a microelectrode. A pair of hook electrodes were placed under the A_3–A_4 connective. These could be used to stimulate the cord while GIs were being penetrated. They could also monitor cord activity as the impaled GI was stimulated through the microelectrodes, thus assuring that each stimulus pulse excited one and only one GI.

Leg motor neurons were monitored by recording with extracellular suction electrodes on nerve branches 5r1 and 6Br4. Activity from depressor motor neurons D_f and D_s could be readily detected in 5r1 records and levator activity could be identified in 6Br4. After recording the motor response to GI stimulation, Procion yellow M4RS or Lucifer yellow CH was injected into the GI iontophoretically so that it could be identified in cross sections of abdominal ganglia.

3.3.4a. Dorsal Giant Interneurons. In terms of selectivity to wind inputs, GI-5 is the most directional GI. It responds exclusively to wind from the ipsilateral rear quadrant. Stimulation of GI-5 consistently excites the ipsilateral slow depressor (D_s) motor neuron (Figure 8A). Levator motor neurons may also be excited but only after D_s. The D_s motor neuron of the contralateral metathoracic leg is also excited by GI-5, but it is much

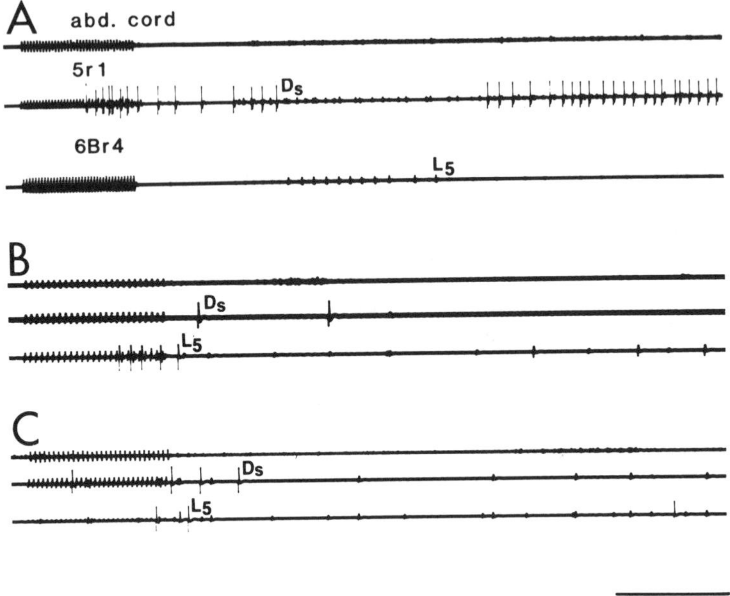

Figure 8. Motor responses recorded extracellularly in metathoracic leg nerves 5rl and 6Br4 in response to trains of current pulses delivered intracellularly to GIs 5, 6, and 7 (A, B, and C, respectively). (A–C) Top trace represents abdominal nerve cord; middle trace represents nerve branch 5rl, which innervates depressor muscle; and bottom trace represents nerve branch 6Br4, which innervates levator muscle. Regular pulses in the abdominal cord are GI action potentials. Pulses associated with these in 5rl and 6Br4 are stimulus artifacts. Action potentials from the slow depressor (D_s) and a levator motor neuron (L_5) are labeled. Calibration represents 75 msec for (A) and 60 msec for (B and C).

weaker and the latency is consistently longer than the activity generated in ipsilateral D_s. The bias toward ipsilateral D_s activation would, in a free-ranging animal, result in a turn away from the ipsilateral rear quadrant, the origin of wind stimuli that excite GI-5.

It should be noted that even in the ipsilateral D_s response to GI-5 activation, the latency from the first GI action potential to the first motor action potential is always quite long (mean of 33.7 ± 13.4 msec reported in Ritzmann and Camhi, 1978). This is also true for all other GI-activated motor outputs and suggests that the GI-to-motor neuron pathway is not monosynaptic. There is a possibility that this time is taken by subthreshold events summing to reach threshold for action potentials. However, motor neurons involved in flight that are also excited by dGIs have been recorded intracellularly in conjunction with dGI stimulation (Ritzmann *et al.*, 1982*b*).

In these motor neurons the initial depolarization is detected only after latencies of 24–34 msec. Similar results have been found in motor neurons with axons in nerve 5 that presumably control leg movements (Ritzmann and Pollack, unpublished data).

Giant interneuron 6 (Figure 8B) is opposite to GI-5 with regard to both sensory input and motor output. It responds only to wind from the front and excites initially ipsilateral levator motor neurons (Ritzmann and Camhi, 1978). The leg movements in response to wind from the front are not as consistent as those in response to wind from the rear. Nevertheless, the most typical initial movement of the ipsilateral rear leg in response to wind from the front is levation. Presumably turns of greater than 90° require more than one leg movement. The turn may be started by another set of legs (perhaps those of the prothorax) while the ipsilateral metathoracic leg is levated and protracted for a subsequent depression that completes the turn.

Giant interneuron 7 combines the inputs of GIs 5 and 6 in that it is excited by wind from the front and ipsilateral rear. The motor output reflects this combination (Figure 8C). Both levator and depressor motor neurons are excited by GI-7 with no consistent bias towards one or the other.

The unbiased input and output of GI-7 makes it an excellent candidate for a general activation neuron that would augment the more biased outputs of GIs 5 or 6. Paired intracellular stimulation of GIs indicates that GI-7 can indeed play this role (Ritzmann and Pollack, 1981). When paired with either GIs 5 or 6, GI-7 increases the motor output and decreases the latency to that output while maintaining the bias towards depressor or levator activity of the other dGI. Thus, paired stimulation of GIs 5 and 7 evokes significantly more D_s action potentials at a shorter latency than either GIs 5 or 7 individually. Similarly, stimulation of GIs 6 and 7 evokes an augmented levator response. GIs 5 and 6 are coactivated only by winds perpendicular to the animal's long axis. Nevertheless, stimulating GIs 5 and 6 together also evokes a stronger response. However, although the levator/depressor bias is consistent in trials of a single preparation, it varies from preparation to preparation.

3.3.4b. Lesioning Single GIs. Since any wind excites 8–12 GIs and summation among GIs occurs, is any single GI essential to the motor output? To answer this question, Westin and Ritzmann (1982) determined the motor response to wind stimulation before and after lesioning a single GI.

Wind stimuli were presented from various different directions while recording extracellularly from metathoracic leg motor neurons. In this preparation, activity in either D_s motor neuron is stronger in response to wind from its ipsilateral rear quadrant than from its contralateral rear quadrant (Figure 9). Levator activity was less consistent.

The stronger response from the ipsilateral rear quadrant could be due to activation of the ipsilateral GI-5. To test this hypothesis, GI-5 was impaled with a microelectrode filled with $CoCl_2$. In addition to its utility in intracellular dye marking, $CoCl_2$ is a strong neurotoxin (Dagan and Sarne, 1979). Injection of $CoCl_2$ into GI-5 created a block for anterior propagation of action potentials, thus selectively eliminating it from the wind-to-motor pathway. As a result of lesioning GI-5, the strong D_s response from the ipsilateral rear was drastically reduced or eliminated (Figure 9). Thus, at least GI-5 plays a critical role in the biased initiation of D_s activity.

3.3.4c. Ventral Giant Interneurons. Giant interneurons 5–7, which consistently evoke motor outputs, are all dGIs. Of the vGIs, only GI-1 has been found to be effective in evoking motor outputs. Stimulating GIs 2 or 3 does not elicit a motor response. Giant interneuron 4 has also been ineffective, but because of its small diameter relative to GIs 2 and 3, which are located on either side of it, GI-4 has been tested in only a few preparations.

In most cases, stimulation of GI-1 evokes action potentials in the widespread common inhibitor (CI) and one to two action potentials to D_s (Ritzmann and Camhi, 1978). Fourtner and Drewes (1977) have reported that tactile stimulation of cerci or electrical stimulation of cercal nerves

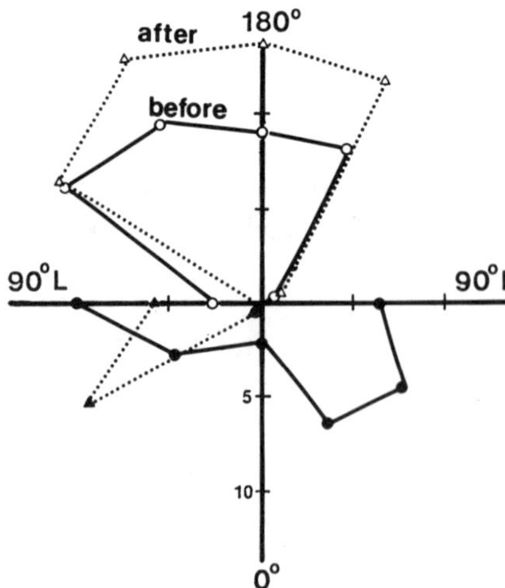

Figure 9. The effect on D_s output of blocking conduction in GI-5. Responses to wind stimulation of the right D_s motor neuron are shown in an intact animal (——) and in the same animal after conduction in the right GI-5 had been blocked (---) by injection of cobalt ions into its axon. The D_s response to wind from the ipsilateral rear was abolished by blocking GI-5, while responses to other wind angles remained. Responses to wind from the front of the animal actually increased slightly following blockage of GI-5, but this was not seen in other animals (from Westin and Ritzmann, 1982).

produces initially a burst of action potential in CI. This is also true for most wind-activated responses (Westin and Ritzmann, 1982) and could be due to activation of GI-1.

Although the D_s response is typically very weak, occasionally the initial response is quite strong (Ritzmann, 1981). However, even in these cases the response wanes quickly in subsequent trials (Figure 10). This suggests that the excitatory pathway between GI-1 and depressor motor neurons contains one or more extremely labile connections. In many preparations, the process of setting up the animal and impaling GIs probably renders this pathway refractory.

In spite of its inability to elicit motor responses on its own, GI-2 activity can enhance the motor output of GI-1. In several preparations, paired stimulation of GIs 1 and 2 evoked significantly more D_s action potentials than did GI-1 alone (Figure 11). An interesting aspect of these pairs is that beyond some threshold, the duration of the GI-2 train has no bearing on the extent of the increase in motor output. This is unlike the case for dGIs and correlates with the phasic nature of the responses to wind stimulation recorded in GIs 2 and 3. Pairing GI-3 and GI-1 has not been effective in altering motor responses. However, its role may be more

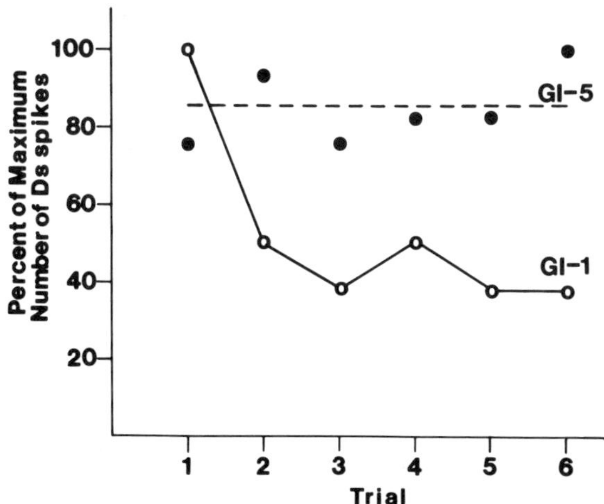

Figure 10. A comparison of the motor response of the first six trials of a preparation in which a ventral GI (○) was stimulated and the first six trials of another preparation in which a dorsal GI (●) was stimulated. The strongest response was taken as 100%. Note that in the ventral GI trials, the response declined sharply after the first trial. However, in the dorsal GI the response stayed at about the same level throughout the six trials. Lines were fitted by eye (from Ritzmann, 1981).

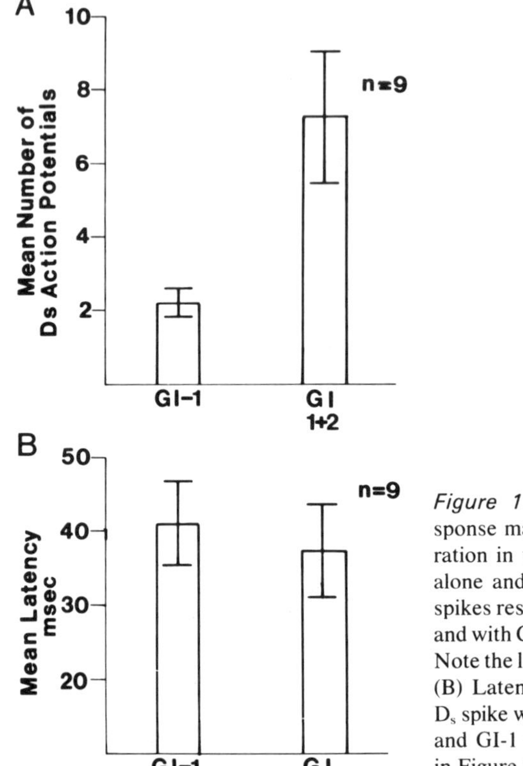

Figure 11. Bar graphs showing motor response magnitudes and latencies for a preparation in which GIs 1 and 2 were stimulated alone and together. (A) Mean number of D_s spikes resulting from stimulation of GI-1 alone and with GI-2. GI-2 alone evoked no response. Note the larger D_s response in the paired trials. (B) Latency from stimulus onset to the first D_s spike was not significantly different in paired and GI-1 trials. Error bars in this figure and in Figure 12 indicate ± 2 SE (from Ritzmann, 1981).

complex. Perhaps GI-3 combines with GIs 1 and 2 and reverses the bias from depression to levation. If so, this effect might only be detected if all three GIs were stimulated simultaneously, a technically demanding experiment.

It should be noted that all of the motor outputs reported for dGIs or vGIs are considerably weaker than that expected in an escape response. They are all limited to slow motor neurons and all have fairly long latencies. Escape, in a relatively freely moving animal, involves much stronger responses including activation of fast motor neurons. Stronger responses with shorter latencies were found in paired experiments. This could be further improved if the full complement of GIs activated by wind could be stimulated simultaneously. Indeed, it is possible that the relatively high threshold for fast motor neurons (Pearson and Iles, 1970) could be surpassed in this manner.

An additional and potentially more important factor is that the restraint and dissection required in intracellular experiments greatly reduces the responsiveness of the system. Restraint has been shown to inhibit the escape system of crayfish (Krasne and Wine, 1975), and Camhi (unpublished data) has noted a similar effect in cockroach; Eaton and Hackett describe a similar effect in goldfish (Chapter 8, this volume). Indeed in an animal that is freely walking on a greased plate, holding one leg can totally eliminate vigorous escape responses (Ritzmann, unpublished data).

3.3.4d. Pairs Involving dGIs and vGIs. Another possibility is that strong motor outputs require activity in both dGIs and vGIs. The motor responses from excitation of dGIs would provide a base of activity on which vGI activation could sum to yield the strong responses observed in escape. Without the dGI-mediated activity, vGIs would be incapable of reaching threshold for any but the weakest motor responses.

As attractive as this hypothesis is, it is not supported by the results of paired experiments involving one dGI and one vGI. Rather than revealing positive summation, the only effect found in dGI/vGI pairs was a decrease in the motor output below that found for the dGI alone. This was true even in pairs including GIs 1 and 5 (Figure 12). Since both of these normally excite the same motor neuron (D_s) this would be the optimal pair for testing the hypothesis.

3.3.5. Functional Separation of dGIs and vGIs

The experiments pairing a dGI and a vGI suggest that these are not only morphologically distinct, but also functionally separate groups of interneurons. This conclusion is supported by several other lines of evidence. In addition to the various physiological differences between dGIs and vGIs that have been noted above, the effects during walking observed in these two groups are totally different. In many systems sensory inputs that are used to evoke locomotor activity are suppressed during locomotion, thus preventing positive feedback loops (Kennedy *et al.*, 1974; Russell, 1976). This is also true for the large GI system of cricket (Murphey and Palka, 1974). In the cockroach, vGI activity is reduced during walking, but dGI activity is actually enhanced (Figure 13) (Delcomyn and Daley, 1979; Daley and Delcomyn, 1980*a*,*b*). Both of these effects appear to arise in large part by central connections. However, the reduction of vGI activity during walking does have a peripheral component, perhaps due to the physical deflection of hairs caused by wind generated by walking movements (Orida and Josephson, 1978; Daley and Delcomyn 1980*b*).

Beyond the physiological and morphological characteristics of dGIs

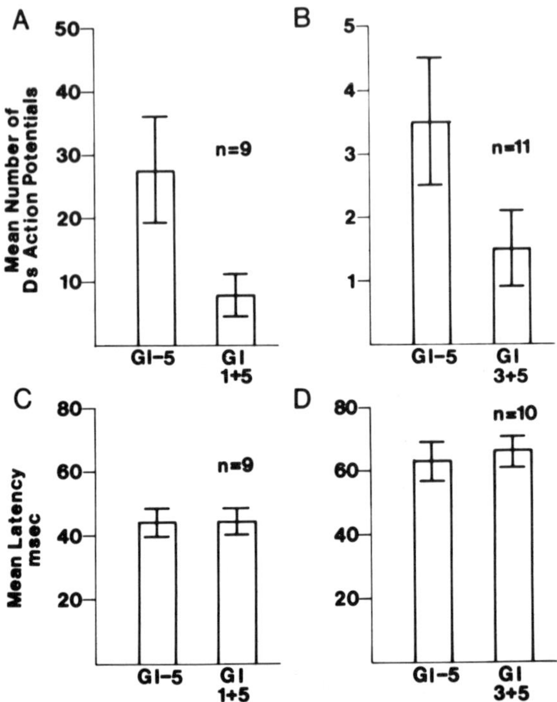

Figure 12. Bar graphs showing the mean responses in paired and unpaired trials involving one dorsal and one ventral GI. Data from two preparations are shown; one (A, C) involving GIs 1 and 5, and one (B, D) involving GIs 3 and 5. In both preparations, the number of D_s spikes in the paired trails was less than that in the GI-5 trials (A, B). However, the mean latency to the first D_s spike was not changed (C, D) (from Ritzmann, 1981).

and vGIs, Westin *et al.* (1977) pointed out that summation between dGIs and vGIs is simply not necessary. Either group could detect wind direction perfectly well independent of the other group. In this discussion, GI-4 (a small omnidirectional vGI) was included with the dGIs due to its similar latency. But it is not required in either group in order to determine wind direction. In the ventral system, wind from the right rear would be detected by activity in right GI-1 coupled with a lack of activity in either GI-3. Wind from the left rear quadrant would be signaled by activity in left GI-1 and again silence in GI-3s. Wind in the front quadrants would be indicated by the same relative activities in GI-1s plus activity in the two GI-3s. Giant interneuron 2 would be excited in all cases. This would enhance the signal and assure against the generation of escape responses due to spurious activity in GI-1 alone.

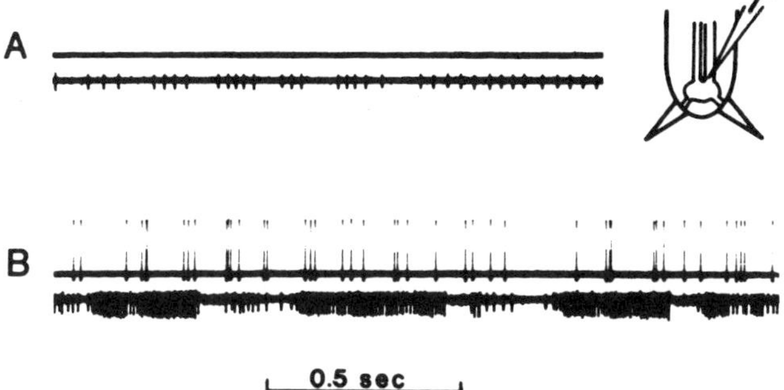

Figure 13. Intracellularly recorded activity from the axon of a dorsal GI-7 during (A) a period of rest and (B) slow walking. Inset represents intracellular recording arrangement. Upper traces are the intracellular record and the lower traces are electrical activity in extensor muscle 177 of the right rear leg (from Dale and Delcomyn, 1980).

In the dorsal system the two GI-7s serve as general excitors. Activity in either right or left GI-5 clearly denotes wind from right or left rear quadrants respectively. Wind from the front excites the two GI-6s. Right or left front is more difficult to surmise with the dorsal system. However, both GIs 6 and 7 respond more to ipsilateral wind. This could provide the left–right bias. In fact, in experiments where the dorsal system is probably employed, motor responses to wind from the front are not as predictable as those from the rear (Westin and Ritzmann, 1982).

3.3.6. Relative Roles of dGIs and vGIs

Why does the cockroach have two separate systems for detecting wind stimuli? At this time we can propose only hypotheses and several are possible. First of all, we must realize that only a small part of the motor consequences of stimulating either system is presently known. The GIs have branches in all abdominal and thoracic ganglia, not just in T_3 (Farley and Milburn, 1969; Harris and Smyth, 1971). Indeed, the vGIs proceed uninterrupted to the supraesophageal ganglion (Spira *et al.*, 1969*b*; Farley and Milburn, 1969). Even in T_3 the motor neurons that have been monitored are those only of the coxal levator and depressor motor neurons. All segments of all six legs are involved in the escape movements. If the entire constellation of motor outputs was known for each system, we might find that the dGIs and vGIs are not at all redundant.

Nevertheless, given that our present knowledge suggests that the two systems are similar, at least two explanations for their existence are possible: (1) the dGIs may serve as a slower, less labile backup system for the vGIs, and (2) the two systems may function sequentially; the vGIs initiating the escape response, and the dGIs completing it.

In several systems employing giant axons for escape, a smaller fiber system is also present in parallel with the giant axons (Krasne, 1965; Schrameck, 1970; Wine and Krasne, 1972; Eaton *et al.*, 1982). The GIs provide more rapid responses to novel threats, but tend to habituate rapidly and perhaps as a result may be actively inhibited if the animal is restrained and escape is hopeless (Krasne and Wine, 1975). The smaller fiber system provides a slower but safer backup system to generate escape movements.

In the cockroach, the vGIs may represent the "true" GI system, whereas the dGIs along with other non-GIs (Westin *et al.*, 1977, Ritzmann and Pollack, 1981) perhaps provide a smaller backup system. The vGI system is significantly faster than the dGIs. The latency for wind-evoked activity reaching the thoracic ganglia is approximately 5 msec less for the vGIs than for the dGIs (Westin *et al.*, 1977). The vGI system is also much more labile than the dGI system (Figure 10B). This could explain the difference in the changes during walking in the two systems. The more labile vGI system must be inhibited to protect the vGI-to-motor pathway for habituation. The dGI system does not readily habituate and can therefore be potentiated by the increase in ongoing activity. The negative influence of vGI activity on motor outputs from dGI stimulation could establish a hierarchy. In a novel situation, vGI activity would not only provide strong motor outputs, but would also effectively prevent the expression of motor outputs from the dGIs. As the motor response habituates, the repression of dGI activity would also habituate, allowing the dGI pathway to function.

An alternate hypothesis has been put forward by Camhi and Nolen (1981). They reason that, at least during slow walking, the vGIs are solely responsible for initiating escape, and the dGIs are responsible for subsequent turning and running movements (Figure 14). During slow walking the mean latency for escape is 14 msec and can be as low as 11 msec (Camhi and Nolen, 1981). According to their calculations, if the vGI pathway is used, at least 8 msec are needed for neural information to flow through the entire escape system. In the shortest latency (11 msec), this would only allow 3 msec more for conduction through GIs. Even the mean latency of 14 msec would allow only 6 msec of additional time. Wind-evoked activity in the dGIs reaches the thoracic ganglia about 5 msec after activity in the vGIs. Thus, under these conditions, the dGIs would probably not be fast enough to evoke the escape movements. During the initiation phase, the vGIs would suppress the dGI system. How-

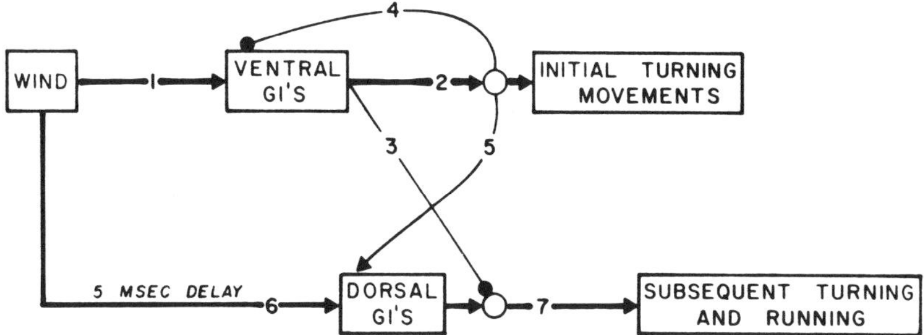

Figure 14. Tentative flow diagram of the activation of escape behavior by the dorsal and ventral GIs. The two circles, one following the ventral GIs (on path 2) and the other following the dorsal GIs (on path 7) represent thoracic GI-to-motor centers. All interactions shown hav been demonstrated. The model reveals that two sets of GIs may work sequentially; first the ventrals would initiate the oriented escape response, then the dorsals would continue to drive the turn during running (from Camhi and Nolen, 1981).

ever, once the cockroach started running, the vGIs would themselves be suppressed, releasing the dGIs to continue the escape movements.

The finding that vGIs are inhibited during walking would seem to argue against this model. However, the extent to which the vGIs are inhibited is correlated with walking speed. Under slow walking the inhibition of vGIs is minimal (Daley and Delcomyn, 1980*a*). With increasing walking speeds, vGI inhibition increases, but so does the latency to escape (Camhi and Nolen, 1981).

Support for this, or any model proposing that vGIs provide the primary input for escape, will require more information on the motor outputs generated by vGI activity. Although the dGI-mediated motor outputs are consistent with behavioral data, the only motor response to vGI stimulation is activation of D_s. How does wind from the front quadrants evoke rearward turns via vGIs? The lability of the vGI system makes this a difficult problem to resolve. Until recently all behavioral analyses have been performed on free-ranging or minimally restrained animals, whereas neurophysiological experiments have been performed on highly dissected and restrained preparations. As is evident from the relatively weak motor outputs recorded in these preparations, one cannot assume that the escape circuitry performs in the same manner under these two conditions. Attempts are presently being made to perform neurophysiological experiments on preparations that are closer to those used in behavioral experiments.

One approach being used by C. M. Comer and J. M. Camhi (personal communication) is to impale a vGI through a small window in the cuticle, inject a toxin into the vGI to kill it, and then allow the wound in one cuticle to heal. Once the cockroach recovers, it can then be tested as a free-ranging animal. A postmortem reveals which GI was lesioned. Preliminary experiments have been encouraging, but more are required. Until more information is available on how the vGIs function, we will not be able to establish what the relative roles of the vGIs and dGIs are.

4. Alterations of the Escape Response

Rapid escape responses are often thought to arise from stable neural circuits. A system that is designed to simply remove the animal from a threatening situation as fast as possible needs little modification. Nevertheless, the circuitry involved in the cockroach escape system does undergo modulation. Two examples have been documented. One is a long-term form of recovery in which activity in the GIs is modified over a period of several days in response to cercal ablation. The other example is a more transient switch in the motor response to dGI activation caused by changes in the momentary conditions the animal encounters.

4.1. Recovery from Cercal Ablation

Quite often individual cockroaches are found with one cercus missing. A missing cercus alters the response properties of GIs to wind stimuli and thus should create serious problems for the directionality of the escape system. Recent experiments by Vardi and Camhi (1982 *a,b*) indicate that the cockroach can in fact adjust to such losses.

Behavioral tests on adult animals 1–5 days after their left cercus was removed confirm that wrong turns are made (Vardi and Camhi, 1982*a*). These animals incorrectly turn left into winds from the left. Wind from the right still evokes correct left turns away from the wind source.

Over a period of 30 days these animals gradually adjust to the absence of a cercus. After 30 days, winds from the left once again evoked right turns, although the mean angle was still less than normal. During this period the cercus was not allowed to regenerate.

Last instar nymphs were also tested. They corrected even better than the adults, indicating a greater potential for plasticity in younger animals.

The mechanism for recovery involves the return of activity patterns in GIs closer to that of a normal animal (Vardi and Camhi, 1982*b*). One day after ablation of the left cercus, left GIs 1, 2, 3, and 6 show little if any response to wind stimuli. These GIs do not normally receive significant inputs from the contralateral cercus (Westin *et al.*, 1977). In contrast, the directionality curves of these GIs 30 days after cercal ablation approach the normal condition. Giant interneurons 4, 5, and 7, which do receive contralateral inputs, have reduced and altered wind response curves immediately after ipsilateral cercal ablation. Their curves are also considerably restored after 30 days.

The source of this correction appears to come from inputs originating in the intact right cercus. However, the cellular mechanism is not as yet understood.

4.2. dGIs as Bifunctional Interneurons

Leg movements associated with running occur only when the cockroach has one or more legs in contact with some surface. In the absence of leg contact, the same wind puffs evoke a totally different behavior, flying. Even though many of the same motor neurons are used in running and flying (Fourtner and Randall, 1982; Ritzmann *et al.*, 1983), the movements are quite distinct. In flight the wings unfold and beat at approximately 42 Hz, legs are held up against the animal's abdomen, and the leg levator motor neurons in nerve 6Br4 burst in phase with both the elevator and depressor cycles on the windbeat (Figure 15A, B).

The switch between running and flying could represent an inhibition of the GI systems allowing a different set of wind-activated interneurons to evoke flight. Alternatively, the GIs could remain active, but have their activity rerouted to the circuitry controlling flight. This latter possibility is in fact what happens (Ritzmann *et al.*, 1980).

Flight can be initiated in a restrained preparation by simply removing all of the legs distal to the coxa-trochantor joint. Under these conditions flight can be evoked with either a wind puff or a train of current pulses in any of the 6 dGIs (Figure 15B, C). The stimulus trains are identical to those used to generate running responses in animals with intact legs. None of the vGIs have been shown to be capable of initiating flight.

The switch between running and flying is provided by campaniform sensilla on the legs (Ritzmann *et al.*, 1980). These have been previously implicated in tarsal contact inhibition of flight (Krämer and Markl, 1978). Axons of the companiform sensilla are located in nerve N_5 of each leg. If only the prothoracic and mesothoracic legs are removed, dGI stimulation evokes a typical running response with no flight-related activity (Figure 16A). Severing N_5 in one of the remaining metathoracic legs may release some flight tendencies, but usually does not. However, when the remaining N_5 is severed, stimulation of a single dGI will evoke flight (Figure 16B). The converse experiment can also be performed. All legs are removed, and dGI stimulation elicits flight activity. However, if stim-

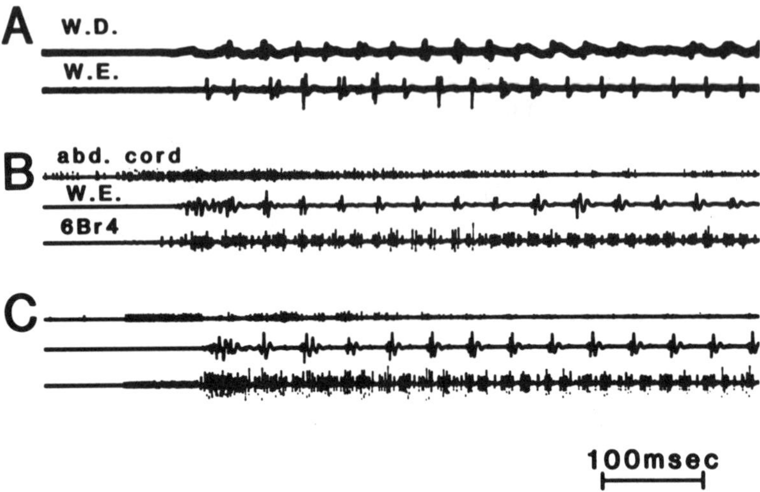

Figure 15. Demonstration of flight in legless restrained animals. (A) For comparison, EMG records from wing depressor (top trace) and elevator (bottom trace) muscles of the metathoracic segment taken during flight in an animal that has been tethered off of the ground. Note the regularly occurring muscle potentials that alternate between elevation and depression. (B) Recording of a flight sequence in an animal that was pinned to a cork but had all legs removed except the right metathoracic leg. In that leg, nerve 5 was severed. Flight was initiated by a brief wind puff directed at the cerci. In (B, C) top trace is the extracellular recording from the abdominal nerve cord, middle trace is an EMG from wing elevator muscles, and bottom trace is from leg nerve 6Br4 (contains levator motor axons of the leg). Note the EMG in the wing elevator muscle is essentialy the same as that in (A). Axons in 6Br4 burst approximately in time with elevator and depressor phases of flight. (C) GI-6 is stimulated intracellularly, in the same preparation as in (B). A train of 73 msec with 2.5 msec interpulse intervals (indicated by dGI spikes in top trace) initiates a flight sequence that is essentially the same as that in (B). Calibration applies to all records (from Ritzmann *et al.*, 1982).

ulation of N_5 in any leg is paired with dGI stimulation, flight is prevented (Figure 16C).

4.2.1. Analysis of Flight Initiation

In dissected and restrained preparations, flight behavior is in fact easier to initiate than a complete running response. This may be because the wings although cut back to stubs, are not hindered from free movement. In any event, dGI activation of flight may be instructive in understanding how dGIs evoke running.

In flight the motor neurons that control wing movements are activated in two parts (Ritzmann *et al.*, 1982). An initial depolarization is followed by a series of rhythmic depolarizations that are in phase with either el-

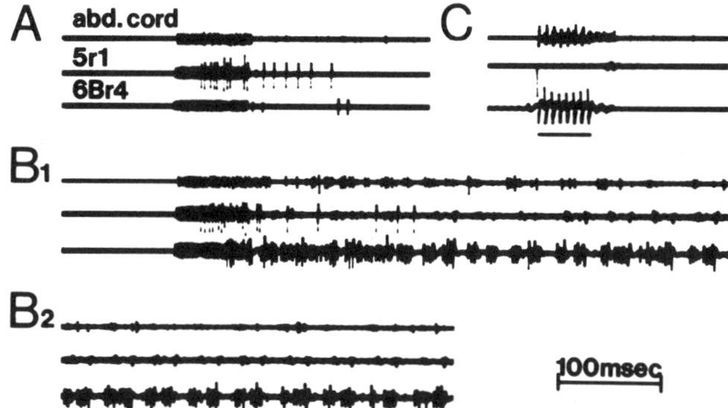

Figure 16. The effect of nerve 5 activity on flight initiation. The order of traces in (A, B): top trace is abdominal cord recording, middle trace is nerve branch 5rl containing leg depressor motor neurons, and bottom trace is nerve branch 6Br4. The prothoracic and mesothoracic legs were removed but both metathoracic nerve 5s were still intact. Under these conditions stimulation of GI-5 (73 msec duration, 2.0 msec interpulse interval) evoked activity in the slow depressor motor neuron D_s but did not evoke flight related activity. Severing the left metathoracic nerve 5 had no effect on the motor output. (B) After the right nerve 5 was also severed, so that no nerve 5 remained intact in any leg, the same stimulus to GI-5 initiated strong flight activity as indicated by the bursts in 6Br4. B_1 and B_2 are continuous records. (C) In the same preparation as in Figure 15, the same stimulus was presented to GI-6 as in Figure 15C. However, at the same time nerve 5 was stimulated with 8 pulses at a 7 msec interval. This was sufficient to prevent flight activity from occurring. A bar under the bottom trace indicates the time of nerve 5 stimulation. The calibration applies to all records (from Ritzmann *et al.*, 1980, copyright 1980 by the American Association for the Advancement of Science).

evation or depression of the wing (Figure 17). In wind-evoked flight, dGI activity is seen at or before the initial depolarization (Figure 17B, C). The activity recorded in dGIs outlasts vGI activity, but does not continue throughout the flight sequence. Thus, dGI activity is sufficient for flight initiation, but not necessary for maintaining the flight sequence.

Nevertheless, the number of dGI action potentials in the input train can influence the duration of the flight sequence. The number of bursts in the flight sequence increases linearly as more dGI action potentials are added to the input (Ritzmann *et al.*, 1982). Flight duration can also be increased by pairing two dGIs. Any two dGIs stimulated together evoke more flight bursts and have a lower threshold duration for the input train than does either dGI alone (Ritzmann *et al.*, 1982). In either case, the input signal eventually gets strong enough that the flight becomes self-sustaining. At that point, flight sequences can last well over several seconds.

All of this suggests that dGIs initiate flight by depolarizing some elements in the flight circuitry, and that the amplitude of the depolarization

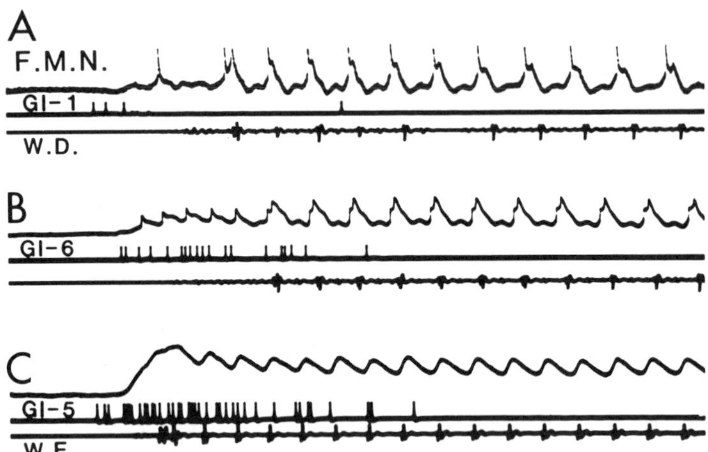

Figure 17. Records showing the relationship between GI action potentials and flight initiation in wind-mediated flight. A flight motor neuron (FMN) and a GI were impaled simultaneously and flight was evoked by a brief wind puff. GI action potentials are recorded before the initial depolarization of the FMN in all cases except (B). In (B), dGI activity occurs slightly after the initial depolarization. Presumably in this case another GI caused the initial depolarization. Even in (B), dGI activity occurs well before the onset of rhythmic activity. In each record the top trace is an intracellular record from the FMN, the middle trace is an intracellular record from a GI, and the bottom trace is an EMG from either a wing depressor (WD) or a wing elevator (WE) muscle. (A, B) From records of the same FMN (a wing depressor motor neuron) but in association with a vGI (A) and dGI (B). (C) From a different preparation in which GI-5 (a dGI) and a wing elevator motor neuron were impaled. The vertical calibration represents 160 mV in all GI traces. In the FMN traces the vertical calibration represents (A) 10 mV, (B, C) 20 mV. The horizontal calibration represents 50 msec in all records.

determines the duration of the flight sequence. This would be similar to the ramp depolarization recorded in dorsal swim interneurons of *Tritonia* (Getting *et al.*, 1980). In *Tritonia* the amplitude of the ramp depolarization affects not only the duration of the swim but also the frequency of motor bursts (Lennard *et al.*, 1980). In cockroach flight the frequency of bursts remains relatively constant throughout the flight sequence (Ritzmann *et al.*, 1982).

5. Summary

Our present evidence points to the dGIs and vGIs as playing important but separate roles in initiating locomotory movements. The larger vGIs may provide the primary initiation signal for most escape responses, especially those that occur while the animal is walking slowly. Information reaches the thoracic ganglia more rapidly via vGIs. However, the vGI system becomes ineffective after a few trials, and the dGIs do not.

The dGIs provide a bifunctional pathway by which wind stimuli can initiate one of two forms of locomotion. In the presence of leg contact, dGIs evoke a running response biased in the proper direction for the escape response. When leg contact is removed, a "switch" directs dGI activity toward the circuitry controlling flight. Thus, the dGIs are capable of initiating at least two totally different behaviors. The result is a very efficient use of a population of interneurons. One would expect to find such multifunctional interneurons in other systems also. Indeed, different kinds of multifunctional interneurons have been located in the cricket (Bentley, 1977) and crayfish (Kramer *et al.*, 1981).

Although the existence of excitatory pathways between GIs and motor neurons has been demonstrated, they appear to be polysynaptic. Does it make sense to use large-diameter axons for rapid escape only to have them connect to motor neurons via one or more interneurons? Perhaps the answer to this lies in the complexity of the cockroach locomotor systems. Unlike systems that do employ monosynaptic connections between GIs and motor neurons (e.g., crayfish and earthworm), the cockroach must coordinate six legs, each of which is composed of five segments. Moreover, the legs of each thoracic segment make movements that are different from the others. A multiplicity of monosynaptic connections between the various GIs and the constellation of motor neurons required for a turning movement may produce an insurmountable problem in coordination. One or more interneurons might be required to collect information from the GIs and direct the motor response. Given this minimal circuit, conduction of the necessary sensory information from the cerci to the motor centers of the thorax as rapidly as possible would still be advantageous. Indeed, even in the crayfish system where giants synapse directly onto fast flexor motor neurons, a parallel polysynaptic pathway now appears to provide most of the excitation (Roberts *et al.*, 1982).

Of course the large axons of the GIs also provide advantages to neurobiologists. The number of large identifiable axons in these two populations is unique. This has allowed detailed experiments to be performed on a variety of neural problems that have been reviewed in this chapter. These include processing of information on wind direction both into and from the various GIs, initiation of motor outputs, switching between different patterned outputs, recovery from sensory ablation, and the relationship of all these factors to a quantified behavior.

Note Added in Proof. Ritzmann and Pollack (in preparation) have recently identified 6 interneurons in the metathoracic ganglion that are monosynaptically excited by GIs. At least one of these can excite leg motor neurons that are excited by GI stimulation. The identification of these interneurons verifies the existence of a polysynaptic pathway between the GIs and the leg motor neurons. Moreover, they are likely to be involved in deciphering the directional information encoded in the GIs and in controlling motor responses.

ACKNOWLEDGMENTS. I thank Drs. C. R. Fourtner and J. Westin for critically reviewing an early draft of this chapter. Preparation of this chapter was supported in part by NIH grant NS 17411-01.

6. References

Bentley, D., 1977, Control of cricket song patterns by descending interneurons, *J. Comp Physiol* **116**:19–38.

Blagburn, J. M., and Beadle, D. J., 1982, Morphology of identified cercal afferents and giant interneurons in the hatchling cockroach *Periplaneta americana, J. Exp. Biol.* **97**:421–426.

Callec, J. J., 1974, Synaptic transmission in the central nervous system of insects, in: *Insect Neurobiology* (J. Treherne, ed.), American Elsevier, New York, pp. 120–185.

Callec, J. J., Guillet, P. C., Pinchon, Y., and Boistel, J., 1971, Further studies on synaptic transmission in insects. II. Relations between sensory information and its synaptic integration at the level of a single giant axon in the cockroach, *J. Exp. Biol.* **55**:123–149.

Camhi, J. M., 1976, Non-rhythmic sensory inputs: Influence of locomotory outputs in arthropods, in: *Neural Control of Locomotion,* (R. M. Herman, S. Grillner, P. S. G. Stein, and D. G. Stuart, eds.), Plenum Press, New York, pp. 561–586.

Camhi, J. M., and Nolen, T. G., 1981, Properties of the escape system of cockroaches during walking, *J. Comp. Physiol.* **142**:339–346.

Camhi, J. M., and Tom, W., 1978, The escape behavior of the cockroach *Periplaneta americana*. I. Turning response to wind puffs, *J. Comp. Physiol.* **128**:193–201.

Camhi, J. M., Tom, W., and Volman, S., 1978, The escape behavior of the cockroach *Periplanets americana*. II. Detection of natural predators by air displacement, *J. Comp. Physiol.* **128**:203–212.

Dagan, D., and Camhi, J. M., 1979, Responses to wind recorded from the cercal nerve of the cockroach. II. Directional selectivity of the sensory neurons innervating single columns of filiform hairs, *J. Comp. Physiol.* **133**:103–110.

Dagan, D., and Parnas, I., 1970, Giant fibre and small fibre pathways involved in evasive response of the cockroach *Periplaneta americana, J. Exp. Biol.* **52**:313–324.

Dagan, D., and Parnas, I., 1974, After effects of spikes in cockroach giant axons, *J. Neurobiol.* **5**:95–105.

Dagan, D., and Sarne, Y., 1978, Evidence for the cholinergic nature of cockroach giant fibers: Use of specific degeneration, *J. Comp Physiol.* **126**:157–160.

Dagan, D., and Volman, S., 1982, Sensory basis for directional wind detection in first instar cockroaches *Periplaneta americana, J. Comp. Physiol.* **147**:471–478.

Daley, D. L., 1982, Neural basis of wind-receptive fields of cockroach giant interneurons, *Brain Res.* **238**:211–216.

Daley, D. L., and Delcomyn, F., 1980*a*, Modulation of excitability of cockroach giant interneurons during walking I. Simultaneous excitation and inhibition, *J. Comp. Physiol.* **138**:231–239.

Daley, D. L., and Delcomyn, F., 1980*b*, Modulation of excitability of cockroach giant interneurons during walking II. Central and peripheral components, *J. Comp. Physiol.* **138**:241–251.

Daley, D. L., Vardi, N., Appignani, B., and Camhi, J. M., 1981, Morphology of the giant interneurons and cercal nerve projections of the American cockroach, *J. Comp. Neurol.* **196**:41–52.

Delcomyn, F., and Daley, D. L., 1979, Central excitation of cockroach giant interneurons during walking, *J. Comp. Physiol.* **130**:39–48.

Eaton, R. C., Lavender, W. A., and Wieland, C. M., 1982, Alternative neural pathways initiate fast-start responses following lesions of the Mauthner neuron in goldfish, *J. Comp. Physiol.* **145**:485–496.

Farley, R. D., and Milburn, N. S., 1969, Structure and function of the giant fiber system in the cockroach, *Periplaneta americana, J. Insect Physiol.* **15**:457–476.

Fourtner, C. R., and Drewes, C. D., 1977, Excitation of the common inhibitory motor neurons: A possible role in the startle reflex of the cockroach, *Periplaneta americana, J. Neurobiol.* **8**:477–489.

Fourtner, C. R., and Randall, J. B., 1982, Studies on cockroach flight: The role of continuous neural activation of non-flight muscles, *J. Exp. Zool.* **221**:143–154.

Getting, P. A., Lennard, P. R., and Hume, R. I., 1980, Central pattern generator mediating swimming in *Tritonia,* I. Identification and synaptic interactions, *J. Neurophysiol.* **44**:151–164.

Gnatzy, W., 1976, The ultrastructure of the thread-hair on the cerci of the cockroach *Periplaneta americana* L.: The intermoult phase, *J. Ultrastruct. Res.* **54**:124–134.

Harris, C. L., and Smyth, T., 1971, Structural details of cockroach giant axons revealed by injected dye, *Comp. Biochem. Physiol.* **40A**:295–303.

Iles, J. F., 1972, Structure and synaptic activation of the fast coxal depressor motoneuron of the cockroach, *Periplaneta americana, J. Exp. Biol.* **56**:647–656.

Kennedy, D., Calabrese, R. L., and Wine, J. J., 1974, Presynaptic inhibition: Primary afferent depolarization in crayfish neurons, *Science* **186**:451–454.

Kramer, A. P., Krasne, F. B., and Wine, J. J., 1981, Interneurons between giant axons and motoneurons in crayfish escape circuitry, *J. Neurophysiol.* **45**:550–573.

Krämer, K., and Markl, H., 1978, Flight-inhibition on ground contact in the American cockroach, *Periplaneta americana.* I. Contact receptors and a model for their central connections, *J. Insect Physiol.* **24**:577–586.

Krasne, F. B., 1965, Escape from recurring tactile stimulation in *Branchioma vesiculosum, J. Exp. Biol.* **42**:307–322.

Krasne, F. B., and Wine, J. J., 1975, Extrinsic modulation of crayfish escape behavior, *J. Exp. Biol.* **63**:433–450.

Lennard, P. R., Getting, P. A., and Hume, R. I., 1980, Central pattern generator mediating swimming in *Tritonia.* II. Initiation, maintenance and termination, *J. Neurophysiol.* **44**:165–173.

Levine, R. B., and Murphey, R. K., 1980, Loss of inhibitory synaptic input to cricket sensory interneurons as a consequence of partial deafferentation, *J. Neurophysiol.* **43**:383–394.

Matsumoto, S. G.. and Murphey, R. K., 1977*a*, The cercus-to-giant interneuron system of crickets. IV. Patterns of connectivity between receptors and the medial giant interneuron, *J. Comp. Physiol.* **119**:319–330.

Matsumoto, S. G., and Murphey, R. K., 1977*b*, Sensory deprivation during development decreases the responsiveness of cricket interneurons, *J. Physiol.* **268**:533–548.

Mendenhall, B., and Murphey, R. K., 1974, The morphology of cricket giant interneurons, *J. Neurobiol.* **5**:565–580.

Murphey, R. K., and Levine, R. B., 1980, Mechanisms responsible for changes observed in response properties of partially deafferented insect interneurons, *J. Neurophysiol.* **43**:367–382.

Murphey, R. K., and Palka, J., 1974, Efferent control of cricket giant fibres, *Nature* **248**:249–251.

Nicklaus, R., 1965, Die Erregung einzelner Fadenhaare von *Periplaneta americana* in Abhängigkeit von der Grösse und Richtung der Auslenkung, *Z. Vergl. Physiol.* **50**:331–362.

Orida, N., and Josephson, R. K., 1978, Peripheral control of responsiveness to auditory stimuli in giant fibres of crickets and cockroaches, *J. Exp. Biol.* **72**:153–164.

Palka, J., and Edwards, J. S., 1974, The cerci and abdominal giant fibres of the house cricket, *Acheta domesticus*. I. Regeneration and effects of chronic deprivation, *Proc. R. Soc. (London) Ser. B.* **185**:105–121.

Palka, J., and Olberg, R., 1977, The cercus-to-giant interneurons system of crickets. III. Receptive field organization, *J. Comp. Physiol.* **119**:301–317.

Palka, J., Levine, R., and Schubiger, M., 1977, The cercus-to-giant interneuron system of crickets. I. Some attributes of the sensory cells, *J. Comp. Physiol.* **119**:267–283.

Parnas, I., and Dagan, D., 1971, Functional organization of giant axons in the central nervous system of insects: New aspects, in: *Advances in Insect Physiology* (J. W. L. Beament, J. Treherne, and V. B. Wigglesworth, eds.), Academic Press, New York and London, pp. 95–143.

Pearson, K. G., and Iles, J. F., 1970, Discharge patterns of coxal levator and depressor motoneurones of the cockroach, *Periplaneta americana, J. Exp. Biol.* **52**:139–165.

Plummer, M. R., and Camhi, J. M., 1981, Discrimination of sensory signals from noise in the escape system of the cockroach: The role of wind acceleration, *J. Comp. Physiol.* **142**:347–357.

Pumphrey, R., and Rawdon-Smith, A., 1936, Synchronized action potentials in the cercal nerve of the cockroach (*Periplaneta americana*) in response to auditory stimuli, *J. Physiol. (London),* **87**:4P–5P.

Ritzmann, R. E., 1981, Motor responses to paired stimulation of giant interneurons in the cockroach, *Periplaneta americana*. II. The ventral interneurons, *J. Comp. Physiol.* **143**:71–80.

Ritzmann, R. E., and Camhi, J. M., 1978, Excitation of leg motor neurons by giant interneurons in the cockroach, *Periplaneta americana, J. Comp. Physiol.* **125**:305–316.

Ritzmann, R. E., and Pollack, A. J., 1981, Motor responses to paired stimulation of giant interneurons in the cockroach, *Periplaneta americana*. I. The dorsal interneurons, *J. Comp. Physiol.* **143**:61–70.

Ritzmann, R. E., Tobias, M. L., and Fourtner, C. R., 1980, Flight activity initiated via giant interneurons of the cockroach: Evidence for bifunctional trigger interneurons, *Science* **210**:443–445.

Ritzmann, R. E., Pollack, A. J., and Tobias, M. L., 1982, Flight activity mediated by intracellular stimulation of dorsal giant interneurons of the cockroach, *Periplaneta americana, J. Comp. Physiol.* **147**:313–322.

Ritzmann, R. E. Fourtner, C. R., and Pollack, A. J., 1983, Morphological and physiological identification of motor neurons innervating flight musculature in the cockroach, *Periplaneta americana, J. Exp. Zool.* **225**:347–356.

Roberts, A., Krasne, F. B., Hagiwara, G., Wine, J. J., and Kramer, A. P., 1982, Segmental giant: Evidence for a driver neuron interposed between command and motor neurons in the crayfish escape system, *J. Neurophysiol.* **47**:761–781.

Roeder, K., 1948, Organization of the ascending giant fiber system in the cockroach (*Periplaneta americana*), *J. Exp. Zool.* **108**:242–261.

Roeder, K., 1967, *Nerve Cells and Insect Behavior,* Harvard University Press, Cambridge, Massachusetts.

Russell, I. J., 1976, Central inhibition of lateral line input in the medulla of the goldfish by neurones which control active body movements, *J. Comp. Physiol.* **111**:335–368.

Schramek, J. E., 1970, Crayfish swimming: Alternating motor output and giant fiber activity, *Science,* **169**:698–700.

Spira, M. E., Parnas, I., and Bergmann, F., 1969*a*, Organization of the giant axons of the cockroach *Periplaneta americana, J. Exp. Biol.* **50**:615–627.

Spira, M. E., Parnas, I., and Bergmann, F., 1969*b*, Histological and electrophysiological studies on the giant axons of the cockroach *Periplaneta americana, J. Exp. Biol.* **50**:629–634.

Spira, M. E., Yarom, Y., and Parnas, I., 1976, Modulation of spike frequency by regions of special axonal geometry and by synaptic inputs, *J. Neurophysiol.* **39**:882–899.

Tobias, M., and Murphey, R. K., 1979, The response of the cercal receptors and identified interneurons in the cricket (*Acheta domesticus*) to airstreams, *J. Comp. Physiol.* **129**:51–59.

Vardi, N., and Camhi, J. M., 1982*a*, Functional recovery from lesions in the escape system of the cockroach. I. Behavioral recovery, *J. Comp. Physiol.* **146**:291–298.

Vardi, N., and Camhi, J. M., 1982*b*, Functional recovery from lesions in the escape system of the cockroach. II. Physiological recovery of the giant interneurons, *J. Comp. Physiol.* **146**:299–310.

Westin, J., 1979, Responses to wind recorded from the cercal nerve of the cockroach *Periplaneta americana.* I. Response properties of single sensory neurons, *J. Comp. Physiol.* **133**:97–102.

Westin, J., and Ritzmann, R. E., 1982, The effect of single giant interneuron lesions on wind-evoked motor responses in the cockroach *Periplaneta americana, J. Neurobiol.* **13**:127–139.

Westin, J., Langberg, J. J., and Camhi, J. M., 1977, Responses of giant interneurons of the cockroach *Periplaneta americana* to wind puffs of different directions and velocities, *J. Comp. Physiol.* **121**:307–324.

Wine, J. J. and Krasne, F. B., 1972, Organization of escape behavior in the crayfish, *J. Exp. Biol.* **56**:1–18.

Yarom, Y., and Spira, M. E., 1982, Extracellular potassium ions mediate specific neuronal interaction, *Science* **216**:80–82.

5

The Drosophila Giant Fiber System

ROBERT J. WYMAN, JOHN B. THOMAS, LAWRENCE SALKOFF, and DAVID G. KING

1. Introduction

Normal functioning of the nervous system depends on the formation of vast numbers of specific connections between neurons. During development, each of the thousands, millions, or billions of cells in a nervous system connects with a specific set of target cells. We currently have no knowledge of the molecular basis of this specificity. The long-term goal of our work is to identify genes that are directly involved in neural connectivity and then to use this knowledge to identify the gene products necessary for proper connectivity. It will be a major advance in neuroscience if the class (or classes) of molecules involved in nerve-cell recognition and connection can be identified.

Our research strategy is to create mutants of *Drosophila melanogaster* in which the specificity of neural connectivity is disrupted. We have developed a system in which we can detect such mutations behaviorally, physiologically, and anatomically. The system comprises 30 identified neurons in *D. melanogaster* that are responsible for the motor aspects of escape jump and flight. We can record from these cells individually, stimulate the cells individually, fill the cells with dye and take electron micrographs of identified synapses between the cells. We have put a lot

---

ROBERT J. WYMAN • Department of Biology, Yale University, New Haven, Connecticut 06511. *JOHN B. THOMAS* • Department of Biological Sciences, Stanford University, Stanford, California 94305. *LAWRENCE SALKOFF* • Department of Neurobiology, Washington University School of Medicine, St. Louis, Missouri 63110. *DAVID G. KING* • Department of Zoology, Southern Illinois University, Carbondale, Illinois 62901.

of effort into learning the anatomy, physiology, and connectivity of these neurons. In particular we know the role of each individual neuron in the generation of the starting jump and the precisely repeated cycle of activation of the flight motoneurons.

It is very easy to screen for mutants of escape behavior. A large number of flies bearing mutagenized chromosomes are generated. When alarmed, the vast majority of these will jump and fly away. Among those few which do not, some may carry mutations that cause defects in the 30 identified neurons. Using simple physiological tests we can identify these mutants.

2. Behavior

Flies will jump to a variety of stimuli. By recording electromyograms of muscle activity during these jump responses, it was found that the pattern of activation of the various muscles involved varied in different responses (Thomas, 1981). Out of this diversity of responses we have focused on one fixed action pattern. This is the jump response elicited by a rapid light-off stimulus. The giant fiber (GF) is a command neuron for this response: a single spike in the GF is sufficient to elicit the stereotypic pattern of muscle activity.

The stimulus in nature that drives the GF is unknown. Wild-type flies infrequently jump in response to a light-off stimulus. In white-eyed flies, however, the GF spike, and an ensuing jump, can be reliably evoked by a light-off stimulus. Typically, no failures occur in a train of 50 consecutive stimuli delivered at 15-sec intervals.

Several possibilities were examined to determine why white-eyed flies respond so much more readily than red-eyed flies. The normal red eye color results from a mix of two types of screening pigments. These are red drosopterin pigments derived from pteridines and brown ommochrome pigments synthesized from tryptophan (Judd, 1976). A number of mutations are known that block the synthesis pathway of either the brown or the red pigment. There are several ways to combine mutations from the two pathways to give white-eyed flies. Thus flies carrying the mutations *brown* and *scarlet,* or *vermilion* and *brown,* or *cinnabar* and *brown* all have white eyes. In addition, the mutation *white,* which may affect a protein component of the pigment granules (Ziegler, 1961), also produces white-eyed flies. All the above genotypes result in flies that jump well to a light-off stimulus (although there are minor differences in jump frequency). This indicates that the lack of an effective light-screening pigment, rather than any pleiotropic effect of a particular mutation, is responsible for the increased tendency to jump. Flies with the genotype *brown; scarlet* (*bw;st*) respond the most reliably and have been used in the mutant screen described below.

It is possible that the high light intensity experienced by white-eyed flies during development causes some developmental change responsible for the increased responsiveness. Consequently, we reared white-eyed flies in continuous dark and red-eyed flies in continuous bright light, but found no change in responsiveness.

The screening pigments are contained in a set of pigment cells (for details of anatomy, see Ready *et al.*, 1976). By preventing light that enters one ommatidium from passing into another ommatidium, they optically isolate the ommatidia. Without screening pigment, a greater fraction of the light that enters the eye is absorbed by the photoreceptor pigment. Thus to a white-eyed fly a light-off stimulus would represent a greater change in absolute intensity of stimulus than for a red-eyed fly. However, we found that white-eyed flies still responded to the switching off of an exceedingly dim spot of light focused on three or four ommatidia in an otherwise dark background. Conversely, a light 4.5 orders of magnitude more intense failed to trigger a response when focused on red-eyed flies.

If an individual red-eyed fly is placed in a diffusely lighted white drum and presented with a light-off stimulus, a jump is elicited. In contrast, if the light is a point source, the jump response is rarely seen. This suggests that activation of the giant fibers requires input from a large number of ommatidia. The white-eyed flies may be more responsive because any stimulus illuminates many ommatidia.

An alternative explanation for the greater reliability of white-eyed flies is based on an observation that motion in a wild-type fly's visual field appears to inhibit the jump response to the light-off stimulus. White-eyed flies, which are not able to perceive pattern due to light scattering within the retina, show no decrement in jumping reliability with the presence of motion in their visual field. To test the effect of motion on the jump response of red-eyed flies, flies that had previously jumped well to the light-off stimulus in a white drum were presented the diffuse light-off stimulus with and without rotation of a striped pattern. During the periods of pattern rotation, the jump response of wild-type flies to the light-off stimulus was abolished. In contrast, rotation of the striped pattern had no appreciable effect on the jumping reliability of white-eyed flies.

Drosophila has a very strong tendency to follow motion of the visual field. When presented with a moving pattern, wild-type flies show this optomotor response, whereas white-eyed flies do not. To check the possibility that the motor activity of the optomotor response is itself responsible for abolishing the jump, red-eyed flies carrying the mutation *opto-*

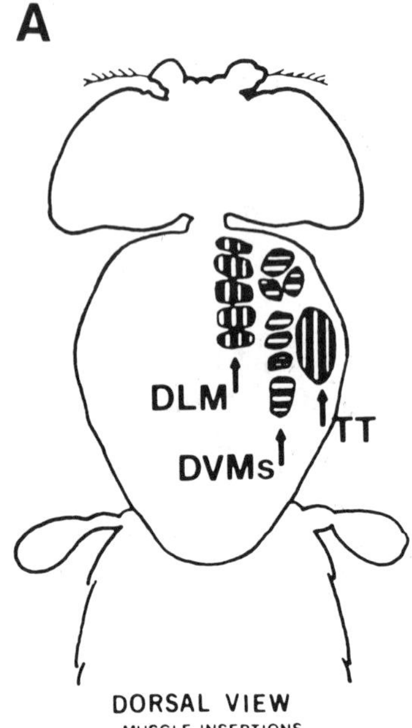

Figure 1. Indirect flight muscles and the jump muscle. A schematic diagram showing the anatomical relationships among the main muscle groups providing power for the escape response. (A) All the muscle cells insert on the dorsal thorax. This allows easy identification of units and penetration of chosen cells with microelectrodes. (B) The main wing depressor is the dorsal longitudinal muscle (DLM), seen here in a parasagittal view. Each solid strip represents a single muscle cell innervated by a single motor axon. (C) The main wing elevators are the dorsoventral muscles (DVMs). The tergotrochanter muscle (TTM) is a large nonfibrillar muscle providing the major thrust for the jump response. Parasagittal view (from Tanouye and Wyman, 1980).

motorblind (*omb*) were tested in the same experimental situation (Heisenberg *et al.*, 1978). Although the *omb* flies used showed no optomotor response to the moving pattern, their jump response to the light-off stimulus, similar to the wild-type, was abolished. Some theories of behavior presume that a critical behavior (such as escape) will take precedence over a less important behavior (optomotoring). It will be difficult to reconcile our finding that a mild optomotor stimulus inhibits the escape response with these ethological theories of behavioral hierarchies.

Each ommatidium in a *Drosophila* eye consists of eight receptor cells. Six of these rhabdomeres (1–6) are arranged on the periphery; two (7 and 8) are in the center. The different cells probably have different functions. To investigate the effect of eliminating certain receptor classes on the light-off induced jump response, the mutations *receptor degeneration B* (*rdgB*, 1:33), *outer rhabdomeres absent* (*ora*, 3:65.3), and *sevenless* (*sev*, 1:32.2) were combined with mutations conferring white eyes and tested for jumping reliability. *Receptor degeneration B* and *ora* each cause absence of the outer rhabdomeres 1–6 (Pak, 1975); *sev* flies lack receptor 7 (Harris *et al.*, 1976). Both *w;rdgB* and *w;ora* flies showed no response to the light-off stimulus; *sev;bw;st*, on the other hand, jumped as well as *bw;st* flies. Thus, the outer rhabdomeres are essential for the jump response.

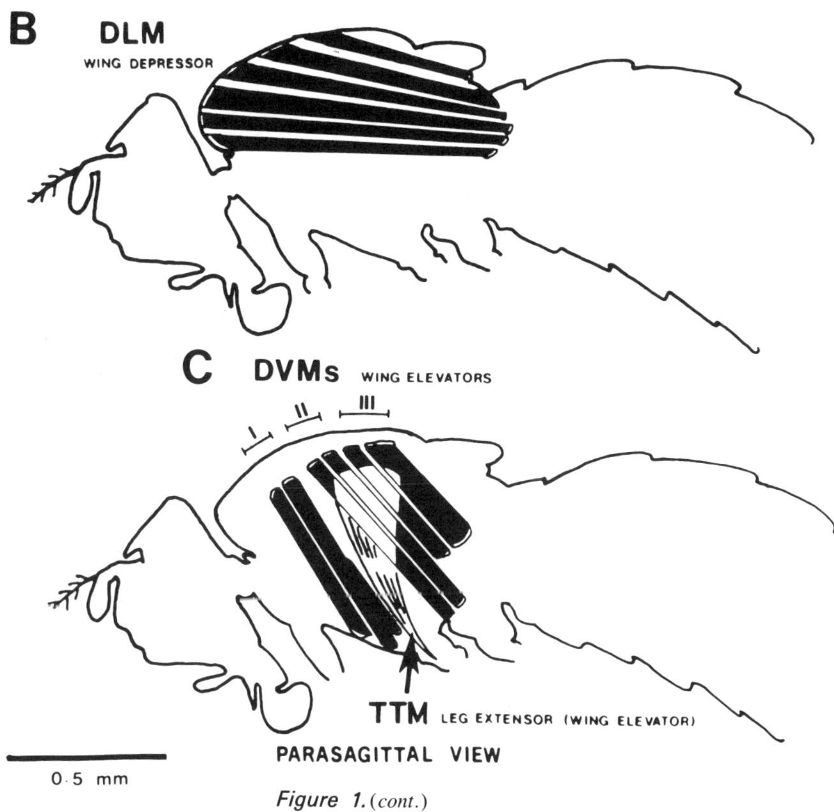

Figure 1. (*cont.*)

3. Normal Anatomy and Physiology

A startled fly jumps and initiates flight. The mesothoracic legs provide the majority of thrust during a jump (Mulloney, 1969). The tergotrochanteral muscle (TTM) is the largest twitch muscle in the fly, and acts as the main extensor muscle of the mesothoracic leg. Flight is maintained by the alternate contraction of two opposing sets of indirect flight muscles. These muscles are shown in Figure 1. The wing elevators (dorsoventral muscles, DVMs) and wing depressors (dorsal-longitudinal muscles, DLMs) are together composed of 13 bilaterally paired giant muscle fibers. Unlike the indirect flight muscles, the TTM is composed of many small muscle fibers.

The GF-mediated escape may be studied in the laboratory by suspending a fly in air and then inserting stimulating and recording electrodes into identified locations (Figure 2). As mentioned above, an effective

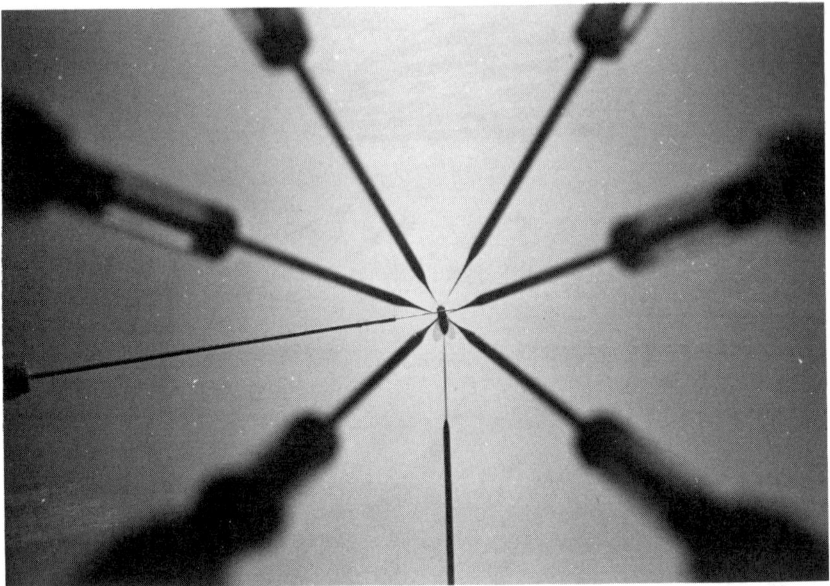

Figure 2. Experimental set up. The fly is glued to a pin (seen entering from left of photo) across the head–neck joint. A ground electrode is placed in the abdomen. Two stimulating electrodes are placed in the head and four recording electrodes are placed into identified muscle cells. The records in Figure 20 were taken with electrodes arranged as above.

stimulus to elicit this response is a sudden light-off. Approximately 24 msec elapse between the light-off stimulus and the GF spike; conduction time from the GF spike to muscle electrical response is less than 2 msec. A single spike is elicited in the TTM muscle (Figure 3) and the resultant twitch causes the jump. Flight then ensues by activity of the DLM and DVM. However, the TTM is thereafter silent. If the initial part of this response is recorded at a high-sweep speed, it can be seen that all the muscles of Figure 1 are driven in a very specific temporal pattern (Figure 4) (Tanouye and Wyman, 1980).

The GFs can be identified in electron micrographs as the largest (about 5 μm) pair of profiles in the cervical connective (Figure 5). When the GF is stimulated by an intracellular electrode, it is seen that a single spike is sufficient to drive the muscle response (Figure 6B). This is the identical response that is elicited by either a light-off stimulus or by extracellularly stimulating the GF in the brain. In both cases the threshold for the response is the threshold for a spike in the GF (as seen by an intracellular electrode).

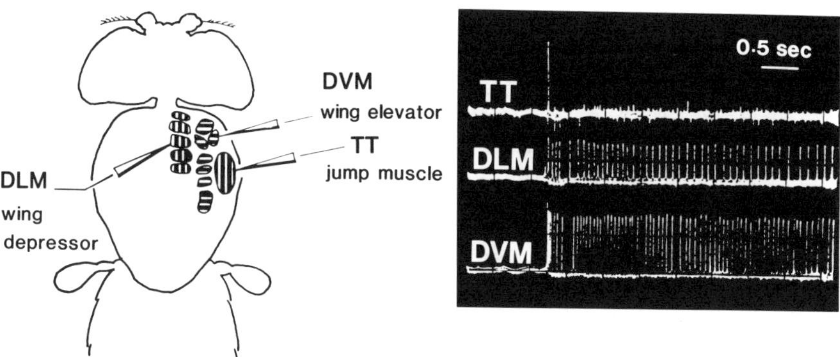

Figure 3. Recording electrodes were placed as shown in the left hand diagram. Escape is initiated by a single spike in all the fibers. Flight then continues with activity in the wing muscles while the jump muscle is inactive.

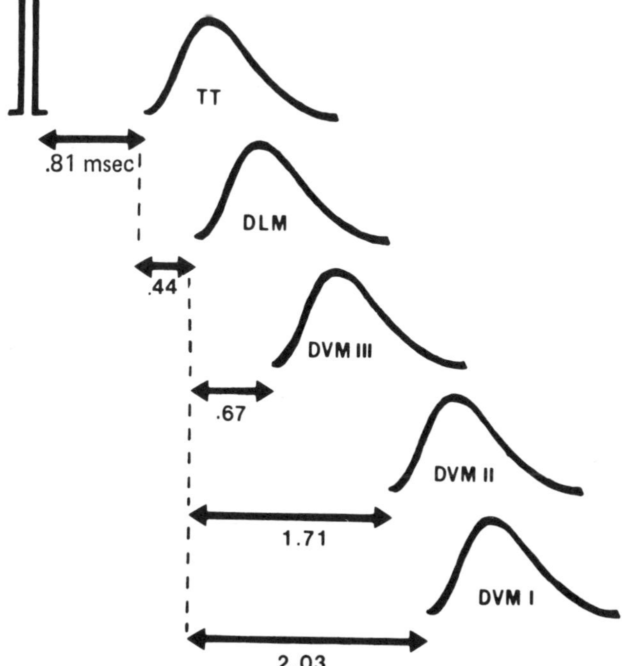

Figure 4. Pattern of muscle potentials following brain stimulation. If the initial events of Figure 3 are recorded at high-sweep speed, the timings shown here can be measured (from Tanouye and Wyman, 1980).

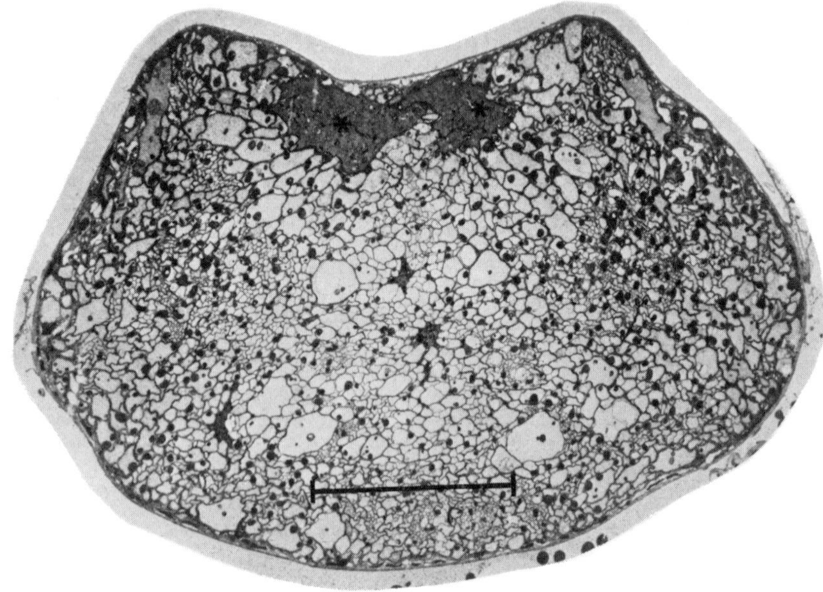

Figure 5. Electron micrograph of a cross section through the cervical connective. The two largest profiles at the dorsal surface (starred) are the GFs Scale bar: 10 μm. (courtesy of Dr. C. Bruce Boschek).

The morphology of the GF can be visualized by filling it with Lucifer Yellow (Figure 7). The GF cell body lies ipsilaterally in the brain and sends an axon through the cervical connective to the thoracic ganglion. The GF is virtually unbranched in its course through the thoracic ganglion. Within the mesothoracic neuromere, it bends laterally and then ends abruptly (Koto *et al.*, 1981). Just before the fiber curves laterally, there are small processes (arrow) that wrap around the peripherally synapsing interneuron (PSI).

When the thoracic bend region of the GF was serially sectioned for the electron microscope, it was found to synapse with the motoneuron for the jump muscle (King and Wyman, 1980). A reconstruction of the region is shown in Figure 8. The total latency from GF spike in the neck to spike in the TTM (jump) muscle is 0.8 msec. Direct stimulation of the TTM motor axon reveals a neuromuscular delay of almost 0.7 msec. This leaves time (0.1 msec) for only one electrical synapse. Thus, starting at the giant fiber, the jump response involves only one interneuronal synapse and a single spike in each element (Tanouye and Wyman, 1980). This beautifully simple circuit has proved very advantageous for finding mu-

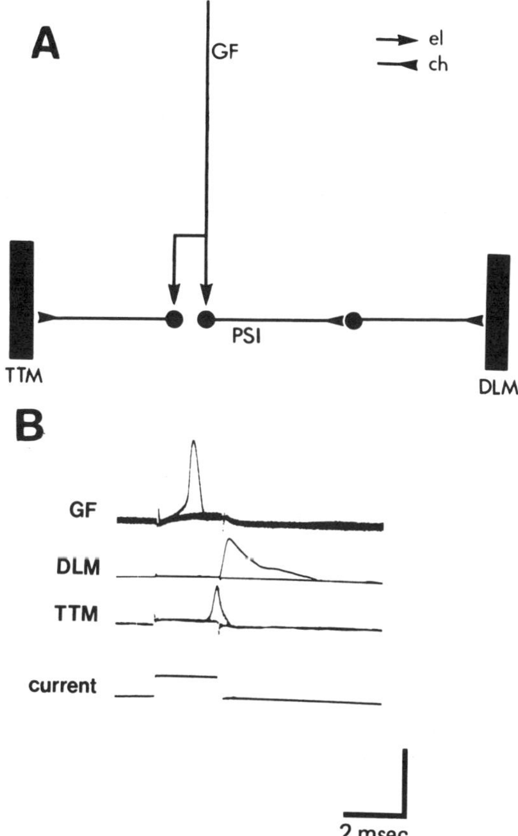

Figure 6. (A) Diagram of the identified components of the GF system. The GF drives the jump muscle (TTM) motoneuron through one electrical synapse and drives the wing depressor (DLM) motoneurons through one central electrical synapse and one peripheral chemical synapse. (B) Intracellular stimulation of the GF. Current passed through an intracellular electrode in the GF elicits a single spike in the GF. This spike in turn elicits spikes in the jump and flight muscles at the same latencies as shown in Figure 4.

tations affecting identified cells. Below, we describe our methods to obtain mutations affecting this synapse.

Just anterior to the region where the GF synapses on the TTM motoneuron it also forms a very large synaptic region with an interneuron, the PSI (King and Wyman, 1980). The electron micrograph in Figure 9 shows how the processes of the GF wrap around the PSI. Each PSI then crosses the ganglion (Figure 8) and exits the ganglion in the contralateral

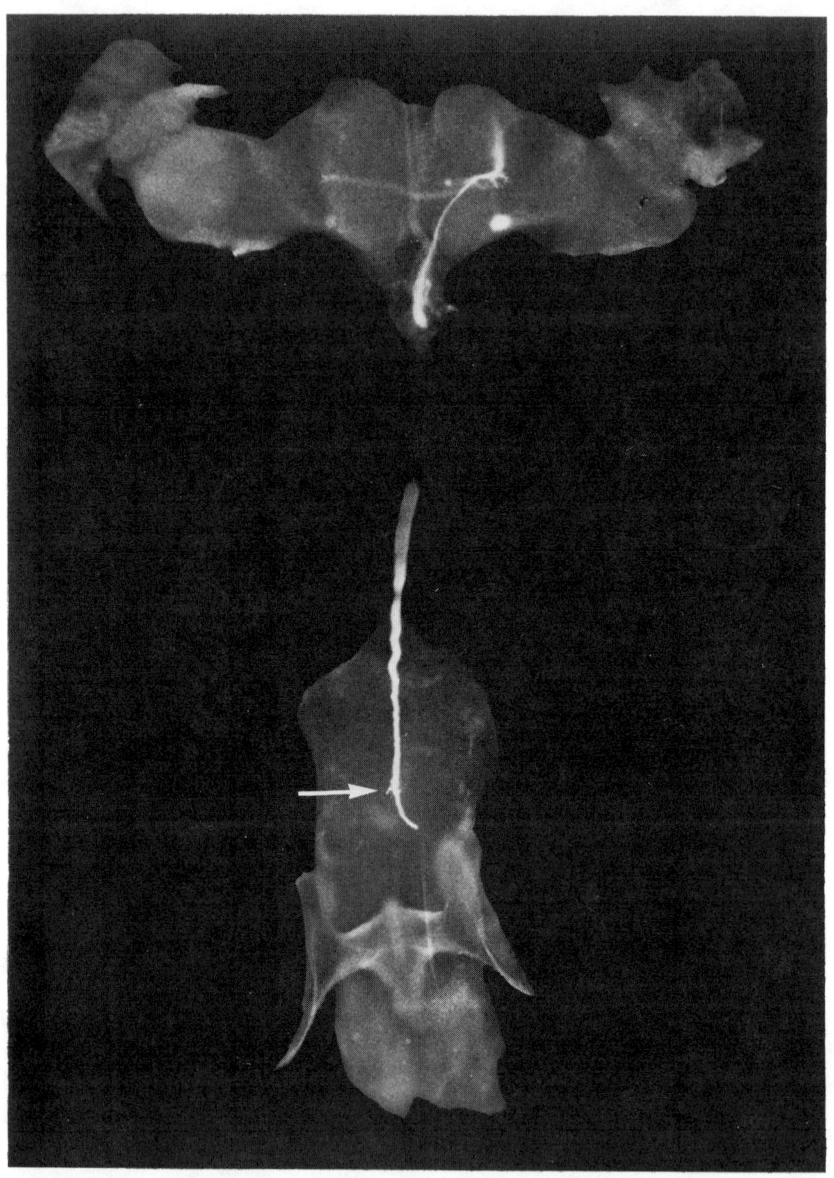

Figure 7. The GF filled with Lucifer Yellow. Wholemounts of the brain (top) and thoracic ganglion (bottom) as seen in dorsal view. The anterior part of the brain is at the top of the picture. The CGF cell body is in the posterior part of the brain. The large ipsilateral branch that ends in the antennal glomerulus runs anteriorly. A contralateral neurite and a cell body, which is probably the contralateral GF cell, are faintly stained. The GF axon turns medially, descends ventrally, and exits the brain in the cervical connective. It then descends (bottom) unbranched into the mesothoracic neuromere of the thoracic ganglion. In the mesothoracic neuromere, it sends out a small tuft of processes (arrow) and then bends laterally.

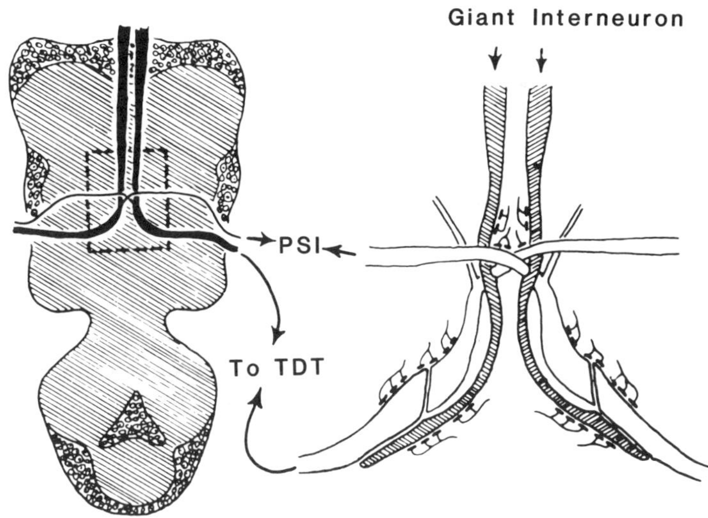

Figure 8. (A) Reconstruction of GFs projected onto a horizontal section through the thoracic ganglion. (B) Enlargement of outlined region (from D. King, 1976).

posterior dorsal mesothoracic nerve (PDMN). Along the way it picks up five axons (Figure 10), which are the motoneurons to the wing depressor muscle (DLM). In the course of the peripheral nerve, the PSI synapses on these five neurons and then terminates. A reconstruction of this is seen in Figure 11. Figure 12 shows an electron micrograph of the ganglionic face of the reconstruction. Note that the nerve is already running outside the ganglion along the muscle and all six elements are present. A few microns further (Figure 13), the PSI has terminated and the five motoneurons have greatly increased in size. A section in between these two shows the PSI synapsing on three of the motoneurons (Figure 14) and shows that the synapses are reciprocal (Figure 15). Note that the vesicles have different shapes on the two sides.

The PSI, to our knowledge, is the only reported example in the literature where two central neurons exit the CNS before making a synapse. We also found that homologous synapses exist in the housefly, *Musca domestica*, a taxonomically distant Dipteran (Valentino, 1977). How widespread the occurrence of peripheral synapses is remains to be discovered. Aside from its heterodoxy, the convenient location of this identified synapse allows us to readily determine whether any mutation that affects synaptic function also alters synaptic morphology.

In Figure 4 it can be seen that the latency from GF to DLM muscle

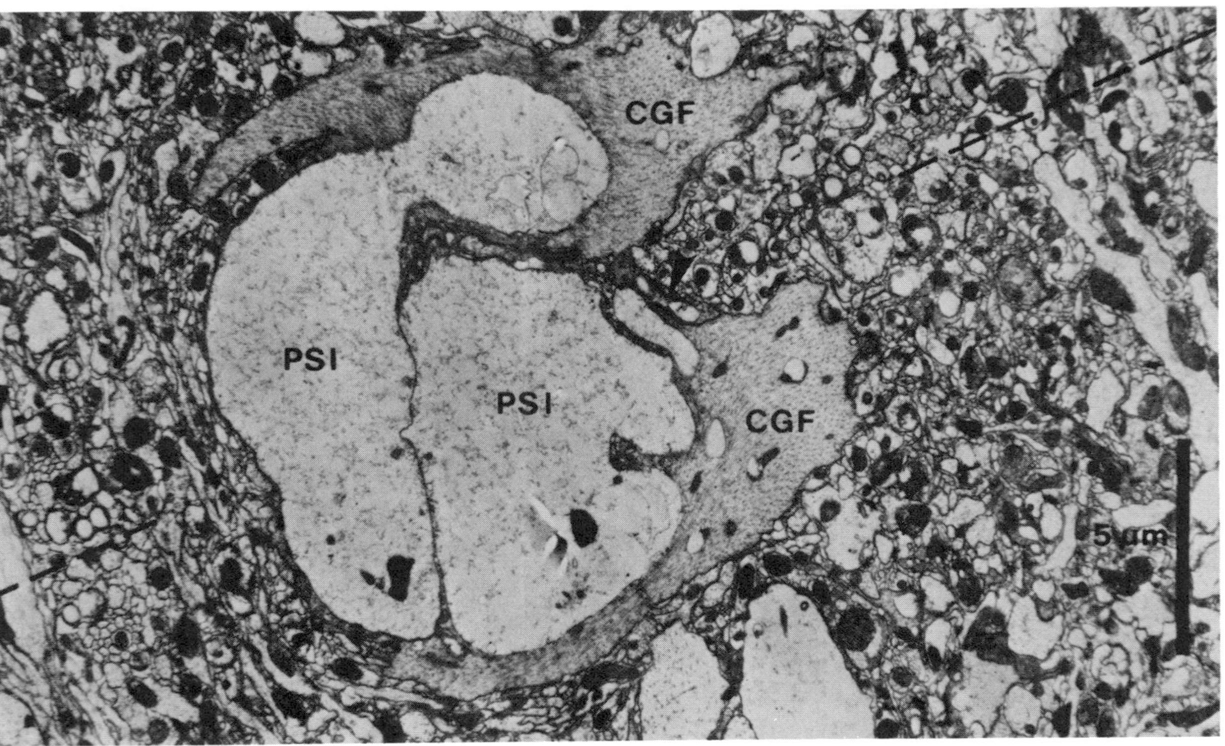

Figure 9. Electron micrograph showing the GF axons contacting the decussating peripherally synapsing interneuron (PSI) axons (from King and Wyman, 1980).

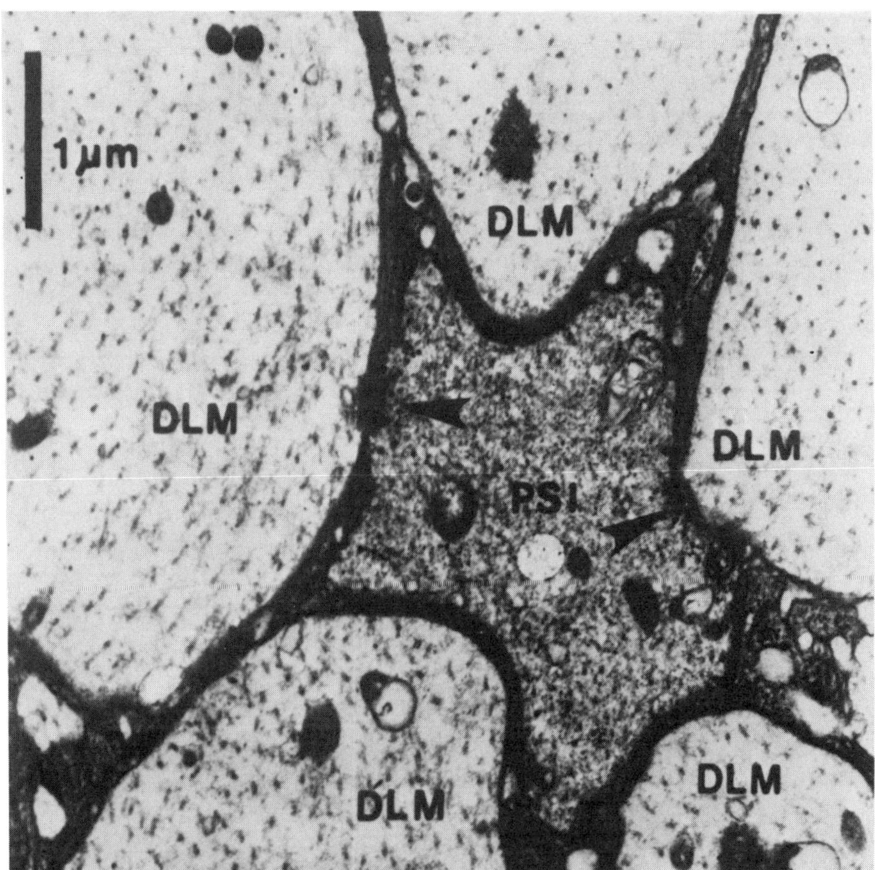

Figure 10. Electron micrograph of a section through the posterior dorsal mesothoracic nerve (PDMN). The PSI is shown in close association with the five DLM axons, making synaptic contacts (arrowheads) with two of them (from King and Wyman, 1980).

is 0.44 msec longer than the latency to the TTM. This extra latency must be the chemical synaptic delay at the peripheral synapses. Thus, the GF driving of the DLM muscles is disynaptic: the GF to PSI electrical synapse (Figure 9) and the peripheral chemical synapse (Figures 14 and 15). These pathways are summarized in Figure 6A (Tanouye and Wyman, 1980).

The five motoneurons seen in Figures 10, 12, and 13 are the motoneurons to the wing depressor muscle. This muscle is composed of six giant (65 × 150 × 300 μm) muscle fibers (Figure 1). Each muscle fiber is singly innervated. One axon innervates both of the top two fibers (a,

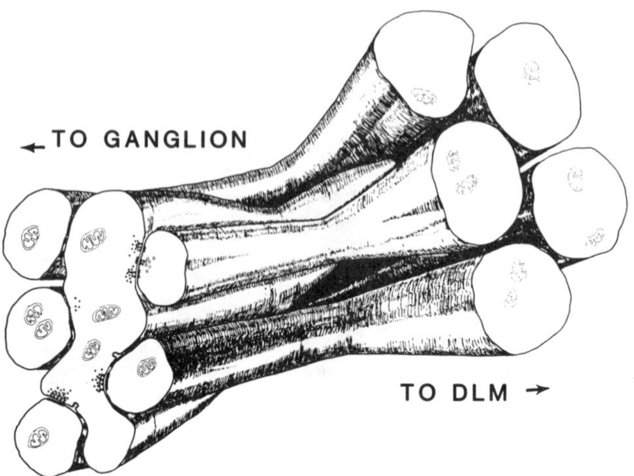

Figure 11. Diagram of the region of termination of the PSI in the PDMN. The five DLM axons and the PSI axon run out of the ganglion into the PDMN. There the PSI synapses with the motor axons just proximal to the termination of the PSI. The DLM axons then increase in size and continue out to the muscle surface. Ganglion face of the diagram reconstructed from micrographs as in Figures 10 and 12; distal face reconstructed from micrographs as in Figure 13; and synaptic region reconstructed from micrographs as in Figures 14 and 15.

b). The four other axons each innervate one of the four ventral fibers (c–f). Peroxidase may be injected into any one of these muscle fibers and it will be taken up by that fiber's motoneuron (Coggshall, 1978). Figure 16A shows the dendritic tree of one of the motoneurons. It lies in a dorsal cap on the thoracic ganglion. The four motoneurons that innervate fibers c–f have their cell bodies ventral and ipsilateral to their muscle fibers (Figure 16B, lower left), but the soma of the motoneuron that innervates the dorsal muscle fibers (a, b) sits dorsally and contralaterally in the ganglion (Figure 16A). This is a very fortuitous arrangement. For example, a bilateral genetic mosaic could be made in which one side of the fly is normal and the other side is mutant. Then, the DLM muscle on one side will have fibers c–f innervated by genotypically mutant neurons, whereas a and b are innervated by a genotypically normal neuron. The converse situation holds on the other side. This can be of great use in studying possible neurotrophic mutants.

Each DLM motor axon branches extensively over the surface of its muscle fiber (Ikeda *et al.*, 1980) and even penetrates into invaginations of the muscle cell (Shafiq, 1964). An electrode placed intracellularly in

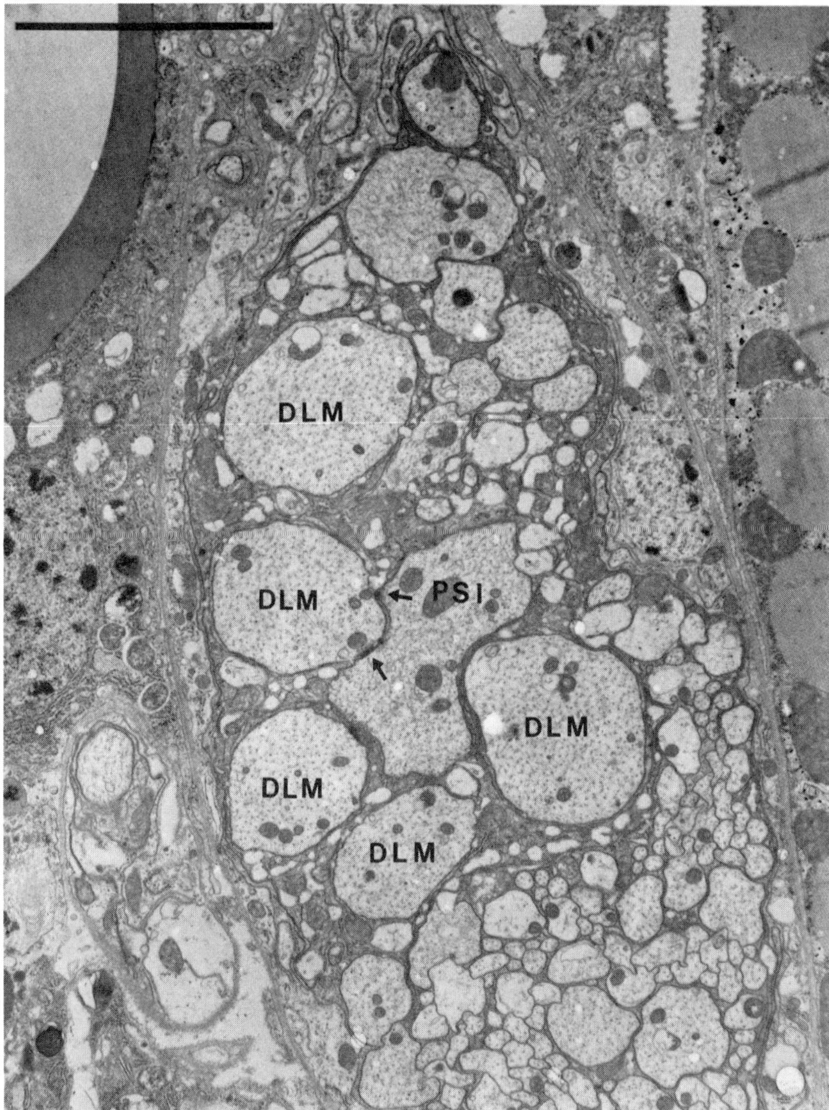

Figure 12. Electron micrograph of PDMN running along flight muscle (right). Nerve is shown with motor component (large profiles), sensory component (small profiles), and neural sheath. Two synapses (arrows) between the PSI and a DLM axon are shown. Scale line (upper left): 5 μm.

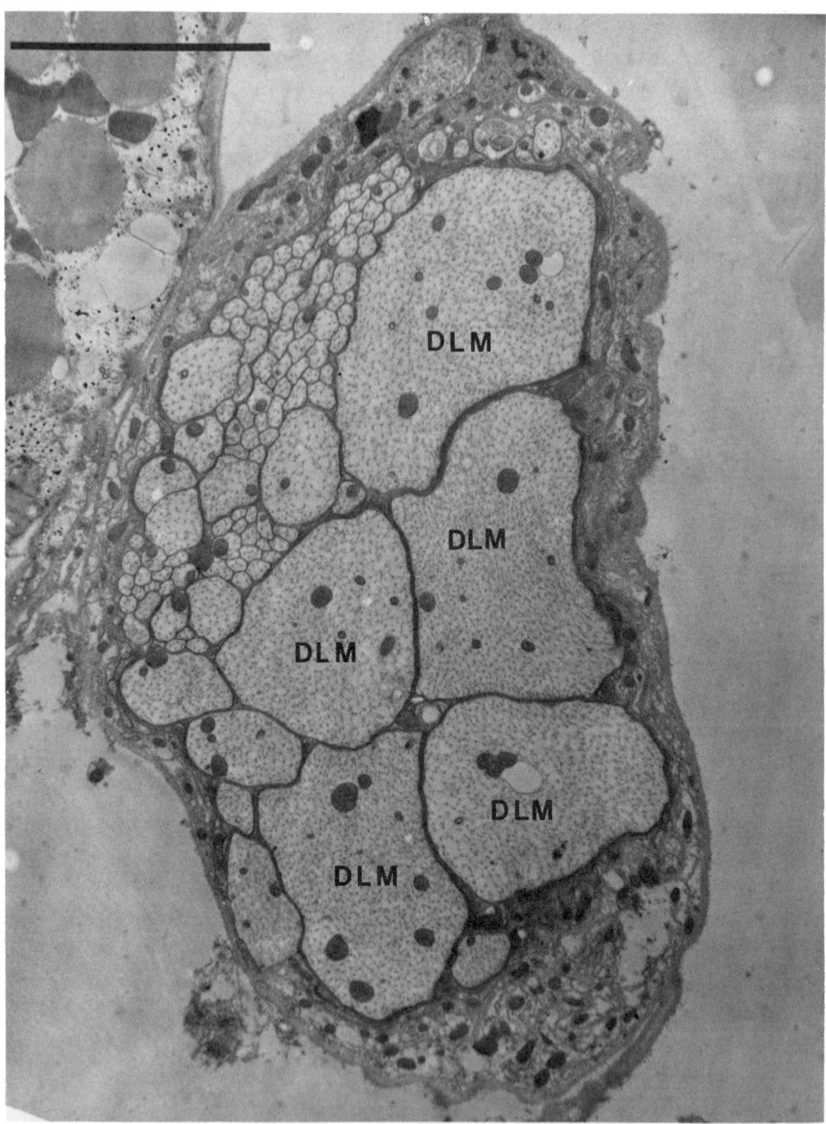

Figure 13. Electron micrograph of a section a few microns further out the PDMN than the section shown in Figure 12. The PSI has terminated and the DLM axons have increased greatly in size. Figures 12 and 13 were taken at the same magnification. Scale line (upper left): 5 μm.

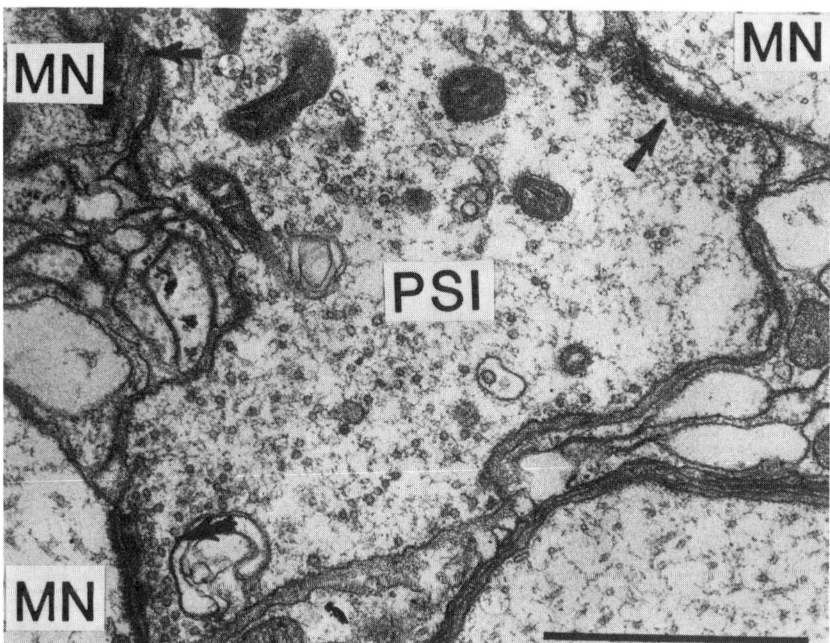

Figure 14. Synaptic region between PSI and DLM motoneurons (MN) diagrammed in Figure 11. A fortuitous section shows the PSI synapsing on three of the DLM motoneurons (arrows). Scale bar = 1 μm.

the muscle sits inside a basket of nerve endings. Current passed out through the electrode will preferentially stimulate that single motoneuron. The antidromic spike propagates back to the neuropil and activates the output synapses of the motoneuron. Using this technique, we have been able to analyze the circuitry that generates the flight motor output pattern. The essence of the circuit, as we see it, is a set of reciprocal inhibitory interactions between all the motoneurons (Figure 17) (Harcombe and Wyman, 1977, 1978; Friesen and Wyman, 1980; Tanouye and Wyman, 1981; Wyman and Tanouye, 1982). This theory would predict that a large, long-lasting hyperpolarization would occur in one motoneuron following the firing of another motoneuron. In intracellular recordings, Koenig and Ikeda (1983) have seen such a hyperpolarization as part of a biphasic (depolarizing, then hyperpolarizing) response. They do not describe the hyperpolarization as an inhibitory interaction, but as a resetting. Depending on how one projects the baseline that would be expected if the driving motoneuron had not fired, either the depolarization or the hyperpolarization can look like the larger effect. They believe that the depolariza-

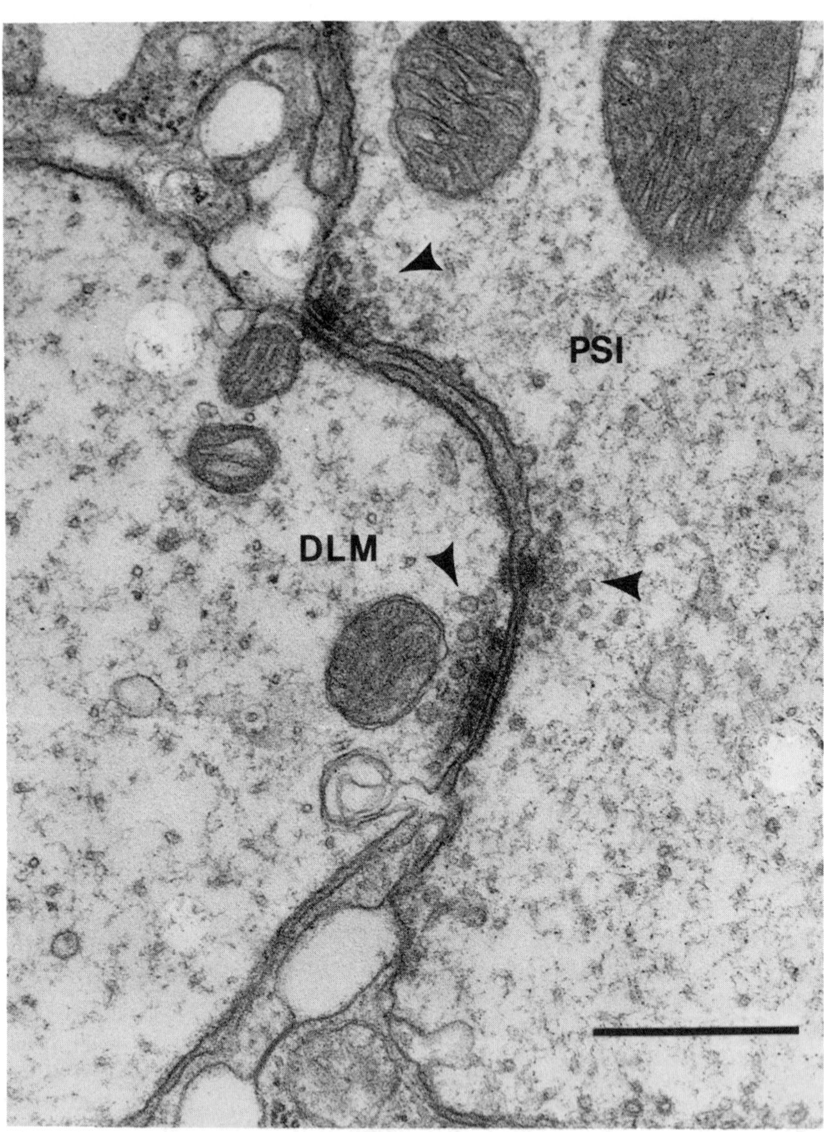

Figure 15. Reciprocal synapse between PSI and a DLM axon in the synaptic region diagrammed in Figure 11. The synaptic vesicles on the two sides are of different size, probably indicating that two different transmitters are used. Scale line (lower right): 0.5 μm (from King and Wyman, 1980).

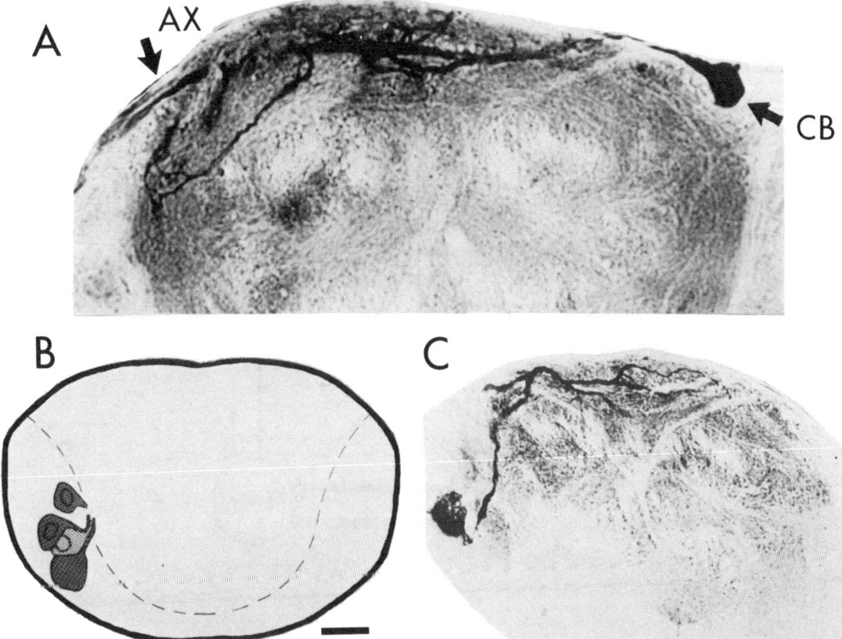

Figure 16. (A) DLM motoneuron that innervates the two most dorsal muscle cells (Figure 1) of the left side. Note that its cell body (CB) is dorsal and contralateral to the muscle innervated. (B) Diagram of location of CBs of the four motoneurons innervating the four ventral DLM fibers. These four CBs are ventral and ipsilateral to the muscle innervated. (C) Dorsal dentritic tree and CB of motoneuron innervation DLM unit e (second most ventral fiber). Cells were filled with peroxidase injections into individual muscle cells. Scale bar: 20 μm (from Coggshall, 1978).

tion/excitation is the main influence on the output pattern, while we believe that the hyperpolarization/inhibition/resetting is the major factor.

The electrophysiology of the flight muscle membrane has also been studied extensively in our laboratory. We now know the channels that are present (Salkoff and Wyman, 1983*a,b*), the sequence of their development (Salkoff and Wyman, 1981*a*), and their role in synaptic plasticity (Salkoff and Wyman, 1980, 1983*a*) and are beginning to identify the genes that code for these channels (Salkoff and Wyman, 1981*b*).

4. Development

The adult nervous system of *Drosophila* develops from larval neurons and from a set of "giant neuroblasts" (Nordlander and Edwards, 1969;

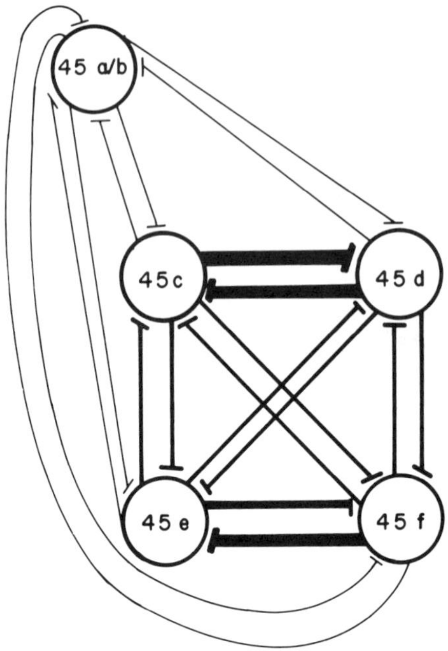

Figure 17. Diagram of the circuitry responsible for the flight motor output pattern. There are six DLM muscle fibers, in a dorsal to ventral stack, named 45 a–f. A single motoneuron, 45 a/b, innervates the top two fibers, a and b. The four ventral muscle fibers, c–f, are each innervated by their own motoneuron. Each of the five DLM motoneurons sends an inhibitory signal to each of the others. The width of the connecting line indicates the strength, or effectiveness, of the inhibitory interaction (from Harcombe and Wyman, 1977).

White and Kankel, 1978) that are born during embryogenesis, but do not proliferate until later. These cells are recognizable by their large size and characteristic position. Some of the daughters of these cells do not differentiate until the pupal period. The GF seems to be one of these late differentiating cells.

In sectioned material (Costello and Wyman, unpublished), the GF cell body can be identified in the brain of 30-hr pupae. The GF cell body can also be identified in a living wholemount under Nomarski optics. We can first locate the axons of the GFs in the cervical connective at 45–50 hr of pupal development. Both the GF and the motoneurons develop concurrently. By 60–65 hr of pupal development, motoneurons can be directly stimulated in the ganglion and are capable of driving the DLM and TTM. The GF contacts the PSI interneuron at about 60 hr of development and then bends laterally. By about 75 hr, the GF makes functional synapses on the jump motoneurons, thus completing the circuit. Hence, it seems that between 45 and 65 hr, the GF is following whatever signals guide it to grow posteriorly and for the next 10 hr or so, it follows cues to the jump motoneurons.

Along with our study of GF development, we have also been studying the development of the fibrillar and jump muscles (W. J. Costello and R. J. Wyman, unpublished). In the early pupa the myocytes proliferate from ventral sites, probably from cells originating in the leg disks (Shatoury, 1956; Deak *et al.*, 1980). They then migrate along the larval nerve to the larval DLMs. At various locations along this path, they branch off and collect to form the adult TTM, DVM, and DLM muscles. The DLM muscle of the adult is built on the framework of a pre-existing larval muscle, the dorsal external oblique muscle. As the larval muscle degenerates, the myocytes collect around the muscle, split its three units into six, fuse, and start making contractile filaments. Cross striations are first visible at about 50 hr. Morphological differentiation is complete at about 84 hr, and the adult emerges at about 105 hr at 25 °C.

The developing electrical properties of the muscle membrane were studied by voltage clamping the DLM cells at various stages of pupal development. Prior to about 60 hr of pupal development, there are no electrically excitable channels in the muscle membrane. At 55 hr the membrane has a small, instantaneously rectifying leak. The first active channel to appear in the membrane is the A-current channel at about 70 hr. The delayed rectification channels appear later and, along with the calcium channels, develop in the next 24 hr (Salkoff and Wyman, 1980, 1981*a,b*).

5. Mutants

5.1. Mutant Selection

From the information summarized above, it can be seen that we have gained a good understanding of the normal anatomy, physiology and development of the escape response neurons. We have therefore started to search for mutants that affect the identified neurons described above. For selecting jumpless mutants, we start with a stock of white-eyed (*bw;st*) flies that also have defective wings due to the mutation *curved*. These flies jump normally, but they cannot fly. We then feed them a mutagen to induce mutations. The progeny of these flies are then screened by a variety of methods for their ability to jump to the light-off stimulus. In one protocol, flies are introduced into the bottom of a vertical glass tube. Those that are capable of walking, crawl up the sides. Severely debilitated flies are left at the bottom and are discarded. The lights are then flicked out and most flies jump into the air; they cannot fly so they fall out the bottom of the tube where they are discarded. Those that remain in the tube do not jump to this stimulus. We have completed the first phase of our search for mutants of the jump system. Over 100,000 F_1 males carrying mutagenized X chromosomes were screened. From these, 104 mutant strains were derived that are defective in jumping, but appear normal in their other behavior.

Mutations in the flight system are detected in a similar way. The selecting arena is a large hollow box. Mutagenized flies are introduced through a funnel at the top. As they are falling, most flies fly away and land on the sides of the box. Flightless flies fall straight down into a collecting tube at the bottom of the box. About 50,000 flies were screened in this way and 40 mutant strains were derived that cannot fly, but appear normal in their other behaviors.

5.2. Physiological Screen

The mutants isolated in the behavioral screens can be defective in either jumping or flying. There are many possibilities for the focus of the defect (e.g., visual system, muscular system, skeletal system, as well as nervous system). Of course we are interested only in those mutants that are defective in the identified neurons previously described. To select these, we have devised simple electrophysiological tests. Some of these are described below.

As can be seen from the circuit of Figure 6A, there are three stages to the jump response: (1) processing of visual information presynaptically to the GF; (2) transmission from the GF to the motoneurons; and (3) transmission from the motoneurons to the muscles. We are primarily interested in defects in stage 2, because these entail defects in the identified neurons. Stage 1 can be bypassed by direct stimulation of the GF while recording from postsynaptic cells. The GF can be easily stimulated extracellularly by passing current across the brain or the cervical connective. Flies that have a normal motor output after GF stimulation are apparently defective in visual reception or integration and thus are not of further interest. Flies with abnormal motor output after brain stimulation are then checked to see if neuromuscular transmission and muscle contraction (stage 3) are normal. This is done by stimulating the motoneurons directly and recording muscle spikes and leg motions. Flies with a normal neuromuscular system must then have defects involving transmission from the GF to the TTM motoneuron, or transmission from the GF to the DLM motoneurons via the PSI.

Three of the mutants that we have isolated will be described as examples of the types of mutants that we find which affect individual identified elements.

5.3. Examples of Mutants

5.3.1. Connection Morphologically Absent

One mutant, called *bendless*, isolated in the screen for jumpless flies, was found by electrophysiology to lack direct transmission from the GF to the TTM motoneuron. The other output connections of the GF appeared to function normally (Figure 18). On filling the GF with Lucifer Yellow, we found that the branch of the GF that makes this synaptic connection was absent (Thomas and Wyman, 1982, 1983).

Dye fills of the mutant GF and the jump motoneuron show that the two do not come into contact. However, excitation of the GF does drive the jump motoneuron; this driving must be via a polysynaptic pathway. Figure 18B shows the GF output to the TTM and DLM of a *bendless*

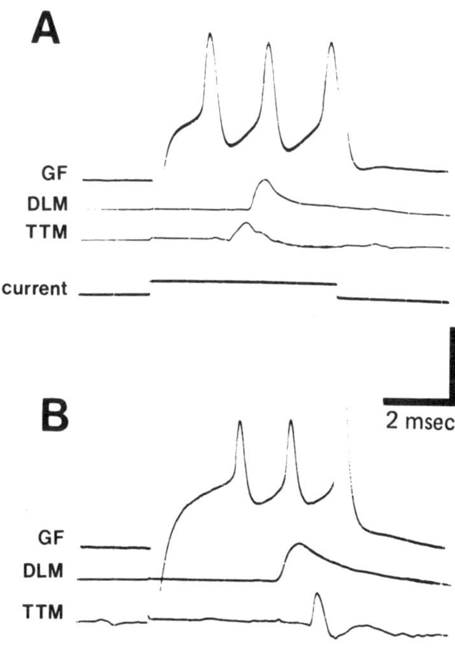

Figure 18. Response in TTM and DLM to intracellular stimulation of the GF in wild type (A) and *ben* (B). Top traces are the intracellular records of the GF. Current is shown in lowest traces. The muscle responses shown in the middle traces are driven by the first GF spike. Vertical calibration is 40 mV for the GF and DLM, 20 mV for the TTM, and 10 nA for the current (from Thomas and Wyman, 1982).

individual. It can be seen that the DLM response latency of the mutant is indistinguishable from that of wild type (Figure 18A). However, the TTM response latency is abnormally long. The mutant's latency is 2.2 msec compared with the normal 0.8 msec. The extra latency is sufficient to allow the interposition of at least one interneuron. It is possible that the polysynaptic pathway is normally present but masked by the fast direct pathway of excitation of the TTM motoneuron. Alternatively, the mutant GF, not making its normal connection, may make aberrant synapses on neurons that are presynaptic to the jump motoneuron. Electron microscopical studies of the distal end of the mutant GFs are planned to resolve this question.

In normal flies, the GF descends into the thoracic ganglion and takes a lateral bend near its termination (Figure 7). From our electron microscopic reconstruction of the normal GF (Figure 8), we know that the GF synapses with the PSI just before the bend; its synapse with the jump motoneuron (TTM) is distal to the bend. In the mutant the GF cell body position and dendritic morphology are indistinguishable from wild type. The GF axon of *bendless* descends unbranched to the thoracic ganglion. It carries spikes normally and its output to the PSI neuron is normal. However, Lucifer Yellow fills in *ben* mutants, which reveals (Figure 19B) that the GF fails to extend beyond the point of its PSI synapse: the complete lateral bend to the jump motoneuron is missing. The jump motoneuron, however, is unaffected; it functions normally when driven by other pathways, and by retrograde labeling with horseradish peroxidase (HRP), we know its morphology is normal (M. Koto, personal communication). Thus, we have made a comprehensive analysis of a single gene mutation and found behavioral defect (no-jump), an electrophysiological defect (absence of the synapse from the GF to the TTM, or jump, motoneuron), and a neuroanatomical defect in a single cell (absence of the lateral extension of the GF).

In the mutant, the GF sometimes terminates in several long tendrils (Figure 19C). One interpretation of this is that these processes are the result of searching attempts by the GF to make contact with the jump motoneuron, but due to the lack of some molecular signal, the connection was never made. In normal flies, we know the time during pupation when the GF grows through the thorax and when functional synaptic contacts are first made. Our next step is to look at the development of the GF using timed injections of Lucifer Yellow. Parallel developmental studies of the wild type and mutant will tell us the time and manner in which the development of the mutant diverges from the normal pattern. We will look for a temperature-sensitive allele so that we can determine the time in development when the normal gene product is needed. We will also

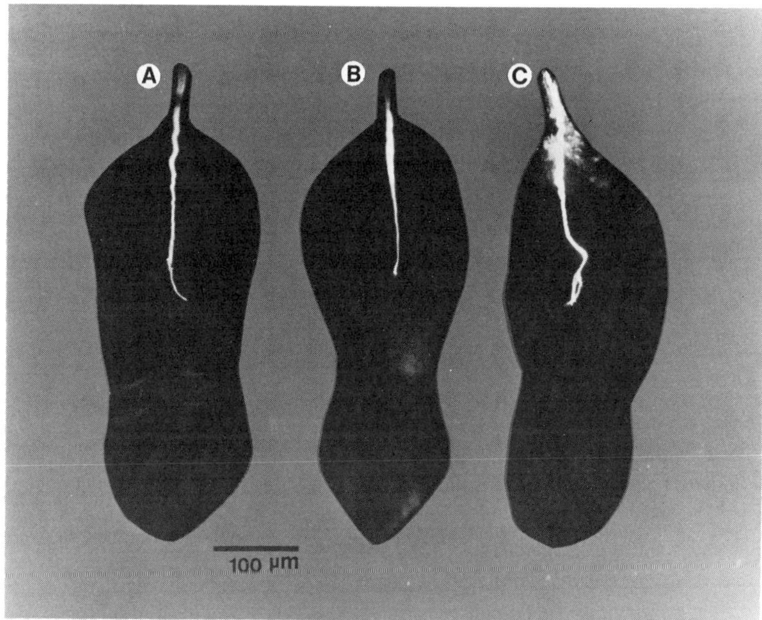

Figure 19. Wholemounts of three thoracic ganglia in which the GF has been filled with Lucifer Yellow. (A) Wild type; (B, C) individuals homozygous for the *bendless* mutation. (A, B) are dorsal views; (C) is a dorsolateral view in which the normal ventral bend and abnormal branches are visible. In the mutant, the GF fails to extend to the region of normal synaptic contact with the TTM motoneuron.

make mosaics to see if the normal gene product is needed by the GF (a brain neuron) or by the postsynaptic element (the thoracic-jump motoneuron) in order for the synapse to form.

5.3.2. Connection Physiologically Absent

We have identified mutations that appear to alter connectivity but in which we cannot yet describe the morphological correlate (if any) of the disruption. In one of these, no normal output at all can be detected from the GF even though the GF is present and drives polysynaptic responses. Another mutant is the reverse of *bendless*. The GF conducts impulses and transmits to the TTM motoneuron, but is defective in its transmission through the PSI pathway. Figure 20A (top four traces) shows the normal timing of the DLM and TTM responses to extracellular stimulation of the GF. The top and bottom traces are the right and left TTM spikes and the middle two traces are muscle spikes in right and left DLMs. The lower

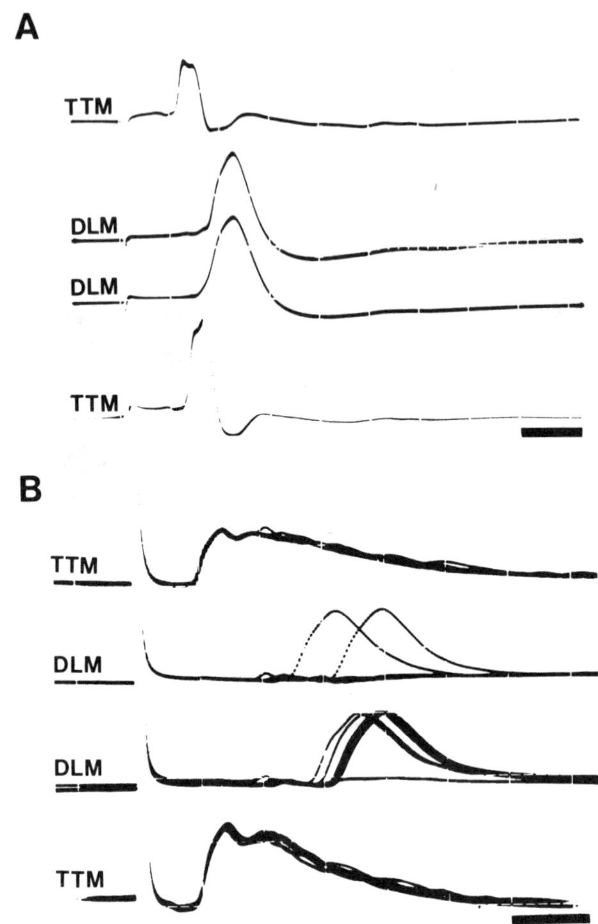

Figure 20. Motor output evoked by extracellular stimulation of the GF in wild-type and *gfA* flies. (A) Muscle recordings from a wild-type fly showing normal timing of responses in right and left TTM muscle (top and bottom traces) and right and left DLM muscles (middle two traces). This pattern is invariably evoked, with no failures, at frequencies up to 50 Hz. (B) Muscle recordings from a *gfA* mutant displayed as in (A). The GF was stimulated at 1 Hz. The driving of the TTM is normal, but the DLMs are driven with a very long and variable latency. There are frequent failures in transmission to the DLM that occur independently in each DLM cell. Scale bar (A, B) 1 msec.

panel (Figure 20B) shows the phenotype of the mutant *gfA*. In this mutant the GF drives the TTM normally but the driving of the DLMs is abnormal. In the middle two traces of the figure, it can be seen that the DLMs are driven at very long and variable latencies and that there are frequent failures where a given DLM unit is not driven at all. Lucifer Yellow fills, however, show normal GF morphology and the PSI is present in light microscopic sections. Since the DLM units each fail independently to GF stimulation, the defects must be at points where the pathways to each unit are separate. The individual motoneurons can all carry spikes, and the neuromuscular junctions can be driven normally at high frequencies. The only point of divergence remaining is the output synapse from the PSI to the DLM motoneurons, i.e., the peripheral synapses. We now presume that this is the affected link.

6. Conclusion

The giant fiber system in *Drosophila* provides an excellent system in which to study the effects of mutations on neuronal connectivity. The small number of neurons and their connections allow both physiological and anatomical identification. The cells can be individually stimulated and recorded from, they can be filled with Lucifer Yellow or peroxidase; and identified synapses can be repeatedly located in electron micrographs. By capitalizing on the finding that this circuit selectively mediates a visually induced escape response, large numbers of flies can readily be screened behaviorally for possible defects in the pathway. Among the behavioral mutants so selected are mutants that have altered connectivity within the GF system. The specificity that is so far apparent in these mutations is nothing short of amazing.

The isolation and phenotypic characterization of mutations such as these represents the first step in probing the question of how genes might be involved in nervous system development. The great store of genetic knowledge available in *Drosophila* provides powerful analytical techniques with which to answer a number of questions. For instance, the site of gene action for each of these mutations can be determined by generating and analyzing mosaic flies containing patches of mutant and normal tissue. The patches can be identified by using a histochemical marking system (Kankel and Hall, 1976). Since the cell body positions of the GF, PSI, and TTM and DLM motoneurons are known, the analysis, in theory, should be straightforward. The acquisition of temperature-sensitive conditional alleles would provide another analytical tool. By analyzing the phenotype of flies carrying a t-s allele, which have been shifted between permissive and nonpermissive temperatures at different times in development, the time of gene action might be estimated.

The use of genetic techniques such as these, coupled with conventional anatomical and physiological investigation of the development of the GF system in both wild type and the mutants will certainly have some surprises in store, and might shed some light on the mechanisms underlying nervous system development. We are continuing work to determine if these genes are indeed involved in the molecular basis of the specificity of neuronal connectivity.

ACKNOWLEDGMENTS. We would like to acknowledge Ms. Catherine Blackmon for her care in processing the manuscript.

7. References

Coggshall, J., 1978, Neurons associated with the dorsal longitudinal flight muscles of *Drosophila melanogaster, J. Comp. Neurol.* **177**:707–720.

Deak, I. I., Rahmi, A., Bellamy, P. R., Bienz, M., Blumer, A., Fenner, E., Gollin, M., Ramp, T., Reinhardt, C., Dubendorfer, A., and Cotton, B., 1980, Developmental and genetic studies of the indirect flight muscles of *Drosophila melanogaster,* in: *Development and Neurobiology of Drosophila,* (O. Siddiqi, P. Babu, L. M. Hall, and J. C. Hall, eds.), Plenum Press, New York, pp. 183–192.

Friesen, W. O., and Wyman, R. J., 1980, Analysis of *Drosophila* motor neuron activity patterns with neural analogs, *Biol. Cybernetics* **38**:41–50.

Harcombe, E. S., and Wyman, R. J., 1977, Output pattern generation by *Drosophila* flight motorneurons, *J. Neurophysiol.* **40**:1066–1077.

Harcombe, E. S., and Wyman, R. J., 1978, The cyclically repetitive firing sequences of *Drosophila* flight motorneurons, *J. Comp. Physiol.* **123**:271–279.

Harris, W. A., Stark, W. S., and Walker, J. A., 1976, Genetic dissection of the photoreceptor system in the compound eye of *Drosophila melanogaster, J. Physiol.* **256**:415–439.

Heisenberg, M., Wonneberger, R., and Wolf, R., 1978, *optomotor-blind*^{H31}—A *Drosophila* mutant of the lobula plate giant neurons, *J. Comp. Physiol.* **124**:287–296.

Ikeda, K., Koenig, J. H., and Tsuruhara, T., 1980, Organization of identified axons innervating the dorsal longitudinal flight muscle of *Drosophila melanogaster, J. Neurocytol,* **9**:799–823.

Judd, B. H., 1976, Genetic units of *Drosophila*—Complex loci, in: *The Genetics and Biology of Drosophila,* Vol. 1b (M. Ashburner and E. Novitski, eds.), Academic Press, London, pp. 767–799.

Kankel, D. R., and Hall, J. C., 1976, Fate mapping of nervous system and other internal tissues in genetic mosaics of *Drosophila melanogaster, Devel. Biol.* **48**:1–24.

King, D. G., 1976, Anatomy of motor neurons in *Drosophila* thoracic ganglion, *Soc. Neurosci. Abstr.* **2**:628.

King, D. G., and Wyman, R. J., 1980, Anatomy of the giant fiber pathway in *Drosophila.* I. Three thoracic components of the pathway, *J. Neurocytol.* **9**:753–770.

Koenig, J. H. and Ikeda, K., 1983, Reciprocal excitation between identified flight motor neurons in *Drosophila* and its effect on pattern generation, *J. Comp. Physiol.* **150**:305–317.

Koto, M., Tanouye, M. A., Ferrus, A., Thomas, J. B., and Wyman, R. J., 1981, The morphology of the cervical giant fiber neuron of *Drosophila, Brain Res.* **221**:213–217.

Mulloney, B., 1969, Interneurons in the central nervous system of flies and the start of flight, *Z. Vergl. Physiol.* **64**:243–253.

Nordlander, R. H., and Edwards, J. S., 1969, Postembryonic brain development in monarch butterfly *Danaus plexippus plexippus,* L. I. Cellular events during brain morphogenesis, *W. Roux' Arch.* **162**:197–217.

Pak, W. L., 1975, Mutations affecting the vision of *Drosophila melanogaster*, in: *Handbook of Genetics* (R. C. King, ed.), Plenum Press, New York, pp. 707–733.

Ready, D. F., Hanson, T. E., and Benzer, S., 1976, Development of the *Drosophila* retina, a neurocrystalline lattice, *Dev. Biol.* **53**:217–240.

Salkoff, L., and Wyman, R. J., 1980, Facilitation of membrane electrical excitability in *Drosophila, Proc. Natl. Acad. Sci. U.S.A.* **77**:6216–6220.

Salkoff, L., and Wyman, R. J., 1981*a*, Outward currents in developing *Drosophila* flight muscle, *Science* **212**:461–463.

Salkoff, L., and Wyman, R. J., 1981*b*, Genetic modification of potassium channels in *Drosophila Shaker* mutants, *Nature* **293**:228–230.

Salkoff, L., and Wyman, R. J., 1983*a*, Ion currents in *Drosophila* flight muscles, *J. Physiol. (London)* **337**:687–709.

Salkoff, L., and Wyman, R. J., 1983*b*, Ion channels in *Drosophila* muscle, *Trends Neurosci.* **6**:128–133.

Shafiq, S. A., 1964, An electron microscopical study of the innervation and sarcoplasmic reticulum of the fibrillar flight muscle of *Drosophila melanogaster, Q. J. Microsc. Sci.* **105**:1–6.

Shatoury, H. H. El., 1956, Developmental interactions in the development of the imaginal muscles of *Drosophila, J. Embryol. Exp. Morph.* **4**:228–239.

Tanouye, M. A., and Wyman, R. J., 1980, Motor outputs of the giant nerve fiber in *Drosophila, J. Neurophysiol.* **44**:405–421.

Tanouye, M. A., and Wyman, R. J., 1981, Inhibition between flight motorneurons in *Drosophila, J. Comp. Physiol.* **144**:345–355.

Thomas, J. B., 1981, Mutations altering the connectivity between identified neurons in *Drosophila melanogaster*, Ph.D. Thesis, Yale University, New Haven, Connecticut.

Thomas, J. B., and Wyman, R. J., 1982, A mutation in *Drosophila* alters normal connectivity between two identified neurons, *Nature* **298**:650–651.

Thomas, J. B., and Wyman, R. J., 1983, Normal and mutant connectivity between identified neurons in *Drosophila, Trends Neurosci.* **6**:214–219.

Valentino, K. L., 1977, Identified neurons make identical synapses in *Musca* and *Drosophila, Soc. Neurosci. Abstr.* **3**:190.

White, K., and Kankel, D. R., 1978, Patterns of cell division and cell movement in the formation of the imaginal nervous system in *Drosophila melanogaster, Dev. Biol.* **65**:296–321.

Wyman, R. J., and Tanouye, M. A., 1982, *Drosophila* flight motor pattern: The evidence from interspike intervals, *J. Exp. Biol.* **96**:413–416.

Ziegler, I., 1961, Genetic aspects of ommochrome and pterin pigments, *Adv. Genet.* **10**:349–403.

6

Escape Behavior of the Locust

The Jump and Its Initiation by Visual Stimuli

KEIR G. PEARSON and MICHAEL O'SHEA

1. Introduction

Within the past decade there has been enormous progress in our understanding of the neural organization of insect nervous systems. Numerous interneurons in a variety of sensory and motor systems have now been identified (Table I), and in some cases the cellular mechanisms for patterning neuronal activity have been determined (O'Shea and Rowell, 1977; Burrows, 1980; Siegler, 1981; Steeves and Pearson, 1982). As yet, however, our knowledge of the neuronal basis for most behaviors in insects is limited; for example, we have no idea how the basic rhythms for walking, respiration, and flight are generated. One clear exception is the jumping system of the locust. Not only have plausible circuits been described for the generation of the motor output for the jump, but the processes by which external sensory signals trigger the jump are also reasonably clear (Pearson *et al.*, 1980; Pearson and Robertson, 1981). In this chapter we describe the neural circuitry patterning motor output for the jump and the integration of visual information for producing signals suitable for initiating a jump.

KEIR G. PEARSON • Department of Physiology, University of Alberta, Edmonton, Alberta Canada, T6G 2H7. *MICHAEL O'SHEA* • Departments of Pharmacology and Physiology, University of Chicago Medical School, Chicago, Illinois 60637. *Present address*: Département de Biologie Animale, Université de Genève, CH-1211, Genève 4, Switzerland.

Table I. Identified Interneurons in Insect Thoracic Nerve Cords

System	Animal	Interneuron(s)	Reference
Motor Systems			
Jump	Locust	M neuron	Pearson *et al.*, 1980
		C neuron	Pearson and Robertson, 1981
Escape	Cockroach	Giant : GI1 to GI7	Daley *et al.*, 1981
Walking/	Cockroach	Nonspiking local	Pearson and Fourtner, 1975
posture	Locust	Nonspiking local	Siegler and Burrows, 1979
	Locust	Spiking local	Burrows and Siegler, 1982
Flight	Locust	Interganglionic	Robertson and Pearson, 1983
Respiration	Locust	Coordinating	Pearson, 1980
		Inhibitory	Burrows, 1982
Sensory Systems			
Visual	Locust	DCMD	O'Shea *et al.*, 1974
		03	Simmons, 1980*a*
Auditory	Locust	G, B1, B2	Rehbein, 1976
	Cricket	ON1, ON2, AN1, AN2 DN1, TN	Wohlers and Huber, 1982
Tactile	Locust	1NL, 1N2	Pfluger and Tautz, 1982
Wind	Locust	TCG	Bacon and Tyrer, 1978

2. The Jump in the Locust

The jump of the locust is a multistaged behavior that is released only after a sequence of events designed to store the energy for the jump has been completed. Prior storage of energy is a commonly used strategy in many jumping insects (Evans, 1973, 1975; Bennet-Clark, 1967, 1975; Frantsevich, 1981; Rothschild *et al.*, 1972), for it has the advantage of allowing the highest possible rate of energy utilization when the jump is triggered.

There are three phases to the locust jump.

1. *Cocking,* locking of both hindleg tibia into full flexion
2. *Co-contraction,* simultaneous contraction of hindleg flexor and extensor tibiae muscles lasting about 0.3 sec during which energy for the jump is stored in elastic elements of the leg
3. *Triggering,* a sudden relaxation of flexor muscles unlocking the tibia and allowing the rapid extension of both hindlegs.

Cocking: When the hindlegs are close to full flexion, cocking is produced by the almost synchronous activation of the single fast motoneuron supplying the extensor tibiae muscle and a variable number of motoneurons supplying the flexor tibiae muscle (Figure 1). The resulting movement of the tibia to full flexion is due to the mechanical advantage enjoyed by the

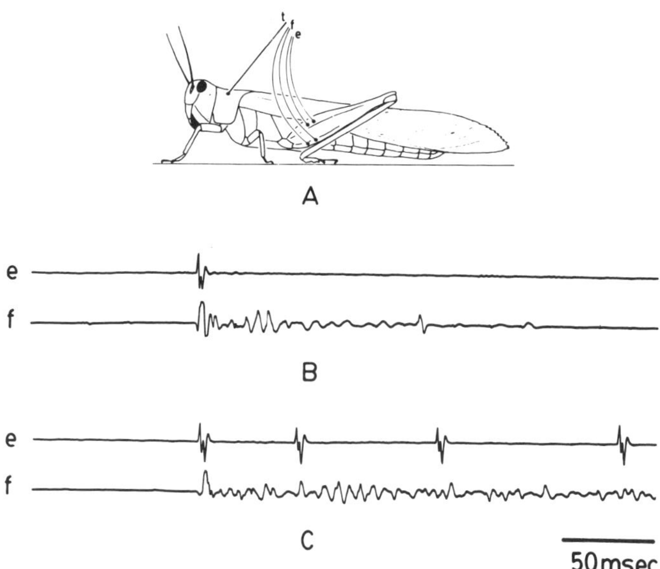

Figure 1. Cocking response in tethered animal. (A) Drawing of preparation showing sites of EMG recording from flexor (f) and extensor (e) tibiae muscles of the hindleg. (B, C) Two examples of cocking response evoked by movement in the animal's visual field. Note the almost simultaneous activation of extensor and flexor motoneurons at the beginning of each response (from Pearson and Robertson, 1981).

flexors when the tibia is close to full flexion (Heitler, 1974). In aroused animals a variety of external stimuli (visual, auditory, tactile, vibration) can initiate this stereotyped cocking response. In particular, the cocking response can often be evoked by slight movements of objects anywhere in the animal's visual field. As we describe later, this is likely due to visual movement-detecting neurons exciting interneurons that coactivate hindleg flexor and extensor tibiae motoneurons.

Co-contraction: The locking of a tibia into full flexion depends on a specialization of the flexor tendon that bifurcates to straddle a cuticular lump in the distal femur when the tibia is flexed. Contractions of the flexor muscle hold this cuticular lump between the two tendon strands thereby preventing extension movements of the tibia. Unlocking requires relaxation of the flexor muscle to release the cuticular lump. In the normal jump this flexor relaxation does not occur until the extensor muscle has had time to generate a large isometric force. This takes from 200 to 500 msec to occur. During this time, energy for the jump is stored as elastic deformations in the cuticle of the distal femur.

Triggering: The triggering of the jump, i.e., relaxation of flexors, can occur either spontaneously, as when the animal uses a jump to move from one place to another, or in response to external stimuli, such as a hissing sound or a sudden movement in the animal's visual field. In both cases flexor activity is terminated abruptly and there are no obvious differences in the overall pattern of motor activity. Thus it appears there is just a single triggering mechanism, and this can be activated by either internal or external sources.

3. Patterning of Motor Activity for the Jump

The neural circuitry for patterning motor activity for the jump is centered around two pairs of identified neurons: the C neurons and the M neurons (Figure 2). Each C neuron forms monosynaptic connections

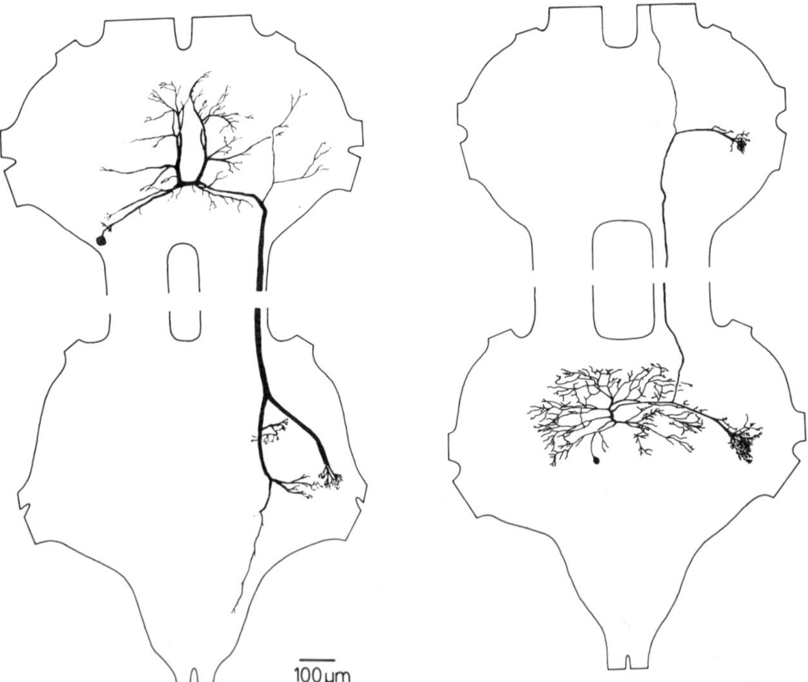

Figure 2. Dorsal view of mesothoracic and metathoracic ganglia showing the structure of (left) a C neuron and (right) an M neuron. The cells were visualized in wholemount following intracellular injection of Lucifer yellow (modified from Pearson and Robertson, 1981; Pearson *et al.*, 1980).

with the fast extensor tibiae and flexor tibiae motoneurons of one hindleg (Pearson and Robertson, 1981). The strength of these connections is such that single action potentials in a C neuron can synchronously coactivate hindleg flexor and extensor tibiae motoneurons with a pattern very similar to that associated with the cocking response. Since the C neurons also receive strong input from a variety of sensory sources (visual, auditory, and tactile) known to evoke cocking, it seems likely these neurons produce the cocking response in normal-behaving animals.

The large lateral process of each M neuron makes inhibitory synaptic connections to hindleg flexor tibiae motoneurons (Pearson *et al.*, 1980). As with the C neurons, the M neurons also receive strong input from the visual, auditory, and tactile sensory systems, but in addition they receive excitatory input from hindleg proprioceptors. The M neurons have a high threshold for spike initiation (12 to 15 mV above resting), which is sufficient to prevent any one sensory input from initiating spike activity. However, the simultaneous presentation of proprioceptive input and any external stimulus will activate the M neurons (Steeves and Pearson, 1982). Recordings from the M neurons during a defensive kick, which has a similar motor program as the jump, shows that the sudden inhibition in flexor activity triggering the kick is preceded by a high-frequency burst in the M neurons (Steeves and Pearson, 1982). This pattern of activity, together with the characteristics mentioned earlier, clearly indicate that the M neurons are responsible for triggering the kick and hence the jump.

The generation of the motor pattern for the jump depends on interactions involving the C and M neurons, the extensor and flexor tibiae motoneurons, and a variety of peripheral receptors in the leg. A schematic diagram summarizing the neural circuitry for the jump is shown in Figure 3. Following coactivation of hindleg extensor and flexor tibiae motoneurons, the flexors are further excited via a strong excitatory pathway from the fast extensor motoneuron to all the ipsilateral hindleg flexor tibiae motoneurons. This central pathway functions to give a high level of flexor activity as soon as the extensor becomes active at the beginning of the co-contraction phase, and thus ensures the tibia is securely locked into the fully flexed position. Activity is maintained in the fast extensor motoneuron by afferent input from cuticular stress receptors in the leg that are activated during the co-contraction phase. This positive feedback pathway may be entirely responsible for generating extensor activity during the co-contraction phase. Excitation of the extensor motoneuron in this manner ensures a rapid build up of the extensor contraction. The mechanism for the maintenance of flexor activity during co-contraction is less clear. It is doubtful that the central excitatory pathway from extensor to flexor motoneurons can maintain flexor activity: transmission

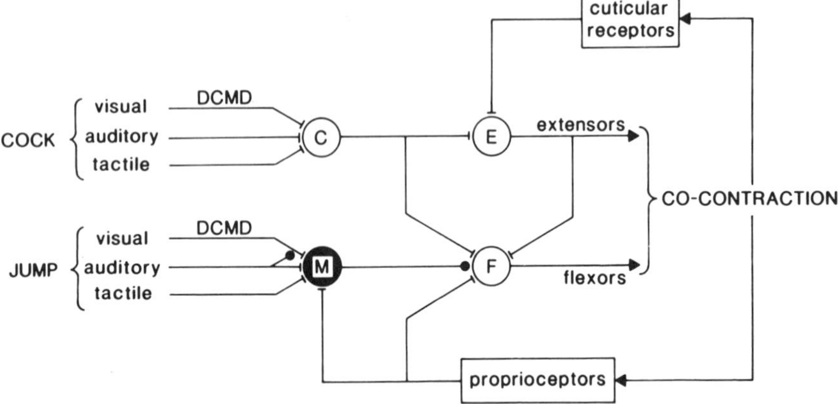

Figure 3. Diagram summarizing the main features of the neural circuits for jumping in the locust. Most aspects of this scheme are discussed in the various sections of this chapter. C, C neurons; M, M neurons; E, fast extensor tibiae motoneuron; F, flexor tibiae motoneurons; and DCMD, descending contralateral movement detector neuron (⊣) Excitatory and (●) inhibitory synaptic connections (from Pearson, 1983).

in this pathway is rapidly depressed with repetitive activity in the extensor motoneuron. On the other hand, flexor motoneurons do receive considerable input from a variety of leg receptors during co-contraction (Heitler and Burrows, 1977). Thus it seems that flexor as well as extensor activity is generated primarily by afferent feedback during the co-contraction phase.

Afferent input from leg proprioceptors during co-contraction also depolarizes the M neurons. This input alone is usually insufficient to activate the M neurons, but it does increase their excitability enough to allow their activation by visual, auditory, and tactile stimuli (Steeves and Pearson, 1982). In the absence of proprioceptive stimuli, these external stimuli rarely activate the M neurons. This proprioceptive feedback to the jump trigger neurons is essential because it ensures that a jump will not be triggered until a substantial co-contraction has been developed. Moreover, the gating of transmission in the inhibitory pathway from the various sensory systems to the flexor motoneurons ensures that flexor motoneurons are inhibited by these stimuli only when the animal is prepared to jump. This is important, for if the M neurons were easily activated by external stimuli, then the strong inhibitory action of the M neurons on flexor motoneurons could severely disrupt any ongoing behavior requiring flexor activity.

4. Movement Detector (MD) Neurons in the Locust

Of the variety of sensory stimuli that can activate both phases of escape behavior in the locust, the visual modality is probably the most striking. Locusts are highly visual animals and it is clear that abrupt, local movements of objects within the animal's extensive visual field can activate cocking and triggering in an aroused animal. The initial cocking phase is almost always elicited by sudden novel movement of small objects in the visual field. Such stimuli can also contribute significantly to triggering the jump. A more complete understanding of escape behavior has been achieved by examining how the activity is derived in visual interneurons that input directly to the cocking and trigger circuitry.

The most prominent excitatory visual input to C and M interneurons is derived from a bilaterally symmetric pair of movement detector (MD) neurons. The axons of these neurons descend to the third thoracic ganglion from the brain and are clearly the most conspicuous in the nerve cord (15–20 μm diameter) (O'Shea *et al.*, 1974). In the brain each axon crosses to the contralateral posterior protocerebrum where the cell body is located and where the input from the visual system is received. Each descending axon is activated by excitatory input from the contralateral eye. These cells are known as the descending contralateral movement detector (DCMD) neurons (Rowell, 1971*a*).

The response features of each DCMD are determined by a single presynaptic neuron, the lobular giant movement detector (LGMD) (Figure 4) (O'Shea and Williams, 1974). This neuron receives its primary input in the lobula, the most proximal region of the optic lobe. A large, fan-shaped dendritic tree provides the input surface for a retinotopic projection of small-field excitatory visual interneurons entering the lobula (O'Shea and Rowell, 1976). Two other dendritic branches of the LGMD are also present in the lobula. One extends dorsally and receives an inhibitory input that is activated by relatively rapid movements of large, patterned objects or by rapid whole-field movements (O'Shea and Rowell, 1975*a*; Rowell *et al.*, 1977). The other is a small fanlike dendritic arborization lying behind the primary fan. We do not yet know what input is received on the secondary fan. A stout axon (15 μm diameter) extends from the peduncle between the optic lobe and protocerebrum to the input segment of the DCMD (Figure 4). Near its synapse with the DCMD, the LGMD receives excitatory input from an ascending auditory interneuron, which is responsive to loud and high-pitched sound (O'Shea, 1975). This input can on occasion initiate spikes that travel out toward the optic lobe on the LGMD axon. This response has a very high threshold, and although

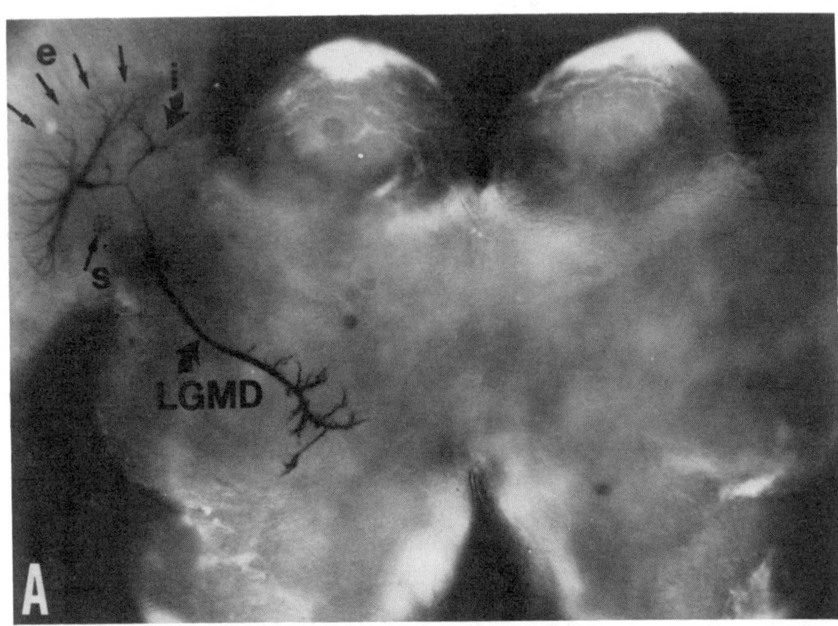
e
i
s
LGMD
A

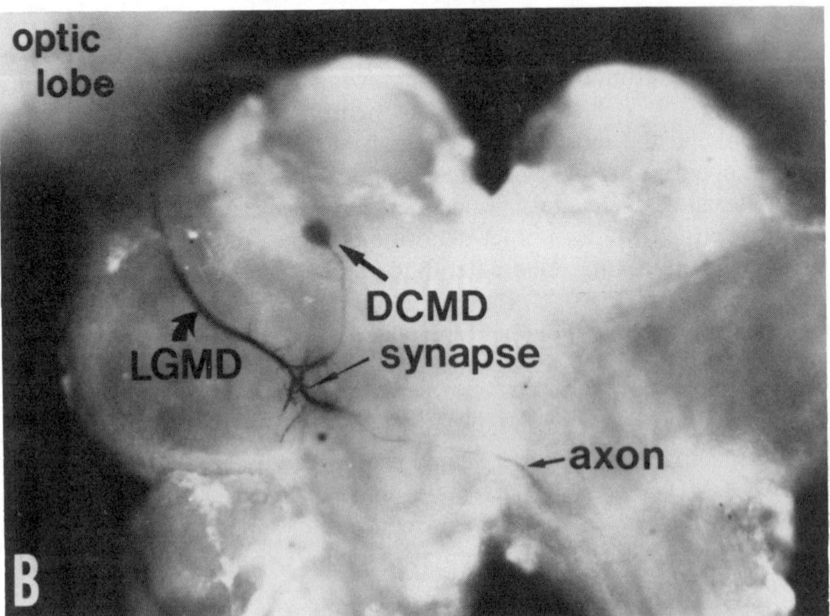
optic
lobe
DCMD
LGMD
synapse
axon
B

the LGMD is bimodal, the auditory sensitivity is very low and not a prominent feature of the response characteristics.

The synapse between LGMD and DCMD is fast, highly reliable (Figure 5A), and has properties suggesting an electrical connection (O'Shea and Rowell, 1975*b*). With rare exceptions, action potentials in the LGMD are transmitted 1:1 to the DCMD (Rowell and O'Shea, 1980). It is clear, therefore, that explaining the visual input to the escape circuitry depends on understanding the LGMD, its responsive features, and antecedent circuitry. What are these response features, how can they be interpreted in the context of escape behavior, and to what extent do we understand how they are generated by the neural machinery?

The visual responsiveness of the LGMD and DCMD neurons can be summarized as follows: (1) highly sensitive detection of abrupt movement of small (5–10°) high-contrast objects anywhere in the 180° visual field, (2) preferential detection of novel movement of such objects, (3) low responsiveness during large- or whole-field movements of visual space resulting either from the animal's own movement or to real external motion, (4) preservation of the ability of the MD neurons to detect novel motion following prolonged large-area movement, and (5) positive correlation between the sensitivity of MD neurons and the state of arousal.

These response features, seen in a behavioral context, combine to produce a neuron that seems ideally suited to an alerting or escape function. Thus it generates brisk bursts of spikes when small, abruptly, and unpredictable moving objects enter anywhere in the visual field. If, however, the object moves smoothly and repeatedly over the same area of the retina, the response decreases (Rowell, 1971*a*). It is as if the neural system interprets repeated movements as nonthreatening, ignores them, and thereby protects the DCMD and the escape circuitry from maladaptive response to them.

The neural mechanism for detecting novel motion, with a preference for small-field stimuli, resides distal to the site of retinotopic convergence

---

Figure 4. (A) Wholemount preparation of the locust brain showing a cobalt-sulfide-stained LGMD neuron. Small arrows (e) indicate the input to the fan-shaped dendritic tree of small-field ON/OFF afferents. The larger arrow (i) points to the small dorsal dendritic field that receives feed-forward inhibition. The soma (s) is lightly stained and lies ventrally in the optic peduncle between the optic lobe and the brain. A stout axon enters the brain and terminates in the ipsilateral protocerebrum where contact is made with the DCMD. (B) Double cobalt labeling of the LGMD and DCMD. The wholemount brain preparation shows contact between the ipsilateral terminations of the LGMD and the integrating segment of the DCMD. The contralaterally descending axon of the DCMD can be seen crossing the midline and descending in the contralateral connective (modified from O'Shea and Williams, 1974; O'Shea and Rowell, 1975*b*).

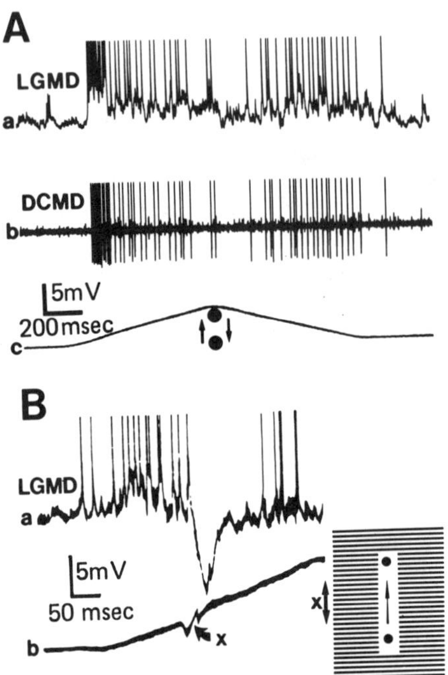

Figure 5. (A) Simultaneous recording of activity in the LGMD (a, intracellular) and the DCMD (b, extracellular) during the movement of a small circular target (●) through the visual field. An analog of the movement is indicated below (c). Note 1:1 correspondence between LGMD and DCMD spikes. Voltage calibration for (a) only. (B) Intracellular recording from the LGMD (a) while a small target is driven through the visual field by a ramp waveform (b). At (x) a striped, large-field background is displaced. This activates the feedforward inhibitory pathway to the dorsal dendritic field of the LGMD. Note the large hyperpolarization and cessation of firing in the LGMD in response to small-object motion (modified from O'Shea and Williams, 1974; O'Shea and Rowell, 1975*a*).

of small-field afferent fibers on the major fanlike dendrites of the LGMD (O'Shea and Rowell, 1976). The neurons presynaptic to the LGMD fan dendrites are a retinotopic array of small-field ON/OFF units (Rowell *et al.*, 1977). Thus the LGMD will generate spikes in response to static, small-field changes in light intensity. The response to movement is derived by the summation of a succession of such local responses produced by a moving object. The preference for small-field moving stimuli is primarily the result of two inhibitory processes. The first is a phasic lateral inhibition operating between small-field afferent pathways to the LGMD. This lateral inhibitory network reduces input to the LGMD during large-field stimulation. A second inhibitory mechanism operates after the point of convergence of the small-field afferents. Intense large-field stimuli produce direct inhibitory potentials in the LGMD (Figure 5B). Part of this feedforward inhibitory pathway has been identified and is shown to form an inhibitory connection with the dorsal dendritic field of the LGMD (Rowell *et al.*, 1977). Both of these mechanisms serve to prevent the LGMD from responding to large-field movement. Such movements are not likely to be threatening (movement of vegetation, for example) and excitation derived from them is not appropriate for an escape circuit.

The sites of decrement, which produce a preference for novel motion, are located prior to the site of convergence of small-field afferents. This was deduced from experiments showing that response decrement is site-specific in the retinotopic array (Rowell and Horn, 1968). Thus decrement occurs only if a stimulus is repeated over the same area of the visual field. Results from intracellular recording in the LGMD suggested that the labile sites are the synapses made by ON/OFF small-field afferent fibers that feed the fan dendrites (O'Shea and Rowell, 1976). It is clear, therefore, that if feed-forward inhibition (inhibition loops around the ON/OFF excitatory afferents) were the only mechanism responsible for small-field preference, these labile sites would be activated during large-field stimulation. This would clearly be maladaptive because after prolonged large-field motion, the sensitivity of the MD system to small-field motion would be depressed. This does not happen due to the action of the lateral inhibitory network located prior to the site of response decrement. Lateral inhibition therefore performs the dual role in this system of producing a preference for small-field stimuli and of protecting the behaviorally important labile novelty-detecting circuit from being activated during large-field motion (O'Shea and Rowell, 1975*a*).

Although retinally generated lateral and feed-forward inhibition do contribute to suppression of the visual system during image movement on the retina caused by eye motion, another mechanism of DCMD response suppression, which does not depend on retinal stimulation, has been discovered by Zaretsky and Rowell (1979). During rapid saccadic head movements, normally potent stimuli do not produce responses in the DCMD. This suppression persists even if the head is immobilized when the animal attempts to make saccadic movements. This shows that suppression of visual responsiveness can be derived centrally during the activation of the motor command that would normally lead to head movement and, of course, retinal stimulation. This centrally derived inhibition or "corollary discharge" and the retinally derived inhibition could act together. In a recent and elegant study, Zaretsky (1982) measured the relative contribution of centrally and retinally generated suppression. In summary, Zaretsky demonstrates that centrally derived inhibition is more powerful than visually derived inhibition. The maximal reduction in DCMD response derived from high-contrast frequency visual background was 0.45 log units. Centrally derived inhibition, which is independent of contrast frequency or pattern in the visual field, was typically 0.94 log units. The significance and necessity of the centrally derived inhibition can best be appreciated under conditions when the visual background is virtually unpatterned (sky or desert sands, for example). Under these circumstances visual input would contribute almost no inhibition. Any small stationary object against such a background would provide a strong stimulus to the DCMD during head movement. Without centrally derived inhibition, such an object might product an inappropriate escape response.

As has been described above, a decremental synaptic mechanism is the key to the novelty-detecting ability of the DCMD. A decreased synapse will slowly, over minutes, recover sensitivity when left unstimulated. The MD system is, however, subject to sudden changes in visual responsiveness (Rowell, 1971*b*). Sensitivity to a repeated stimulus to which the system has become unresponsive can therefore be re-established rapidly if necessary, without a long period of no stimulation. This incremental process is correlated with what can best be described as the animal's state of arousal. An increase in arousal level is accompanied by an increased MD sensitivity and a rapid re-establishment of response to stimuli that had previously been ignored. This incremental system provides an adaptive flexibility to a highly selective sensory system that generally ignores much of the visual information potentially available to it. By this system an animal in an aroused state allows a wider range of visual stimuli access to the escape circuitry. We do not yet understand the cellular basis of the incremental process or how the sensitivity of the MD system is linked to a general behavior-arousal system. Because the sites of response decrement lie more peripherally in the visual system than the LGMD, we can predict the existence of a neural pathway that carries centrally generated instructions causing increment at these sites. Activity in this pathway must be correlated with high arousal state. In the MD system, some preliminary experiments have implicated octopamine as a candidate neurotransmitter that can elevate visual responsiveness when perfused into the optic lobe distal to the LGMD (M. O'Shea, unpublished data). This and other modulatory transmitters may function widely in the CNS to control levels of responsiveness and synaptic efficacy to make them appropriate for the particular behaviors that are to be initiated or suppressed. We are far from understanding how this is achieved.

5. Initiation of the Jump by Movement-Detecting Neurons

As we have seen in the previous section, locusts have evolved an interneuronal system for rapidly transmitting information about novel movements anywhere in the visual field to thoracic ganglia. What function, or functions, does this movement-detecting system serve? There is now considerable evidence that one function of the DCMDs is to initiate the jump. Not only are the discharge characteristics of these neurons ideally suited for signally information about potentially threatening situations, but their synaptic connections to neurons in thoracic ganglia are appropriate for initiating a jump. The most important of these connections are to the C and M neurons. Each of these neurons receives a strong monosynaptic excitatory input from both DCMDs. The amplitude of the individual EPSPs (excitatory post-synaptic potentials) is usually in the range of 4–6 mV but can be as large as 9 mV. Despite these large amplitudes, high-frequency bursts in one or both DCMDs rarely evoke action potentials in the C and M neurons in dissected preparations. For the M neurons, this is understandable because it would be inappropriate for these to be activated by the DCMDs alone (see Section 2). Activation of the M neurons by the DCMDs requires summation with synaptic input from proprioceptors (activated during co-contraction). On the other hand, the C neurons might be expected to be easily activated by the DCMDs, since the cocking response can often be evoked by movements in the visual field. However, the cocking response occurs only in aroused animals and never in the dissected preparations in which measurements of the DCMD input to the C neurons are made. Thus it seems that the threshold of the C neurons is regulated depending on the level of arousal. How this occurs is not known, but from a functional viewpoint it is sensible. Obviously it would not be appropriate for the C neurons to be activated every time the DCMDs discharged, for this would cause considerable disruption to normal ongoing behavior (feeding, grooming, sexual activity, etcetera). Only when the animal is aroused by a threatening situation should the C neurons be activated. Thus we expect the C neurons should have a high threshold, which indeed they do, and also that the excitability of these neurons is regulated by arousal.

The sequence of events by which movements in the visual field lead to a jump is as follows: in an aroused animal any novel movement in the visual field evokes a high frequency discharge in a DCMD and this activates the C neurons but not the M neurons. This leads to cocking and the initiation of the co-contraction phase. The excitability of the M neurons is then increased by proprioceptive feedback, thereby permitting further activity in the DCMDs (evoked by any additional movements) to activate the M neurons and thus trigger a jump. In the event that additional movements do not occur, only the cocking response will be elicited. If this sequence of events is correct, then we see that the DCMDs have two distinct functions in the jump: the first is to evoke cocking and the second is to trigger the jump.

6. Conclusions

An obvious feature of the neuronal systems controlling the jump and signaling information about movements in the visual field is that they contain a small number of large identifiable interneurons. Each pair is uniquely characterized by its structure and physiological properties, and each is specialized for a specific purpose: the LGMDs for collecting information about movements from the entire retina, the DCMDs for relaying movement information rapidly to the thoracic ganglia, the C neurons for eliciting the cocking response, and the M neurons for triggering the jump. There are, however, some interesting functional differences between the interneurons in the jump circuitry (the C and M neurons) and the interneurons in the movement-detecting system (LGMD and DCMD). The C and M neurons receive input from a variety of sensory sources, have a high threshold for spike initiation, and each appears to have only one behavioral function. By contrast, the LGMD–DCMD neuron (taken together) primarily respond to a single sensory modality, are very easily activated, and have multiple behavioral functions [it should be noted here that the DCMDs are also involved in the control of flight, but their precise function in this behavior has not been established (Simmons, 1980*b*)]. These functional differences between the C and M neurons and the LGMD–DCMD neurons may illustrate an important organizational feature of sensory-motor integration in insects. On one hand, interneurons in motor systems are committed to generate single components of a behavior and are activated only under specific conditions, whereas sensory systems are designed to extract a number of features from the environment and the higher-order neurons signaling each feature have the capacity to evoke a variety of behaviors depending on the animal's state. If this is generally true, then the important question arises about what directs feature–detecting neurons to produce different behaviors under different conditions. In the movement-detecting system this control depends to a large extent on sensory input from leg receptors. The elicitation of the cocking response by the DCMDs requires the tibia to be close to full flexion, the triggering of the jump by the DCMDs requires feedback from leg proprioceptors during co-contraction, and any influence of the DCMDs on flight requires the loss of leg contact with the ground so flight activity can be generated. Similarly in the cockroach, single giant interneurons can have multiple functions (escape or flight) depending on sensory input from the legs (Chapter 4, this volume). Thus, in insects at least, we are beginning to get a glimpse of the mechanisms involved in linking sensory to motor systems and how this linkage is controlled by the state of the animal. So far this analysis has been confined to the largest interneurons in insect nervous systems, many of which are involved in escape behavior. Whether the organization of these interneurons reflects general features of interneuronal systems in insects and other animals remains to be seen.

7. References

Bacon, J., and Tyrer, N. M., 1978, The tritocerebral commissure giant (TCG): A bimodal interneuron in the locust, *Schistocerca gregaria*, *J. Comp. Physiol.* **126:**317–325.

Bennet-Clark, H. C., 1967, The jump of the flea: A study of energetics and a model of the mechanism, *J. Exp. Biol.* **47:**59–76.

Bennet-Clark, H. C., 1975, The energetics of the jump in the locust *Schistocerca gregaria*, *J. Exp. Biol.* **63:**53–83.

Burrows, M., 1980, The control of sets of motoneurons by local interneurons in the locust, *J. Physiol.* **298:**213–234.

Burrows, M., 1982, Interneurones co-ordinating the ventilatory movements of the thoracic spiracles in the locust, *J. Exp. Biol.* **97:**385–400.

Burrows, M., and Siegler, M. V. S., 1982, Spiking local interneurons mediate local reflexes, *Science* **217:**650–652.

Daley, D. L., Vardi, N., Appignani, B., and Camhi, J. M., 1981, Morphology of giant interneurons and cercal nerve projections of the american cockroach, *J. Comp. Neurol.* **196:**41–52.

Evans, M. E. G., 1973, The jump of the click beetle (Coleoptera, Elateridae)—Energetics and mechanics, *J. Zool. (London)* **169:**181–194.

Evans, M. E. G., 1975, The jump of *Petrobius* (Thysanura, Machilidae), *J. Zool. (London)* **176:**49–65.

Frantsevich, L. I., 1981, The jump of the black beetle (Coleoptera, Histeridae), *Zool. Jb. Anat.* **106:**333–348.

Heitler, W. J., 1974, The locust jump. Specializations of the metathoracic femoral-tibial joint, *J. Comp. Physiol.* **89:**93–104.

Heitler, W. J., and Burrows, M., 1977, The locust jump. II. Neural circuits of the motor programme, *J. Exp. Biol.* **66:**221–242.

O'Shea, M., 1975, Two sites of axonal spike initiation in a bimodal interneuron, *Brain Res.* **96:**93–98.

O'Shea, M., and Rowell, C. H. F., 1975*a*, Protection from habituation by lateral inhibition, *Nature* **254:**53–55.

O'Shea, M., and Rowell, C. H. F., 1975*b*, A spike-transmitting electrical synapse between visual interneurons in the locust movement detector system, *J. Comp. Physiol.* **97:**143–158.

O'Shea, M., and Rowell, C. H. F., 1976, Neuronal basis of a sensory analyser, the acridid movement detector system. II. Response decrement, convergence, the nature of the excitatory afferents to the LGMD, *J. Exp. Biol.* **65:**289–308.

O'Shea, M., and Rowell, C. H. F., 1977, Complex neural integration and identified interneurons in the locust brain, in: *Identified Neurons and Behavior in Arthropods* (G. Hoyle, ed.), Plenum Press, New York. pp. 307–328.

O'Shea, M., and Williams, J. L. D., 1974, Anatomy and output connections of the lobular giant movement detector neuron (LGMD) of the locust, *J. Comp. Physiol.* **41:**257–266.

O'Shea, M., Rowell, C. H. F., and Williams, J. L. D., 1974, The anatomy of a locust visual interneuron; the descending contralateral movement detector, *J. Exp. Biol.* **60:**1–12.

Pearson, K. G., 1980, Burst generation in coordinating interneurons of the ventilatory system of the locust, *J. Comp. Physiol.* **137**:305–313.

Pearson, K. G., 1983, Neural circuits for jumping in the locust, *J. Physiol. (Paris)* **78**:765–771.

Pearson, K. G., and Fourtner, C. R., 1975, Nonspiking interneurons in the walking system of the cockroach, *J. Neurophysiol.* **38**:33–52.

Pearson, K. G., and Robertson, R. M., 1981, Interneurons coactivating hindleg flexor and extensor motoneurons in the locust, *J. Comp. Physiol.* **144**:391–400.

Pearson, K. G., Heitler, W. J., and Steeves, J. D., 1980, Triggering of locust jump by multimodal inhibitory interneurons, *J. Neurophysiol.* **43**:257–278.

Pfluger, H. J., and Tautz, J., 1982, Air movement sensitive hairs and interneurons in *Locusta migratoria*, *J. Comp. Physiol.* **145**:369–380.

Rehbein, H., 1976, Auditory neurons in the ventral cord of the locust: Morphological and functional properties, *J. Comp. Physiol.* **110**:233–250.

Robertson, R. M., and Pearson, K. G., 1983, Interneurons in the flight system of the locust: distribution, connections and resetting properties, *J. Comp. Neurol.* **215**:33–50.

Rothschild, M., Schlein, Y., Parker, K., and Steinberg, S., 1972, Jump of the oriental rat flee, *Xenopsylla cheopsis* Roths, *Nature* **239**:45–48.

Rowell, C. H. F., 1971*a*, The orthopteran descending movement detector (DMD) neurones: A characterisation and review, *Z. Vergl. Physiol.* **73**:167–194.

Rowell, C. H. F., 1971*b*, Variable responsiveness of a visual interneuron in the free-moving locust, and its relation to behaviour and arousal, *J. Exp. Biol.* **55**:727–748.

Rowell, C. H. F., and Horn, G., 1968, Dishabituation and arousal in the response of single nerve cells in an insect brain, *J. Exp. Biol.* **49**:171–183.

Rowell, C. H. F., and O'Shea, M., 1980, Modulation of transmission at an electrical synapse in the locust movement detector system, *J. Comp. Physiol.* **137**:233–241.

Rowell, C. H. F., O'Shea, M., and Williams, J. L. D., 1977, The neuronal basis of a sensory analyser, the acridid movement detector system, IV. The preference for small field stimuli, *J. Exp. Biol.* **68**:157–185.

Siegler, M. V. S., 1981, Posture and history of movement determine membrane potential and synaptic events in non-spiking interneurons and motor neurons of locust, *J. Neurophysiol.* **46**:296–309.

Siegler, M. V. S., and Burrows, M., 1979, The morphology of local non-spiking interneurons in the metathoracic ganglion of the locust, *J. Comp. Neurol.* **183**:121–148.

Simmons, P., 1980*a*, A locust wind and ocellar brain neurone, *J. Exp. Biol.* **85**:281–294.

Simmons, P., 1980*b*, Connexions between a movement-detecting visual interneurone and flight motoneurones of a locust, *J. Exp. Biol.* **86**:87–98.

Steeves, J. D., and Pearson, K. G., 1982, Proprioceptive gating of inhibitory pathways to hindleg flexor motoneurons in the locust, *J. Comp. Physiol.* **146**:507–515.

Wohlers, D. W., and Huber, F., 1982, Processing of sound signals by six types of neurons in the prothoracic ganglion of the cricket, *Gryllus campestris* L., *J. Comp. Physiol.* **146**:161–174.

Zaretsky, M., 1982, Quantitative measurements of centrally and retinally generated saccadic suppression in a locust movement detector neurone, *J. Physiol.* **328**:521–533.

Zaretsky, M., and Rowell, C. H. F., 1979, Saccadic suppression by corollary discharge in the locust, *Nature* **280**:583–585.

7

The Production of Crayfish Tailflip Escape Responses

FRANKLIN B. KRASNE and JEFFREY J. WINE

1. Introduction

1.1. Multiple Systems for Escape

Crayfish escape from potentially serious threats to life and limb by means of powerful flexions of their abdomens, which thrust the animal through the water away from danger. Aspects of the behavioral and neural analysis of this tailflip escape behavior are the subject of this essay. Since a comprehensive and systematic treatment of the same topic has recently been published in *The Biology of Crustacea*, Vol. 4 (Wine and Krasne, 1982), we here emphasize recent advances and discuss interpretations of them.

The crayfish ventral nerve cord is traversed by two bilaterally paired, dorsally situated giant axons, the medial and lateral giant axons (MGs and LGs, or collectively simply "giants"). Classical work of Johnson (1924, 1926) and Wiersma (1947) established that these giant axons produce tailflip escape responses when directly stimulated and showed that the giants could be fired by peripheral stimuli, the receptive field for the MGs being mainly the head and thorax, whereas that of the LGs was the abdomen. These conclusions have been amply verified by more modern anatomical and physiological techniques. From this it was presumed that the giants mediate all tailflip escape. However, in 1970 it was first reported that tailflip escape could occur without giant activity (Schrameck, 1970),

FRANKLIN B. KRASNE • Department of Psychology, University of California at Los Angeles, Los Angeles, California 90024. *JEFFREY J. WINE* • Department of Psychology, Stanford University, Stanford, California 94305.

thus establishing the existence, in addition to LG- and MG-mediated escape, of a category of nongiant-mediated tailflip responses.

Subsequent analysis of the circumstances under which each category (LG, MG, and nongiant) of mediation was employed and the characteristics of the responses produced by each mediational system led to a parsimonious picture (Wine and Krasne, 1972).

Giant-mediated tailflip responses have properties that make them well suited for emergency escape maneuvers. The giants are fired only by stimuli of abrupt onset, and their latency of firing is always short (usually about 3–7 msec; see Figure 7). Escape movements invariably follow giant axon spikes at very short latency (2–3 msec to muscle potentials and another 10 to the start of movement). These latencies provide little time for evaluation of the nature and location of the stimulus beyond the coding inherent in the receptive fields of the giant axons. Consistent with this, the tailflip movements produced are quite stereotyped. The MGs, which have a rostral receptive field, are associated with flexions about each segmental hinge of the abdomen, which drives the animal directly backward away from the rostral threat; the LGs, which have a caudal receptive field, are associated with flexions only about rostral abdominal joints, and this pattern of flexion causes a substantial lift and pivoting upward and forward of the abdomen that removes the hind end of the animal from the threatening object (Figure 1).

In contrast to the LG and MG escape systems, nongiant escape production circuitry can generate responses that are flexibly structured to provide for accurate, probably visually guided trajectories away from threats and toward known places of safety. This has long been known from informal observation (Wine and Krasne, 1972), but formal analysis has only recently begun (Figure 2) (Reichert and Wine, 1983). Although the nongiant response generation system is thus much more sophisticated in its capabilities than that of the giant system, a serious price is paid in terms of rapidity of reaction. Whereas flexor muscle potentials for giant-mediated responses start about 3–10 msec after an abrupt stimulus, nongiant responses require some 50–500 msec for their initiation (Wine and Krasne, 1972; Reichert and Wine, 1983; Figure 7). Giant-mediated responses occur only to abrupt stimuli and only once per stimulus, but nongiant-mediated responses can occur to any kind of threat and are often repeated in bouts called "swimming." Thus, a stimulus can lead to a single giant-mediated response or to one or more nongiant responses or to a giant-mediated response followed by one or more nongiant responses.

Which of these patterns will occur depends on the strength and abruptness of the stimulus. It also depends on the past history and situation of the animal, since both giant- and nongiant-mediated escape sys-

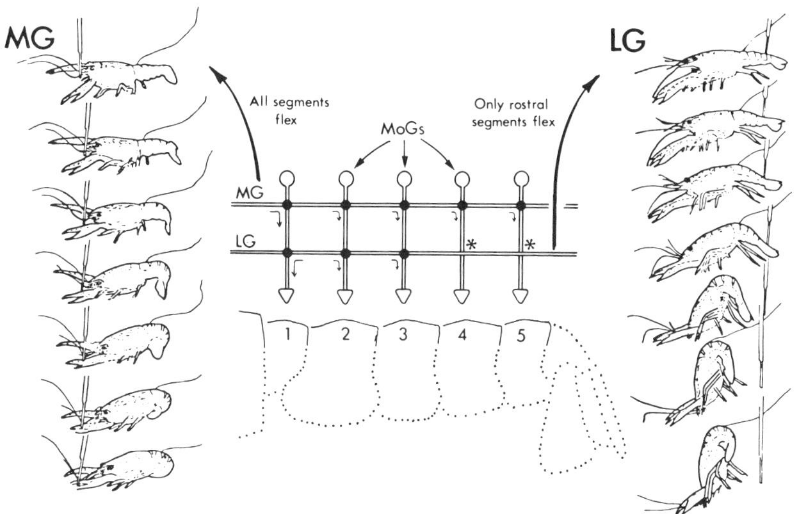

Figure 1. Forms of giant-mediated tailflips. When the MGs fire, all segments flex and the abdomen curls and propels the animal backward. When LGs fire, caudal segments remain straight and cause the thrust to be directed mainly down, thus pitching the animal forward. Since MGs respond to rostral inputs and LGs to caudal ones, tailflips always remove the animal from the source of stimulus. Consistent with the difference in form of MG and LG flips, the MGs excite MoGs in every abdominal segment, whereas the LGs excite MoGs only in more rostral segments (circuit of center top) (based on Wine and Krasne, 1972; Mittenthal and Wine, 1973; and taken from Wine and Krasne, 1982).

tems are subject to habituation, sensitization, and to control by motivational variables (see Krasne and Wine, 1977).

Strong, abrupt threats most often cause a giant-mediated reaction followed by a nongiant-mediated swimming sequence, and this is an optimally adaptive strategy for escape given the limitations of each control system. The giant-mediated response gets the animal moving away from the stimulus almost immediately, and by the time flexion and re-extension have occurred (about 100–200 msec), enough time has passed to prepare for well-directed nongiant-mediated tailflipping.

2. The Roles of the Giant Axons

When giant axons are made to fire by directly depolarizing them, vigorous tailflips are produced (Wiersma, 1947); moreover, under normal circumstances there is a perfect correlation between the natural firing of

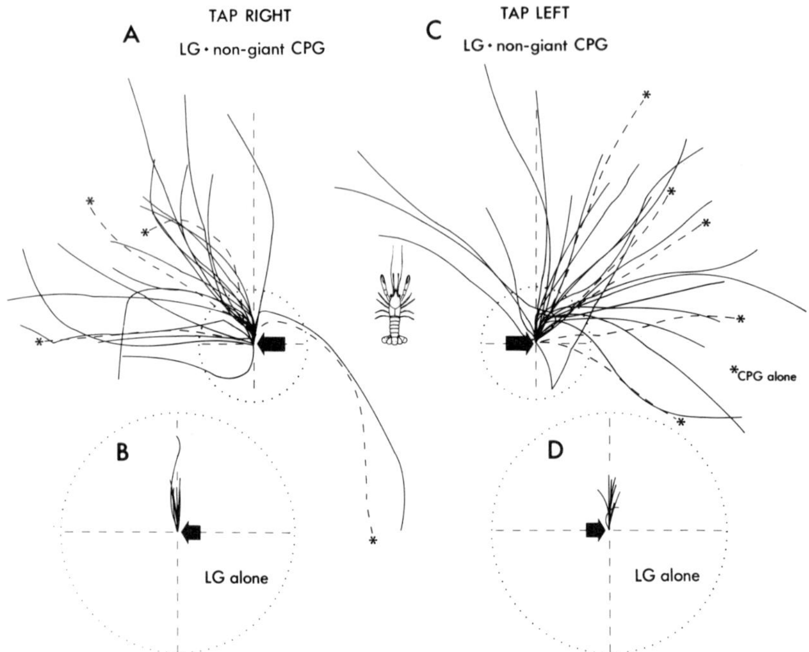

Figure 2. Orientation during nongiant-mediated swimming has a lateral component that on average removes the animal from the stimulus source. Based on video tapes filmed from above. (A, C) Taps were administered to the right and left side of unrestrained animals; the trajectories show that animals react to the laterality of the stimulus when they swim. (B, D) In contrast, the initial, giant-mediated tailflips (shown at enlarged scale) are laterally symmetric; their only significant vector is a forward movement. Starred trajectories are ones produced by nongiant swimming that was not preceded by firing of the giants. Diameter of dotted circles is 6 cm; inset shows orientation and size of crayfish. Origins and trajectories were plotted from a point on the center of the posterior edge of the telson (from Riechert and Wine, 1983).

the giant axons and the occurrence of short latency stereotyped tailflips (Wine and Krasne, 1972). It has therefore been natural to presume that the giants are an obligatory link mediating between receptors and motor neurons in the production of short latency responses. However, normal (nongiant) escape responses, albeit ones of long latency, can be generated without the operation of the giants. This, as well as recognition of the possibility that the giants might enhance intersegmental synchrony of short latency reactions without actually being essential to their generation (see Kupfermann and Weiss, 1978) has raised the question of precisely what role the giants do play in the various events that ordinarily follow their firing. These events are phasic flexion itself, phasic re-extension (most often starting about 70 msec after LG firing), and sometimes one or more cycles of nongiant-mediated swimming, the first flexion of which most often begins about 140 msec after firing of the LGs (Figure 3). In the last few years the role of the LGs in these events has been examined in some detail; this section summarizes what has been found.

2.1. The LGs Are Necessary for the Short Latency Phasic Flexions That Follow Them

It has recently been demonstrated that when the LGs are hyperpolarized sufficiently to prevent their firing, stimuli that would otherwise cause short latency tailflip responses no longer do so (Figure 4). Moreover, for any given stimulus strength, the amount of hyperpolarization needed to abolish motor responses is precisely that which is needed to prevent the LGs from firing. This establishes the necessity of LG firing in the chain of command leading to "LG" tailflips (Olson and Krasne, 1981).

It leaves open, however, their precise function in that chain of command. Their input regions might be the site where information derived from sensory receptors is integrated and the decision to respond is made, or the decision might be made at an earlier station and the LGs might be mere followers that serve to distribute excitation to the extensive, segementally distributed population of motor neurons whose firing produces the response. In either case the firing of the LGs would always precede a short latency tailflip, but the jobs done by the LGs in the two cases are very different.

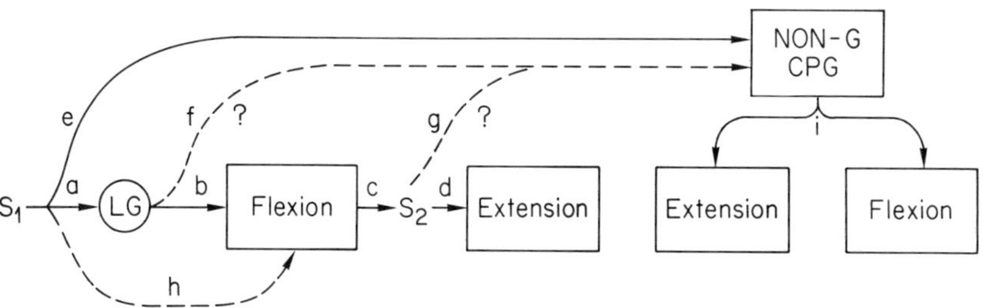

Figure 3. Causality of components of tailflip escape behavior. (——) Major causes and (---) minor or contributory ones. S_1 denotes the stimulus initiating escape behavior; S_2 denotes stimuli produced *by* the LG-mediated fast flexion. Causal pathways f and g have question marks because it is not clear whether the facilitating effects of LG firing on subsequent nongiant tailflipping are mediated centrally (causal pathway f) or indirectly via flexion-produced stimuli (causal pathway g).

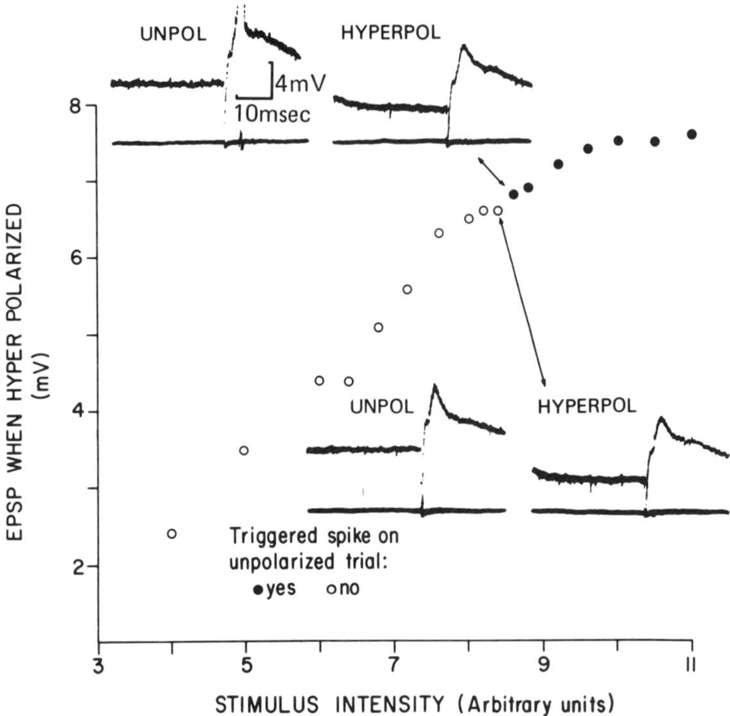

Figure 4. The LGs are the decision neurons of the LG circuit. This experiment rules against the possibility that a neuron prior to the LG is the essential convergence point for sensory information and determines the firing of the LGs. If such a hypothetical decision neuron fired, we would expect to see a large increase in the sensory evoked PSP in the LG when the stimulus exceeded threshold for a tailflip. The graph of this figure shows that there is no increase in the PSP, which is instead a smooth function of stimulus intensity. PSPs in the LG were evoked by an electrical stimulus of the afferents through the 2nd nerve root of the 3rd ganglion. The stimulus intensity was increased step-wise as the LG was alternatively hyperpolarized (Hyperpol) or left at normal resting potential level (Unpol). The hyperpolarization kept the cell from spiking and obscuring any change in PSP amplitude. There was no discontinuous increase in PSP amplitude as the cell went from a subthreshold level (○) to suprathreshold (•). This is also indicated by the four inserted traces. The lower two traces show the PSP just before the firing level was reached; the upper traces show the spike and PSP just after the firing level (from Olson and Krasne, 1981).

2.2. The LGs Are the "Decision" and "Trigger" Neurons for LG Tailflips

It has been known for more than a decade that sensory stimuli too weak to cause an LG escape response nevertheless produce both monosynaptic and disynaptic EPSPs in the LG dendrites and that the amplitude of the resultant compound EPSP increases as stimulus intensity increases, until, with a sufficiently strong stimulus, a spike is initiated on the ascending limb of the EPSP near its peak (Krasne, 1969). Thus the LG dendrites have been presumed to be the site where sensory-derived excitation is integrated. However, it is very common in this system for neurons to receive convergent excitatory input deriving from a given source via several routes of different length and directness. Thus, one must consider the possibility that the sources of input seen with subthreshold stimuli merely "prime" the LGs and bring them near to firing level, whereas firing per se depends on an independent source of input that is recruited only by stimuli that are strong enough to cause escape. Of course, such a source of input would be singled out as special only if it produced exceptionally large EPSPs in the LGs; however, if it did, it would have to be considered as playing a special triggering role in LG firing (even if not alone sufficient to produce such firing), and the LG itself could not then be considered *the* decision neuron of the circuit. Although recruitment of a large EPSP in association with the firing of the LGs has never been noticed, in an LG already near firing threshold such an EPSP could easily merge so smoothly with the spike it triggered that it would go entirely unnoticed. A series of experiments were therefore performed in which EPSPs were evoked in hyperpolarized LGs by a range of stimuli with intensities near the threshold intensity for causing firing of the unpolarized LG. Under such conditions the recruitment of a strong source of input at the threshold stimulus intensity for producing an escape response would be conspicuous if it were present. However, as shown in Figure 4, these experiments provided no evidence whatsoever for recruitment of a special input at the response threshold (Olson and Krasne, 1981). Therefore, the LGs themselves do appear to be the crucial site at which the decision to make an LG-mediated escape response is taken.

2.3. *The LGs Are Sufficient for Phasic Flexion but Do Not Produce Fully Normal Responses*

Although the LGs (and by analogy the MGs) are necessary for short-latency escape responses, it is entirely conceivable that such firing might trigger escape only when accompanied by sensory-derived activity that is independent of the giants. However, it has long been known that LG firing produced by direct depolarization of the LGs does produce a robust tailflip response. On the other hand, it is now also known that, independent of the giants, sensory afferent stimulation can cause both excitatory and inhibitory input to the motor neurons (Sherwood and Wine, 1979) and premotor interneurons (Kramer *et al.*, 1981*a*; Section 3.2, this chapter) that participate in generation of the tailflip. This raised the possibility that the tailflip produced by directly elicited firing, although robust and seemingly well formed (see Figure 2 of Wine and Krasne, 1982), is in some way abnormal, and this possibility has motivated a careful comparison of directly and naturally evoked LG tailflips (G. Hagiwara, L. Miller, and J. J. Wine, unpublished data). Differences have in fact emerged from this analysis; both the precise direction of the animal's movement (G. Hagiwara, L. Miller, and J. J. Wine, unpublished data) and the duration of fast flexor ("tailflip"—see Section 2.4) muscle EMGs (Reichert *et al.*, 1981) differ between naturally and directly evoked LG-mediated responses (Figure 5). Thus, LG firing is necessary for short-latency flips, and direct stimulation of the LGs produces a fair approximation to a normal flip. However, sensory information that bypasses the LGs must exert an influence in order for the LGs to trigger an entirely normal LG-type tailflip.

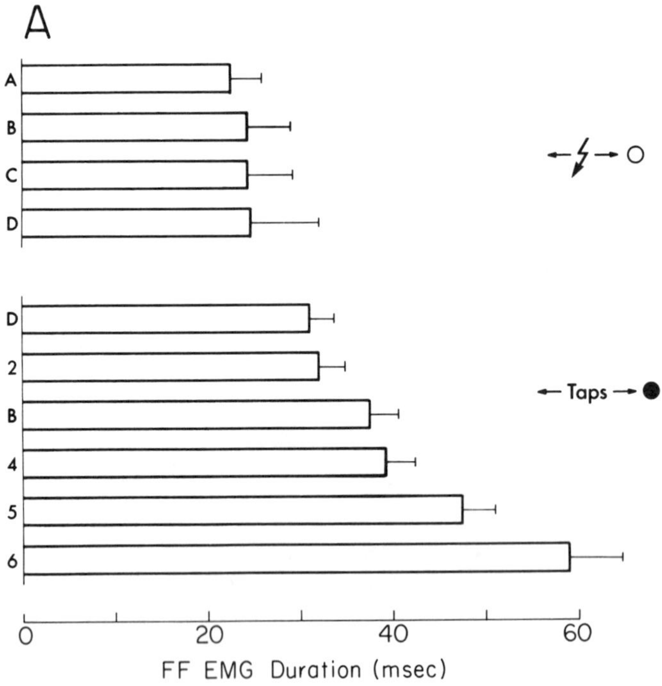

Figure 5. Comparison of tailflips produced by naturally and directly (electrically) elicited LG firing. (A) Duration of FF EMGs. Each bar gives results for the individual animal whose identification label appears at the left. (B) Vertical and horizontal distances covered as a

2.4. *The Central Consequences of LG Firing Do Not Produce Postflexion Re-extension, Which Is Instead a Chain Reflex*

In a normal, free, rested animal the flexion that is produced by either natural or directly evoked LG firings is almost always followed, 50–175 msec after the LG firing, by a rapid re-extension of the abdomen (Figure 6A,B) (Reichert *et al.*, 1981). It would thus be natural to presume that re-extension, as with the flexion response itself, is in large measure a direct consequence of central events set in motion by the firing of the LGs. However, directly elicited firing of the LGs in restrained acute preparations of isolated abdomens causes no excitation of phasic extensor motor neurons whatever, whereas firing of either extensor muscle stretch receptors or tactile afferents causes substantial excitation (Wine, 1977*a,b*). Moreover, interfering with rapid flexion in intact animals either by flexor de-efferentation or by restraint prevents postgiant extension (Figure 6C,D), whereas augmenting flexion-produced exteroceptive input by providing a substrate for the tail to flap against augments the vigor and shortens the latency of re-extension (Reichert *et al.*, 1981). It is not yet known whether the giants might make some centrally mediated contribution to excitation of phasic extensors in animals with intact nervous systems, but it seems

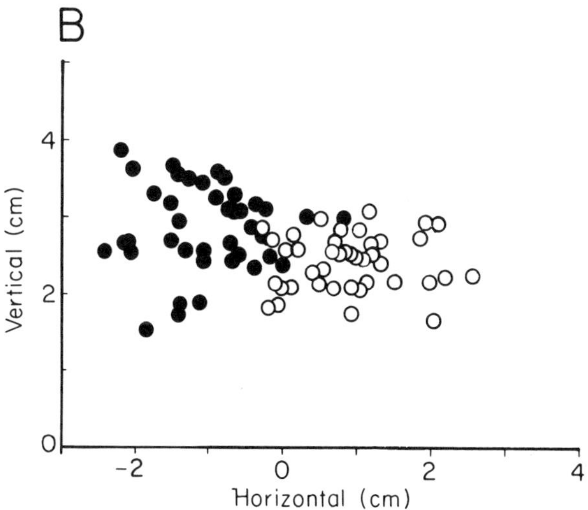

result of each type of flip; naturally evoked flips (taps) produce more vertical trajectories than do flips evoked by direct electrical stimulation of LGs (based on Reichert, Wine and Hagiwara, 1981 and unpublished data of G. Hagiwara, L. Miller, and J. J. Wine).

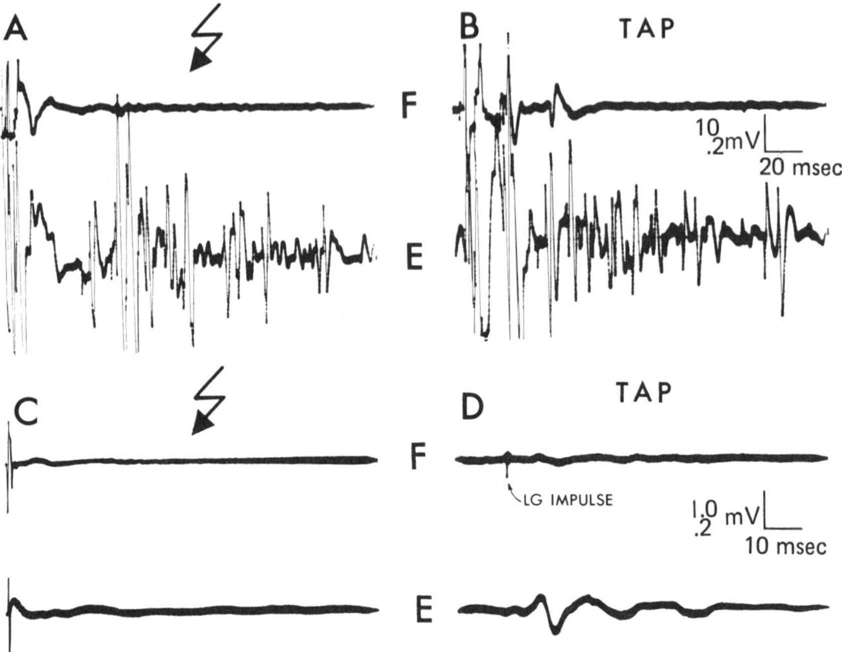

Figure 6. Postgiant extension requires flexion. (A, B) Flexor and extensor EMGs of an intact, unrestrained animal resulting from electrical (A) and natural tactile (B) stimulation of the LG axons. In each case LG axon stimulation resulted in only a single tailflip. In these and the following EMG records, the upper trace of each pair (labeled F) shows the summed extracellular potentials recorded by the electrodes implanted in the fast flexor muscles just dorsal to the nerve cord. The lower trace of each pair (labeled E) shows the summed extracellular potentials recorded by the electrodes implanted in the fast extensor muscles. All EMG recordings are from the muscles of the 3rd abdominal segment. (C, D) Results of stimulating the LG axons following transection of all abdominal fast flexor efferent roots. Neither electrical (C) nor natural tactile stimulation (D) of the LG axon evoked phasic flexor or phasic extensory EMG activity. The extracellularly recorded LG axon spike appears in the flexor trace of D (in C it is partly masked by stimulus artifact), and a small amount of *tonic* extensor activity is seen on the extensor trace (note fast sweep speed in C, D) (from Reichert *et al.*, 1981).

clear from the above facts that re-extension is in substantial measure a chain reflex response to reafference produced by the fast flexion response itself. Whereas the giants play no direct central role in the excitation of postgiant re-extension, they do drive abdominal circuitry that inhibits extensor stretch receptors and the phasic extensors themselves until shortly after peak flexor tension is developed; without these inhibitory actions, reafference and proprioception might well cause premature reflex re-extension while flexion was still in progress.

Although it is not understood in detail how postgiant re-extension is triggered and organized, it seems clear that reafference plays a dominant role in its production, and no central excitatory role for the giants has been found. This comes as something of a surprise at a time in scientific history when fixed progression patterns such as the LG flip and re-extension are commonly found to be centrally generated or to depend on cooperative interactions between central pattern generators and proprioceptive or exteroceptive reflexes.

2.5. LG Firing Promotes but Does Not Drive Subsequent Swimming

Lateral giant tailflips evoked by natural stimuli are very commonly succeeded by bouts of one or more nongiant-mediated tailflips (called here "swimming"), the first nongiant flexion of which most often starts about 150 msec after the stimulus.

In the original study of the stimulus dependence of the two types of escape, Winc and Krasne found that giant-mediated escape was caused by abrupt stimuli, whereas nongiant reactions that occurred on their own were usually due to gradual stimulation (Wine and Krasne, 1972). Thus, it was natural to suppose that nongiant responses that follow giant ones were in some way triggered by the giants rather than by the stimulus.

However, it has recently been shown that tailflips and re-extension caused by direct stimulation of the LGs, even in intact, unrestrained, unhabituated animals, is virtually never followed by nongiant-mediated swimming. On the other hand, in well-rested animals, abdominal taps that are just subthreshold for firing the giants often cause nongiant-mediated tailflips, and when this happens, these nongiant responses occur at about the same time that they would have occurred had the giants fired (Figure 7A). Therefore, postgiant swimming is produced by circuitry that is at least in part independent of and parallel to giant circuitry (Reichert and Wine, 1982; 1983).

However, if there has been any habituation to taps, the likelihood of postgiant swimming is much greater than can be accounted for by the intrinsic responsiveness of the nongiant system to phasic stimulation; the likelihood of nongiant swimming is commonly augmented some two to ten times when a stimulus fires the LGs compared with when it does not (based on data of Reichert and Wine, 1982). Figure 7B demonstrates the augmentation produced by the giants more directly by showing the effect of electrically firing the giants just following a gentle tap (Reichert and Wine, 1983). Thus, in most circumstances sensory activity from the initial

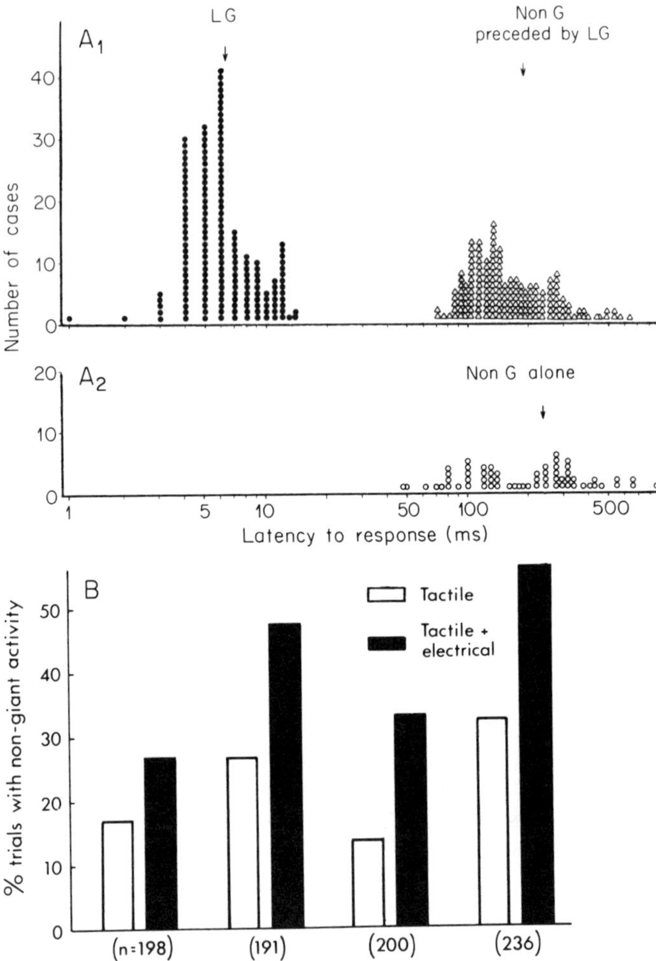

Figure 7. Relationship between LG firing and subsequent nongiant-mediated tailflips. (A) Histograms of latencies to tailflip flexions for three categories of tailflips. All responses were produced by lightly tapping the abdomen. (A_1) Latencies of LG-mediated responses and the nongiant responses that followed them in cases where stimuli caused both LG and subsequent nongiant responding. (A_2) Latencies of nongiant flips when only nongiant responding occurred. Arrows indicate means that are 6.3 and 184 msec in (A_1) and 240 msec in (A_2); corresponding medians are 6, 150, and 210 msec, respectively. (B) Giant axon activity primes the nongiant system. Each pair of columns shows the data for one animal. The number of trials is indicated beneath the columns. Trials were at 1-min intervals in blocks of approximately 40 trials separated by 24 hr of rest. Half the trials (▭) were taps alone; the other half were taps followed 8 msec later by a suprathreshold electrical pulse directly to the giant fibers (▬). Thus for black bars, giants fired on all trials, whereas for white bars they fired on only a fraction of trials. The difference between the probability of nongiant responding for the two types of trials indicates that the effects of natural stimulation and the firing of giants summate in the production of nongiant responding (from Reichert and Wine, 1983).

stimulus plus giant firing summate to trigger postgiant nongiant swimming. It is not yet known whether the facilitation by the giants is mediated entirely centrally or, as with the production of post-LG re-extension, indirectly via peripheral feedback from the LG flip (see Figure 3).

2.6. Commentary

In their extensive evaluation of the command neuron concept, Kupfermann and Weiss (1978) proposed that a command neuron should be defined as "a neuron that is both necessary and sufficient for the initiation of a given behavior" and put forward the giant axons of the crayfish as presently being the best candidates for such neurons.

The above discussion, however, shows how complicated the real application of Kupfermann and Weiss' definition can be. The greatest difficulties arise from the terms "sufficient" and "given behavior." The LGs are necessary and sufficient for the production of short latency phasic flexions that *approximate* the flexion phases of natural LG responses. However, sensory-derived information that bypasses the LGs is needed to make the responses entirely normal. Is it then acceptable to refer to the LGs as "command neurons" for short latency fast flexions? We think it probably is. Animals must always be tested in particular environmental and motivational contexts, and it would be a very restrictive definition that would require a putative command neuron's output effects to be completely independent of these contexts. We grant that the framing of a suitable definition that would include a role for contextual variables presents serious problems. However, the lesson provided by LG escape responses is that dependence on context, in this case the state produced by preceding stimulation, is a reality that must be contended with. Similar problems also arise when Kupfermann and Weiss' definition is applied to the Mauthner neuron system (see Chapter 8, this volume).

If the LGs are to be considered "command neurons," what behavior is it that they command? They do surely command one class of short latency phasic flexions and they certainly do not command the nongiant-mediated swimming that often follows these short latency flexions, since they promote but are neither necessary nor sufficient for its initiation. In the normal course of things their firing is operationally speaking both necessary and sufficient for the production of the re-extension that follows LG-produced flexion—but only because the stimuli that cause the apparently independent re-extension reflex are produced by the LG-commanded flexion. Thus, in one sense the LGs should perhaps be considered merely as command and decision neurons for one type of phasic flexion rather than for the corresponding escape reactions as a whole. On the other hand, whatever the terminology, it is clear that it is the LGs that set the entire flexion/extension pattern into motion. And one viewpoint might well be that the giants should be considered as command neurons for a complex flexion/extension motor pattern in the production of which reflexes play some important roles; major roles for peripheral reflexes are also known in the production of both vertebrate and invertebrate quadrapedal progression patterns, although as a whole they are considered to be centrally generated (see Pearson, 1979).

Issues of definition aside, it is also known that firing of the LGs has a host of central actions that help coordinate the muscle groups that produce phasic flexion and re-extension with each other and with other muscles that are important for the overall response. These include inhibition of postural muscles whose contraction would interfere with both phasic flexion and re-extension (Kuwada and Wine, 1979; Kuwada *et al.*, 1980), prolonged inhibition of sensory circuitry whose activity during both flexion and re-extension would be maladaptive (Krasne and Bryan, 1973), delayed driving of peripheral inhibitors of fast flexor muscles concomitant with re-extension (Wine and Mistick, 1977), and extensor motor neuron inhibition, the cessation of which may determine the timing of re-extension (see Wine and Hagiwara, 1978; Reichert *et al.*, 1981). These actions imply a pivotal status for the giants in short-latency escape response production as a whole even though its direct production of movement is somewhat more limited than might have been anticipated. In time we will probably come to understand why, on functional and/or evolutionary grounds, the organism employs the particular mix of central and peripheral control that we see here.

3. The Circuitry for Tailflip Production

Until recently the known facts about the motor circuitry that generates tailflips seemed straightforward (Figure 8A). There were known to be two classes of motor neurons, "motor giants" (MoGs) and the nongiant "fast flexor" motor neurons (FFs). Each FF motor neuron (from five to nine per hemiganglion depending on segment number (Mittenthal and Wine, 1978) was known to project to a different portion of the phasic flexor musculature of its hemisegment, whereas each MoG (one per hemiganglion) projected to almost all the phasic flexor muscles of its hemisegment, thereby overlapping the distribution of the FF pool. Contacts from the giants to both the MoGs and FFs were believed to be monosynaptic (Selverston and Remler, 1972; Atwood and Pomeranz, 1974).

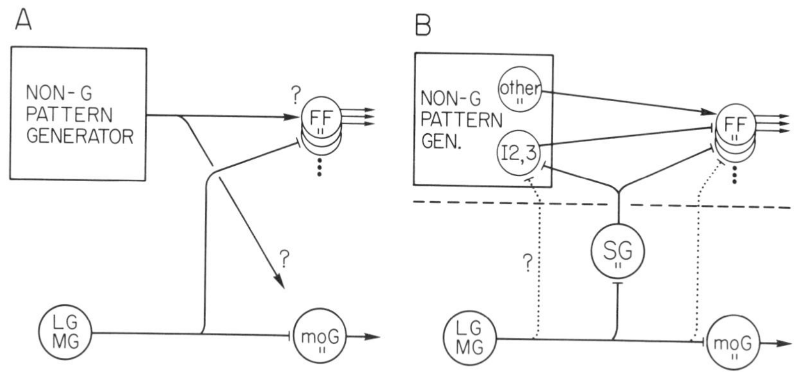

Figure 8. (A) Previous and (B) current versions of the motor circuitry that produces tailflips. Ditto marks indicate that there are multiple neurons of this category; in the case of the SGs and MoGs this is merely segmental repetition. Arrowheads indicate excitatory synapses of unspecified type; bars indicate synapses that are believed to be electrical. Dotted lines indicate very weak connections. Inhibitory circuitry, which is extensive, has been omitted for simplicity.

Moreover, a number of differential connections of LGs and MGs to motor neurons were known that at least in part accounted for the different movement patterns produced by the two sets of giants (e.g., see Figure 1). Nothing was known about the circuitry responsible for producing nongiant-mediated tailflips.

Recently, however, the picture has altered rather dramatically (Figures 8B and 9). In this section we will first review new findings and then attempt to make some sense out of the major features of the circuitry as it is now conceived.

3.1. New Findings

3.1.1. Giants Excite FFs Indirectly

Contrary to previous beliefs, the direct, monosynaptic connection between giants and FFs are extremely, and in some cases vanishingly, weak (Roberts *et al.*, 1982). The giants excite the FFs primarily via electrotonic connections from neurons called the segmental giants (SGs) of which there is one on each side of every abdominal ganglion (except perhaps the sixth, which has not been evaluated) (Figure 10). The SGs are coupled to the LGs by unusually effective rectifying electrical synapses that allow the giants to bring the SGs to firing level within a few tenths of a msec (Figure 10A). It is because they follow the giants so

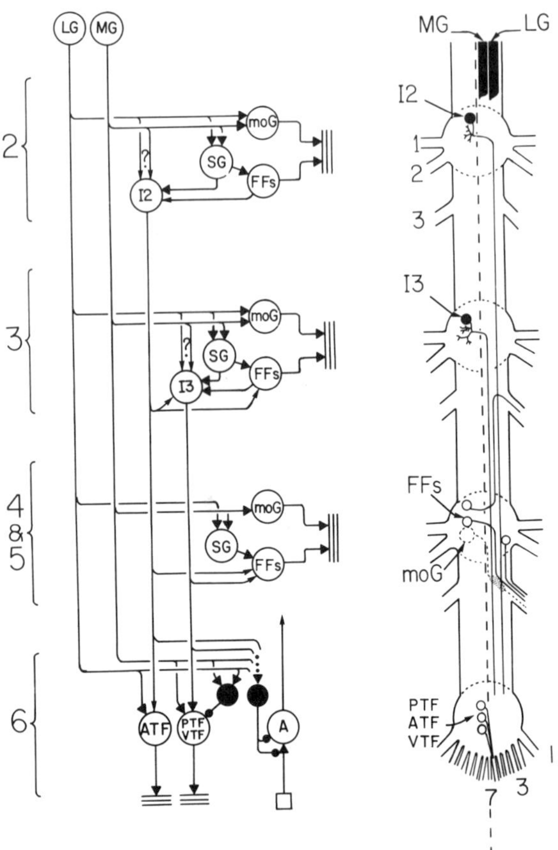

Figure 9. Layout of motor circuitry for tailflip production. Known connections and geometry of neurons involved in tailflip production. Abdominal ganglion number is shown at the left; only one ganglion has been shown for segments four and five, as the relevant connections in these ganglia are similar. Ganglionic root numbers (1–3) are indicated at the left of the 2nd abdominal ganglion. The arrows in the left hand portrayal represent excitatory synapses; the large arrowheads indicate synapses strong enough to fire their targets, whereas the small ones indicate that summation is necessary. The solidly filled neurons are inhibitors. Fast flexors are pluralized in the diagram because there are about nine per hemisegment. Note that the MoGs of caudal ganglia are not excited by the LGs. I2 and I3 are intersegmental neurons involved in both giant and nongiant tailflip production. ATF, PTF, and VTF are motor neurons to tailfan muscles discussed in the text. Interneuron A is a sensory interneuron transmission to which is inhibited during tailflips (to prevent reafference); the inhibitors of interneuron A are driven by I2 and I3, which may provide a crucial nonmotor function for I2 and I3. At the right the locations of some of the neurons are indicated. The MGs and LGs are only portrayed rostrally. The FFs and MoGs, which are similar in ganglia 1–5, are shown in ganglion 4/5 for convenience. The three FFs (solidly drawn neurons) shown illustrate the three types of FFs (with regard to location of soma and course of axon) that occur in each ganglion. The dendrites of the FFs (not shown) are within the ganglion whereas those of the moGs are near the third (motor) roots (based on Kramer *et al.*, 1981*a*; Roberts *et al.*, 1982; Selverston and Remler, 1972).

reliably and introduce so little synaptic delay that the SGs so long escaped detection. They were recently discovered and their important role uncovered only because of the fortuitous fact that they send into the first segmental root of their hemiganglion an axon that can be used to activate them antidromically (in Figure 10 they are activated through electrode Rld). Figure 11 shows the results of experiments that take advantage of the SG's peripheral axon. The figure shows that when antidromic stimulation of SG axons is used to fire the SGs of a ganglion bilaterally (the giants ordinarily drive both SGs to fire), the EPSPs produced in FFs are just as large as if a giant axon were fired. On the other hand, if a giant is fired while the SGs are refractory (MG without SG in Figure 11), almost no EPSP is produced in FFs; even the small EPSP that is produced is at least in part due to passive conduction from giants to FFs via the nonfiring SGs and does not prove the existence of direct giant-to-FF synapses. Thus, the giant-to-FF pathway is very much weaker than the giant-SG-FF pathway.

The function and destination of the efferent axon is unknown, but we believe that it probably affects the swimmerets, which retract during each tailflip. The possibility must be considered that neurons analogous to the SGs lurk in many supposedly monosynaptic connections.

3.1.2. Only FFs Are Used during Nongiant Tailflips

Whereas both MoGs and FFs operate synergistically during giant-mediated tailflips, nongiant-mediated phasic flexions are produced entirely by the FFs; the MoGs are in fact powerfully inhibited during nongiant tailflips (Figure 12; Kramer and Krasne, 1984; F. B. Krasne, unpublished data).

3.1.3. During Nongiant Flips the FFs Are Excited by Large Intersegmental Neurons That Lie Under the Giants

The SGs are not utilized to excite the FFs during nongiant flips. However, several other premotor interneurons involved in producing nongiant responses have been found, and two, named I2 and I3, out of what is surely a much larger population, have been identified and well characterized (Figures 8B and 9) (Kramer *et al.*, 1981*a*; Kramer and Krasne, 1984). The I2s and I3s both span a number of segments, and their axons, which are large, lie just under the MGs. I2 originates and receives all of its input in the second abdominal ganglion and sends its axon to the sixth (last) abdominal ganglion, making excitatory connections to FFs in all ganglia caudal to that of its origin. I3 originates and receives input in the third abdominal ganglion and runs to the sixth exciting FFs caudal to the

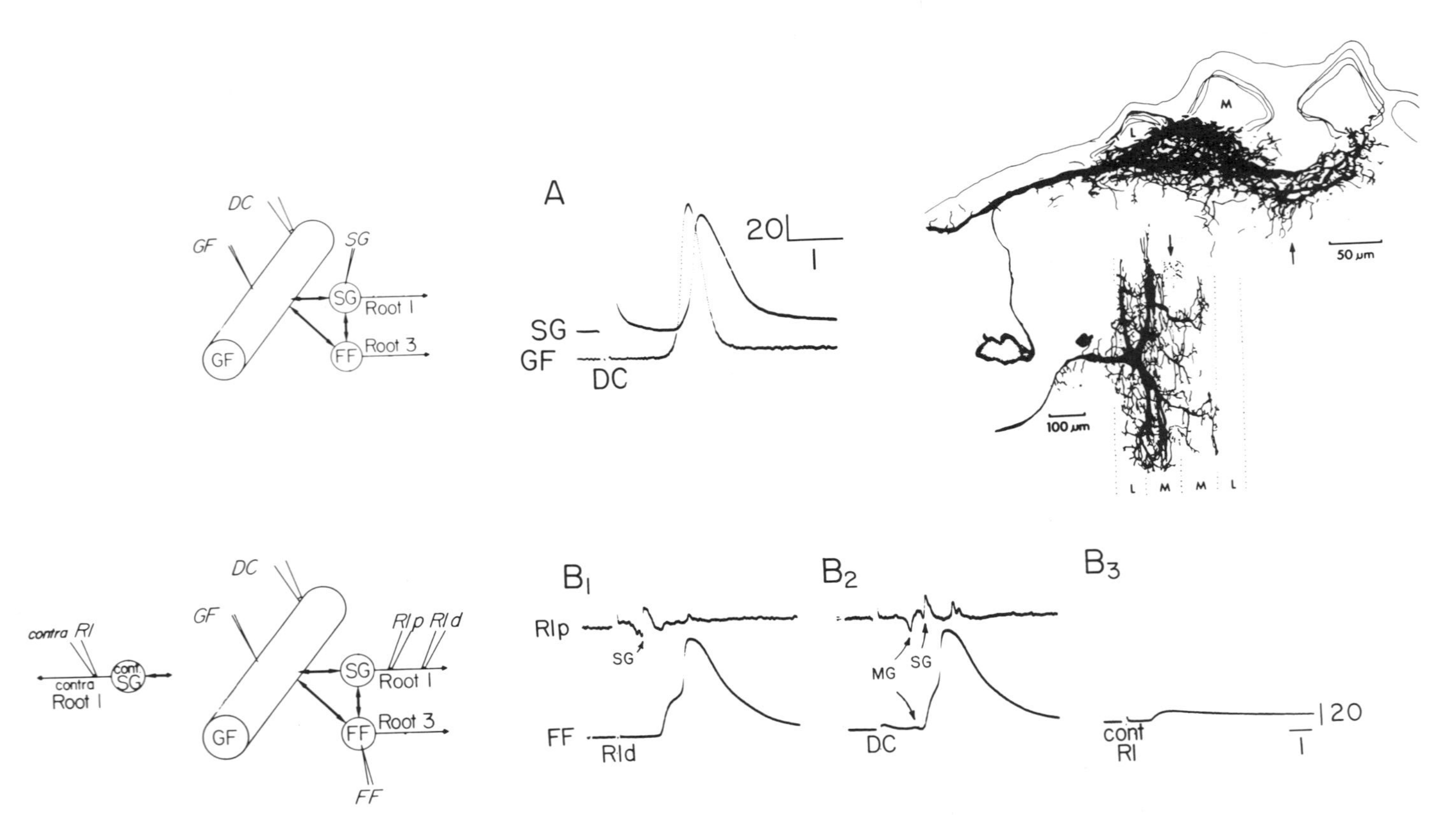

DC
GF
SG
SG
Root 1
FF
Root 3
GF
A
20
1
SG
GF
DC
M
L
50 μm
100 μm
L
M
M
L
contra RI
contra
Root 1
cont
SG
DC
GF
RIp
RId
SG
Root 1
FF
Root 3
GF
FF
B1
RIp
SG
FF
RId
B2
MG
SG
DC
B3
cont
RI
20
1

Figure 10. The segmental giants. Inset, upper right, shows anatomy of an SG neuron in the 3rd abdominal ganglion based on cobalt-filled, intensified preparations. Top shows cross sections; bottom shows horizontal views. Arrows indicate midline; L and M are lateral and medial giant axons whose outlines are dotted in the horizontal view. Note dense connections to ipsilateral giant axons and substantial but weaker contact with the contralateral MG. (A) Upper trace, intracellular response of a 3rd ganglion SG to stimulation of an ipsilateral LG axon (labeled "GF" for "giant fiber"). Stimulus was delivered through a suction electrode, DC, placed on the dorsal surface of the nerve cord over the LG. Lower trace, intracellular response of the LG. Calibrations are in mV and msec. The short SG latencies to LG firing suggests electrotonic coupling. Many such delays are even shorter than the one illustrated here. (B) FF responses to giant axon and SG input to show relative sizes of EPSPs from major routes of input. Upper trace, suction electrode recording from root 1 (Rlp:root 1, proximal). Lower trace, intracellular recording from FF. (B_1) FF response to antidromic firing of the ipsilateral SG by means of a stimulus through distal root 1 suction electrode (Rld). (B_2) FF response to activation of the ipsilateral giant axon (an MG) through DC electrode. (B_3) Response to antidromic firing of the contralateral SG stimulated via root 1 (cont R1). Medial giants ordinarily drive both ipsilateral and contralateral SGs, although excitation of the contralateral SGs is weaker than that of the ipsilateral ones. Note that the EPSP produced by the MG is approximately the sum of that produced by the ipsilateral and contralateral SGs (from Roberts *et al.*, 1982).

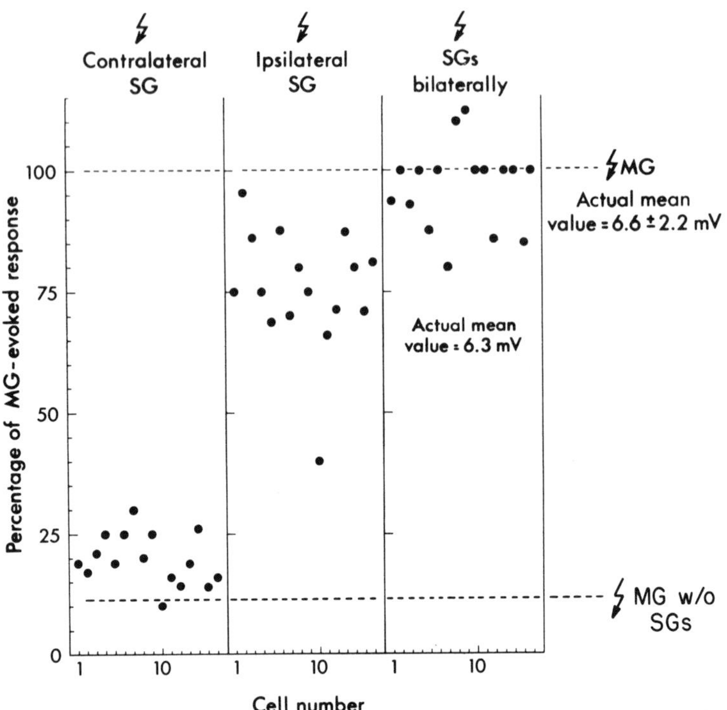

Figure 11. Comparison of giant- and SG-evoked depolarizations in FF motor neurons. For each of 16 FF motor neurons from ten isolated abdominal nerve cords, responses were recorded to stimulation of the ipsilateral MG and to antidromic stimulation of the contralateral SG, the ipsilateral SG, and both ipsilateral and contralateral SGs together (i.e., "SGs bilaterally"). Responses to MG firings were set to 100 and responses to SG firings were expressed as a percentage of the MG-evoked response. In all these preparations, one MG spike fired both SGs, and the FF response was subthreshold. Results show that the response to bilateral SG stimulation often equals the MG-evoked response and is on the average 95% of the MG response. The dashed line at the bottom of the figure shows the relative size of EPSPs produced (in different preparations) when SGs failed to fire (due to refractoriness or deterioration of the preparation); see text (based on Roberts *et al.*, 1982).

third enroute. For the most part the outputs of I2 and I3 to FFs are quite weak compared with those of the giant-SG input route to the FFs; whereas the firing of the giants leads to FF EPSPs that are commonly large enough to cause firing, I2, I3, and other unidentified interneurons of the same class produce rather small EPSPs in most FFs and so must operate cooperatively to bring FFs to firing level during nongiant flips. The different nongiant premotor interneurons are also a heterogenous class; in addition to spanning and providing outputs in different sets of abdominal segments,

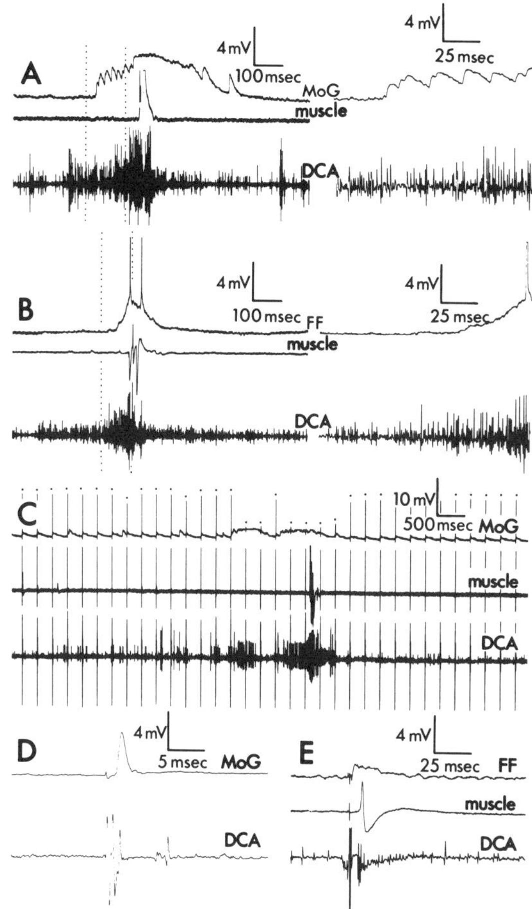

Figure 12. Comparison of MoG and FF activity during nongiant responses. Traces labeled "muscle" are thoracic flexor EMG's; MoG and FF activity was recorded intracellularly at origin of third roots; traces labeled DCA (dorsal cord axons) were recorded with a large suction electrode. (A) Depolarizing IPSPs in a fourth ganglion MoG preceding and during a nongiant tailflip produced by manipulating exopodite of tailfan. Note that both dorsal cord activity and IPSPs start well before motor response. First part of response (between dotted lines) is shown expanded on right. (B) EPSP leading to spikes in an FF motoneuron during a nongiant response evoked by brushing the ventrum between abdomen and thorax. Again, note gradual development of EPSP and DCA activity. (C) In a different preparation a nongiant response was evoked in a fifth ganglion MoG by pinching an appendage of the tailfan (just before the start of the segment shown) during continuous 4-Hz stimulation of cord giants in order to examine the extent of inhibition of MG-produced EPSP's. Note that nongiant responses are commonly poorly time locked to their initiating stimuli. Dots indicate tops of EPSP's in MoG. Note decrease in EPSPs preceding and during response. (D) A giant-evoked EPSP in the above MoG neuron at fast sweep. (E) A giant-evoked EPSP in the FF motoneuron shown in (B). Note much more rapid rise and short duration of the EPSP in (E) in comparison with the EPSP in (B). Duration of such EPSPs is often augmented by late-arriving input from nongiant premotor interneurons that are driven by the giants and the EPSPs sometimes show a second rise from such input (from Wine and Krasne, 1982).

they can also excite motor neurons of a given ganglion differentially. For example, in the last abdominal ganglion, I3 strongly excites posterior and ventral telson flexor muscle motor neurons (this is one of the few cases we have found of strong motor outputs of the nongiant premotor interneurons), which are typically utilized in producing flips that propel the animal directly backward. By contrast, I2 does not excite these telson flexor muscles, but does produce fairly large EPSPs in anterior telson flexor muscle motor neurons, which are commonly utilized during flips that lift the hind end of the abdomen. Consonant with their differential effects on motor neurons, given nongiant premotor interneurons are fired differentially in different types of nongiant responses. Thus, I3 often (but not always) fires during nongiant responses produced by rostral stimulation but not ones produced by caudal stimulation, whereas I2 does not show such selectivity.

3.1.4. Some Nongiant Premotor Interneurons Are Utilized during Giant-Mediated Tailflips

Whereas the SGs are not utilized during nongiant mediated tailflips (F. B. Krasne, unpublished data), a number of nongiant premotor interneurons, among them I2 and I3, fire during LG- and MG-mediated tailflips (Kramer *et al.*, 1981*a*). The excitatory input to FFs thus recruited adds to that produced by the giant-SG pathway and reinforces the excitatory effects of that pathway. In the case of I2 and I3 a pathway from the giants to SGs to I2 and I3 is sufficient to mediate I2/I3 firing; we suspect, but are not sure, that direct giant-to-I2/I3 connections, if any, are weak.

3.2. The Relationship between Giant and Nongiant Tailflip Pattern Generating Circuitry

We see from the above facts that contrary to what was formerly thought, the MoGs are not just extra large, extra widely projecting FFs. Rather, the MoGs and FFs are the final common output pathways for two somewhat separate systems, the giant system and the nongiant system. The large, highly specialized MoGs are the private pathways of the giant axons (and in fact are inhibited deeply during nongiant responses). The FFs, on the other hand, are the sole output route of nongiant circuitry. However, they are *also* used by the giants. There is a similar separation to be seen at the premotor interneuronal level: The SGs belong solely to the giant axon system. Nongiant premotor interneurons such as I2 and I3 distribute excitation to the FFs during nongiant flips, but as with the FFs themselves, they are also commandeered during giant-mediated flips.

A similar arrangement has also been found in the hydrozoan, *Aglantha*, where the same motor neurons are activated by motor giants involved in escape and by another slow-conducting pathway (see Chapter 2, this volume).

Thus there is an asymmetrical sharing of motor and premotor elements by giant and nongiant systems. Nongiant motor circuitry is recruited by the giant systems, but giant motor circuitry is not used by the nongiant system. Good reasons for this can be seen when we compare the organization and mode of operation of giant and nongiant motor circuitry, as inferred from the facts of the previous section.

The giant systems are designed for rapid production of two (MG and LG) fixed action patterns. The axons of their neurons are large; neurons of these circuits drive their followers with large, suprathreshold EPSPs; transmission delays are short; and virtually all the flexor musculature of a segment is caused to contract by the firing of the major segmental output neuron of the circuit, the MoG.

The nongiant system is designed for finesse of control. The premotor interneurons of this system are large but not giant neurons spanning a number of segments, which, as with the giants, presumably serve to promote nearly synchronous contractions in the segments they span. However, because their spans are limited, they are, as a group, in a position to exert more flexible control over abdominal flexions than are the giants. Moreover, since the EPSPs produced in most FFs by nongiant premotor interneurons are small, these neurons must work in groups to produce FF firing; this makes possible much more subtle and variable control over the firing of the nine or so FF muscles in each segment. This is precisely the sort of organization that would seem to be needed to provide the wide range of flexor patterns that we know from behavior the nongiant system can generate.

Given this, it would make poor sense for the nongiant system to recruit SGs or MoGs, for if it did, the powerful, highly generalized outputs of these neurons would largely override any possible differential recruitment of different muscles by the nongiant circuitry. However, it might make excellent sense for giant fibers to recruit nongiant elements that can reinforce the giants' own actions. The SGs function in fact appears to be to effect precisely this sort of recruitment.

In location, form, and outputs neurons such as I2 and I3 are very similar to the giants, and there are still further similarities. Lateral giants are actually a chain of neurons, one bilateral pair of which originates in each segment, where its dendrites ramify and receive input from sensory circuits. Its axons are sent forward one segment to make a powerful electrical synapse with the next member of the chain. As with members of the LG chain, I2 and 3 dendrites in fact receive input both from sensory circuitry of their segments and from nongiant premotor interneurons originating in other segments. We have thus conjectured that the giant system actually evolved under selection pressures for speed from elements of the nongiant system. We believe that the giants themselves are modified nongiant premotor interneurons, and the MoGs are probably evolved from an exceptionally widely distributing FF. In this view, those connections of the ancestral precursors to giant elements that caused consequences that reinforced giant actions were selectively maintained and strengthened, whereas other connections were lost or weakened.

3.3. Why Are the SGs Interposed between Giants and FFs?

We have seen above that, contrary to previous belief, the giant axons probably excite all of their nongiant subsystem targets *indirectly* via the SGs. Why? Driver neurons are also interposed between command neurons and power stroke motor neurons in the Mauthner cell system (see Chapter 8, this volume) and the *Drosophila* escape flight reflex (Tanouye and Wyman, 1980; see Chapter 5, this volume), but the reason for their presence is not clear.

In the crayfish we can conceive of at least three possibilities.

1. Electrical load reduction. In a network where a single intersegmental command neuron must drive a large number of followers (certainly more than a dozen per hemisegment) via electrical synapses, the followers could place a severe load (current sink) on the command neuron and cause slowing or even blockage of intersegmental conduction. A spiking driver interposed between the command neuron and the followers in each segment could greatly reduce the severity of this problem.
2. Evolutionary and/or developmental convenience. Perhaps in order to minimize electrical loading, the giant axons themselves do not have efferent arborizations. Rather, the axis cylinder itself is the efferent presynaptic element, and thus it is the dendrites of targets that must grow to meet the giant axons. If, as we believe, the giant system evolved out of elements of the nongiant system, then it may have been simpler to restrict developmental changes needed to provide contacts between the giants and its multiple followers to a single neuron per hemiganglion rather than try to alter the form of more than a dozen FF motor neurons and premotor interneurons in each ganglion in such a way that they would retain their old nongiant connections and yet also make suitable new

ones with the giants. If, as seems conceivable, there was a segmental neuron capable of exciting fairly large groups of FFs prior to the evolution of giant axons, then its assumption of a driver role would have been particularly plausible.

3. To provide a point of control over giant-to-fast flexor transmission. Although the synapses between giants and SGs have a very high safety factor for transmission, those between SGs and FFs do not; in acute preparations EPSPs produced in FFs by SGs are usually marginally above or marginally below FF firing levels. Therefore, variation in the amplitude of giant-evoked depolarizations in SG would produce large effects on probability of FF firing. The SGs could thus provide an effective and unitary control point for modulation of giant-to-FF transmission.

An experiment done more than a decade ago by Schrameck (1970) suggests a remarkable use to which such control via the SGs might be put. The recurrent tailflips called swimming are ordinarily generated by nongiant circuitry; however, Schrameck reported that sometimes the giants did fire during swimming (Figure 13). She wrote

> . . . often during low frequency swimming . . . lateral giant fiber impulses occurred in . . . later cycles. . . . When lateral giant fiber potentials were not involved . . . anterior abdominal flexor potentials preceded or succeeded . . . posterior potentials by up to 5 msec. But when the lateral giant fiber impulses did occur, the anterior potentials preceded the posterior potentials, most often by only 2 msec. This evidence suggests that the lateral giant fibers might function during slow swimming to synchronize the flexor motor neuron activity more exactly. . . .

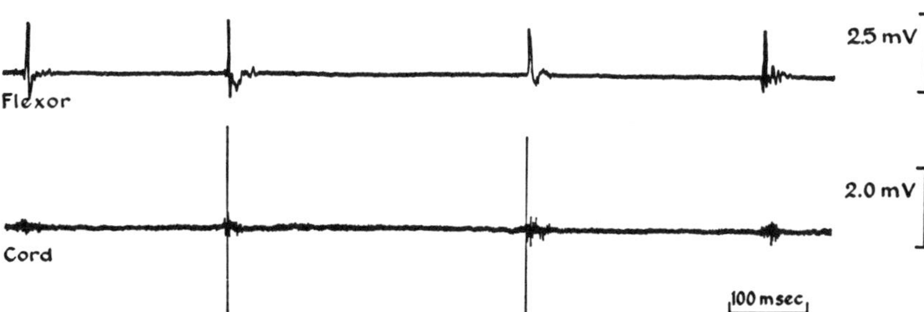

Figure 13. Lateral giant firings during swimming. The top trace shows flexor muscle EMGs and the bottom trace nerve cord potentials. Four flips of a swimming sequence are illustrated. The large spikes seen in the nerve cord trace on the second and third flips of the swim are LG spikes. The occurrence of such giant spikes during swimming is rare and is believed to occur only under special circumstances (see text) (from Schrameck, 1970).

Although our laboratories have never been able to find circumstances that would make this happen, Schrameck's results clearly indicate that the nongiant circuitry can under some perhaps very unusual circumstances co-opt the LGs, presumably to enhance flexor synchrony. However, if firing of the giants during swimming were to lead to full-sized EPSPs in FFs, the nongiant circuitry would then lose control over the precise form of the swimming power strokes—unless the effectiveness of transmission from giants to FFs via SG were attenuated by inhibition of the SGs. By inhibiting the SGs during nongiant tailflips the coupling of the giant axons to the FFs could be reduced, and the giants could be made to operate not as command neurons but rather as extrafast-conducting extralong premotor interneurons of the nongiant system, producing FF firing in cooperation with a number of other shorter slower nongiant premotor interneurons. That is, by partially inhibiting the SGs, the giants might be transiently restored to something resembling their ancestral role. A similar result could not readily be achieved by directly inhibiting FFs, because this would tend to attenuate EPSPs from nongiant sources as well as from the giants. If this notion were correct, the giants would be bifunctional neurons operating in one circumstance as decision and command neurons for producing stereotyped tailflips and under other circumstances as one of many premotor neurons of a complex pattern generator.

There is in fact some evidence that, as this hypothesis would require, the SGs can be inhibited during nongiant tailflips. Schrameck herself showed that shortly after the individual flips of a swimming episode the giants can become unable to cause phasic flexor muscle activity. In support of this, we have found that both after *and* just before nongiant mediated tailflips, the EPSPs produced in FFs by firing of giants are commonly reduced by 20% or more (Figure 14A); it is difficult to be certain what is happening *during* the flip because nongiant sources are themselves providing extensive input to FFs and causing firing of them during this interval. As illustrated in Figure 14D and F, direct recordings from SG during nongiant flips have shown that at the recording site in the giant part of the SG dendrite, giant-produced EPSPs can be substantially reduced and spike amplitudes reduced by about 5% for a period, before, during, and after nongiant flips. Thus, there is considerable evidence to support the idea that inhibition of the SGs is used to reduce, without abolishing, the influence of the giants during nongiant tailflips. Missing, however, is knowledge of the circumstances that caused Schrameck's animals to utilize the giants during some swimming sequences and evidence that during nongiant flips that actually utilize the giants the inhibition of SGs is deep enough to really prevent excessive influence of the giants on targets of the SGs.

Thus, the idea that the giants can play rather different roles, depending on the context of their usage, should presently be viewed as speculation.

3.4. Switching of Outputs of the Nongiant Premotor Neuron, I3

Economy requires that nervous systems be organized so that whenever possible circuitry that is capable of performing a particular kind of task should be utilized whenever such a task must be carried out. In motor systems this requirement is believed to lead to modular construction and systems of hierarchical control in which lower level building blocks are selected for use, as necessary, by higher level controllers (see, e.g., Gallistel, 1980). Still greater economy could be attained if the operation of the basic modules were themselves flexible, with higher-level controllers being able to bias or "program" them to produce a variety of related but not identical movements. The various modes of quadrapedal locomotion and also perhaps other rhythmic activities such as scratching are commonly believed to provide examples of such flexible operation of a single pattern-generating network, but evidence that this is really so has mostly been quite indirect, and details of how variations in circuit operation are achieved have been quite speculative (see, e.g., Greene, 1972; Bernstein, 1967; Szentagothai and Arbib, 1974; Grillner, 1975; Berkinblit *et al.*, 1978*a,b*). The switching of roles of the giant axons hypothesized above, if confirmed, would provide a direct and elegant example of flexible operation of motor circuits. But whatever the outcome of those speculations, the circuitry generating tailflips provides another definitive example of flexible operation and provides an unusually direct glimpse of the details of how this flexibility is achieved.

The nongiant premotor interneurons called the I3s are known to fire under three circumstances: (1) When a rostrally located stimulus causes a nongiant response having a low, directly backward trajectory, (2) when a rostral stimulus of abrupt onset fires the MGs (which also results in a low, directly backward trajectory), and (3) when an abrupt caudal stimulus causes the LGs to fire (which results in a flip that lifts the abdomen). The I3s have a variety of output effects. There are some, such as the driving of inhibitory circuits that prevent reafference from tailflips from having undesirable effects, which are presumably adaptive during all tailflips. However, there are others, specifically the strong excitation of the ventral and posterior telson flexor muscle motor neurons that force the animal to move directly backward, which are adaptive during MG- and nongiant-mediated backward-directed flips, but would be disastrous during LG-mediated forward pitches evoked by caudal threats. Since I3 *is* utilized,

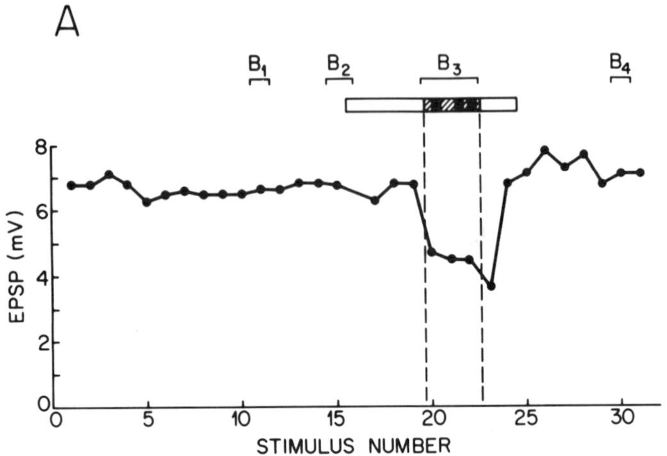

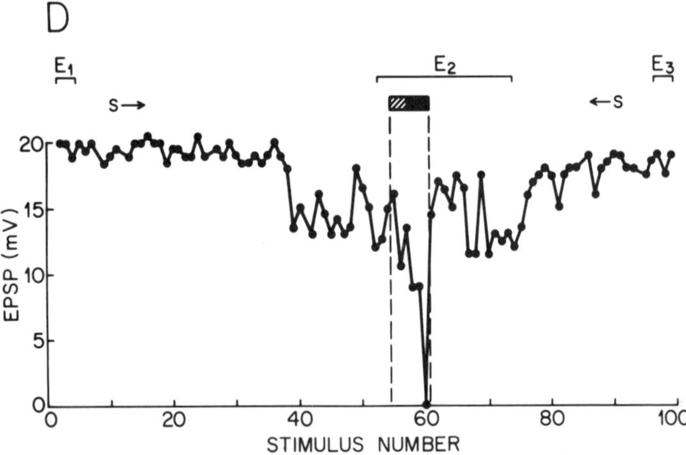

Figure 14. Decrements of giant-to-FF and giant-to-SG transmission during nongiant responses. In each experiment medial giant axons were stimulated repetitively before, during, and after elicitation of a nongiant-mediated escape response. (A–C) The EPSPs produced in an FF soma were recorded intracellularly; (D–F) recordings were from an SG dendrite. (A, D) Graphs of the amplitudes of each MG-evoked EPSP from a period during which a nongiant response was elicited by manipulating the tailfan. In each case bars above the graph indicate heightened dorsal nerve cord activity that reflects operation of the circuitry generating nongiant responses: solid black indicates that phasic flexon is in progress; cross hatching indicates strong activity of the kind that just precedes flips; an open bar indicates some increase of cord activity that might be merely the firing of sensory interneurons. (D) The approximate beginning and end of the period of tailfan stimulation are marked by Ss.

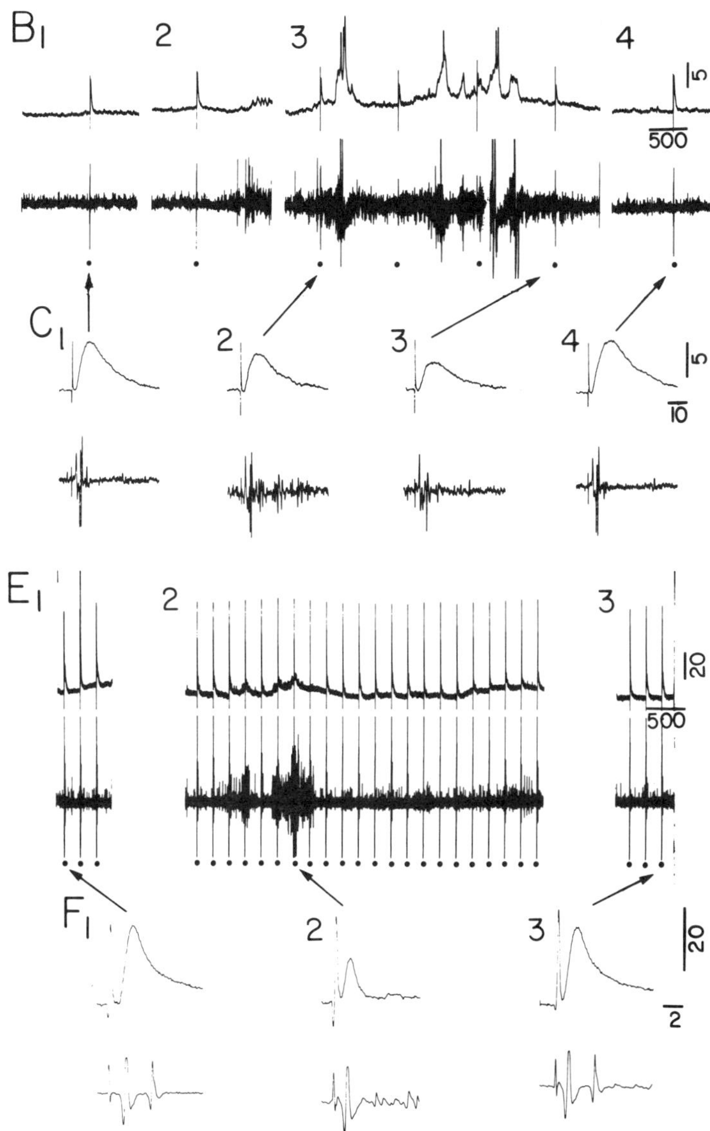

(B, E) Segments of record from the periods indicated at the top of A and C, respectively. In each record the top trace is intracellular and the bottom one is from a large surface electrode on the dorsal surface of the nerve cord; dots mark MG stimulation. (C, F) High-speed records of the response to selected MG firings as indicated by the arrows; MG spikes are the first spikes after the stimulus artifact in the lower traces. Calibrations are in mV and msec (from F. B. Krasne, unpublished data).

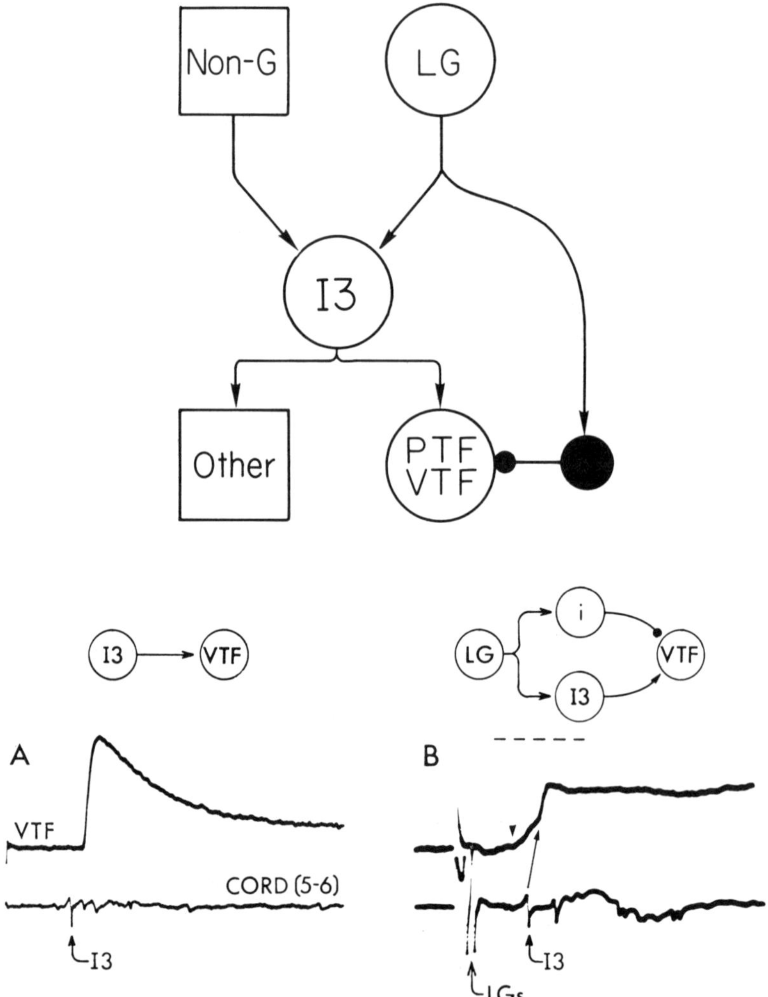

Figure 15. Selection of I3 outputs according to type of tailflip. As indicated at the top of the figure, I3 can be driven by both LGs and nongiant circuitry and in turn causes a variety of effects. However, some of these effects (firing of certain telson flexor motoneurons), which would be maladaptive during LG-mediated flips, are prevented by LG-evoked inhibition (produced via the neuron drawn in solid black). Arrows indicate excitatory synapses. VTF and PTF indicate ventral and posterior telson flexor motoneurons. Below, the LG-evoked inhibition is illustrated. Recordings are intracellularly from a VTF motoneuron. (A) The effects of firing I3 alone. A large, just subthreshold EPSP is produced. (B) The LGs are fired, and they recruit I3. However, they also recruit an inhibitory neuron (i) whose depolarizing IPSP (down-pointing arrowhead in B) starts before excitatory input from I3 (indicated by full arrow) arrives. The inhibition dramatically reduces the EPSP produced by I3 (based on Kramer *et al.* 1981*a*,*b*).

presumably for those of its output effects that *are* adaptive, during LG flips as well as during backward nongiant and MG flips, the excitation of ventral and posterior telson flexor muscles must be, and in fact is, nullified by inhibition during LG flips (Figure 15B). This is accomplished by a fast inhibitory pathway driven by the LGs that inhibits the ventral and posterior telson flexor motor neurons just before the onset of I3-produced EPSPs. Thus, in this case higher level controllers in effect select possible output effects of the I3s by inhibiting some target neurons and permitting others to fire in response to the intrinsically suprathreshold inputs they receive from the I3s.

4. Concluding Remarks

The tailflip behavior of crayfish is obviously a highly specialized object of study. We chose to study it because we believed the performance requirements placed on it would maximize its simplicity and hence our chances of achieving some understanding of it. As our study of it continues, we do feel a growing sense of understanding, but we also become increasingly aware of its subtleties; indeed, it is the relative simplicity of this system that has allowed us to discover its fine points. We also wish to emphasize that although our object of study is highly specialized, it nevertheless can provide what we feel are very general insights into organizational features of nervous systems. This point is illustrated by the influence the analysis has had on concepts of command neuron function, roles for chain reflexes, and flexible operation of "hard-wired" motor pattern generating circuitry.

5. References

Atwood, H. L., and Pomeranz, B., 1974, Crustacean motor neuron connections traced by backfilling for electron microscopy, *J. Cell Biol.* **63**:329–334.

Berkinblit, M. B., Deliagina, T. G., Feldman, A. G., Gelfand, I. M., and Orlovsky, G. N., 1978*a*, Generation of scratching. I. Activity of spinal interneurons during scratching, *J. Neurophysiol.* **41**:1040–1057.

Berkinblit, M. B., Deliagina, T. G., Feldman, A. G., Gelfand, I. M., and Orlovsky, G. N., 1978*b*, Generation of scratching. II. Nonregular regimes of generation, *J. Neurophysiol.* **41**:1058–1069.

Bernstein, N., 1967, *The Co-ordination and Regulation of Movements,* Pergamon Press, Oxford.

Gallistel, C. R., 1980, *The Organization of Action: A New Synthesis,* Lawrence Erlbaum Associates, New Jersey.

Greene, P. H., 1972, Problems of organization of motor systems, in: *Progress in Theoretical*

Biology, Vol. 2 (R. Rosen and F. M. Snell, eds.), Academic Press, New York, pp. 303–338.

Grillner, S., 1975, Locomotion in vertebrates: Central mechanisms and reflex interaction, *Physiol. Rev.* **55**:247–306.

Johnson, G. E., 1924, Giant nerve fibers in crustaceans with special reference to Cambarus and Palaemonetes, *J. Comp. Neurol.* **36**:323–373.

Johnson, G. E., 1926, Studies on the functions of the giant nerve fibers of crustaceans, with special reference to Cambarus and Palaemonetes, *J. Comp. Neurol.* **42**:19–33.

Kramer, A. P., and Krasne, F. B., 1984, The production of crayfish tailflip escape responses by circuitry that does not utilize the giant fibers, *J. Neurophysiol.* **52**:189–211.

Kramer, A. P., Krasne, F. B., and Wine, J. J., 1981*a*, Interneurons between giant axons and motoneurons in the crayfish escape circuitry, *J. Neurophysiol.* **45**:550–573.

Kramer, A. P., Krasne, F. B., and Bellman, K. L., 1981*b*, Different command neurons select different outputs from a shared premotor interneuron of crayfish tailflip circuitry, *Science* **214**:810–812.

Krasne, F. B., 1969, Excitation and habituation of the crayfish escape reflex: The depolarizing response in lateral giant fibres of the isolated abdomen, *J. Exp. Biol.* **50**:29–46.

Krasne, F. B., and Bryan, J. S., 1973, Habituation: Regulation through presynaptic inhibition, *Science* **182**:590–592.

Krasne, F. B., and Wine, J. J., 1977, Control of crayfish escape behavior, in: *Identified Neurons and Behavior of Arthropods* (G. Hoyle, ed.), Plenum Press, New York, pp. 275–292.

Kupfermann, I., and Weiss, K. R., 1978, The command neuron concept, *Behav. Brain Sci.* **1**:3–39.

Kuwada, J. Y., and Wine, J. J., 1979, Crayfish escape behaviour: Commands for fast movement inhibit postural tone and reflexes, and prevent habituation of slow reflexes, *J. Exp. Biol.* **79**:205–224.

Kuwada, J. Y., Hagiwara, G., and Wine, J. J., 1980, Postsynaptic inhibition of crayfish tonic flexor motor neurones by escape commands, *J. Exp. Biol.* **85**:344–347.

Mittenthal, J. E., and Wine, J. J., 1973, Connectivity patterns of crayfish giant interneurons: Visualization of synaptic regions with cobalt dye, *Science* **179**:182–184.

Mittenthal, J. E., and Wine, J. J., 1978, Segmental homology and variation in flexor motoneurons of the crayfish abdomen, *J. Comp. Neurol.* **177**:311–334.

Olson, G. C., and Krasne, F. B., 1981, The crayfish lateral giants as command neurons for escape behavior, *Brain Res.* **214**:89–100.

Pearson, K. G., 1979, Modulation of reflex pathways and sensory control of central pattern generators, *Neurosci. Res. Program Bull.* **17**:610–614.

Reichert, H., and Wine, J. J., 1982, Neural mechanisms for serial order in stereotyped behaviour sequence, *Nature* **296**:86–87.

Reichert, H., and Wine, J. J., 1983, Coordination of lateral giant and nongiant systems in crayfish escape behavior, *J. Comp. Physiol.* **153**:3–15.

Reichert, H., Wine, J. J., and Hagiwara, G., 1981, Crayfish escape behavior: Neurobehavioral analysis of phasic extension reveals dual systems for motor control, *J. Comp. Physiol.* **142**:281–294.

Roberts, A. M., Krasne, F. B., Hagiwara, G., Wine, J. J., and Kramer, A. P., 1982, The segmental giant: Evidence for a driver neuron interposed between command and motor neurons in the crayfish escape system, *J. Neurophysiol.* **47**:761–781.

Schrameck, J. E., 1970, Crayfish swimming: Alternating motor output and giant fiber activity, *Science* **169**:698–700.

Selverston, A. I., and Remler, M. P., 1972, Neural geometry and activation of crayfish fast flexor motoneurons, *J. Neurophysiol.* **35:**797–814.

Sherwood, D. N., and Wine, J. J., 1979, Orderly sequence of polysynaptic sensory inputs to crayfish tailflip motoneurons, *Soc. Neurosci. Abstr.* **5:**261.

Szentagothai, J., and Arbib, M. A., 1974, Conceptual models of neural organization, *Neurosci. Res. Program Bull.* **12**(3)**:**313–510.

Tanouye, M. A., and Wyman, R. J., 1980, Motor outputs of giant fiber in *Drosophila, J. Neurophysiol.* **44:**405–421.

Wiersma, C. A. G., 1947, Giant nerve fiber system of the crayfish. A contribution to comparative physiology of synapse, *J. Neurophysiol.* **10:**23–38.

Wine, J. J., 1977*a*, Neuronal organization of crayfish escape behavior: Inhibition of the giant motoneuron via a disynaptic pathway form other motoneurons, *J. Neurophysiol.* **40:**1078–1097.

Wine, J. J., 1977*b*, Crayfish escape behavior. II. Command-derived inhibition of abdominal extension, *J. Comp. Physiol.* **121:**173–186.

Wine, J. J., and Hagiwara, G., 1978, Durations of unitary synaptic potentials help time a behavioral sequence, *Science* **199:**557–559.

Wine, J. J., and Krasne, F. B., 1972, The organization of escape behavior in the crayfish, *J. Exp. Biol.* **56:**1–18.

Wine, J. J., and Krasne, F. B., 1982, The cellular organization of crayfish escape behavior, in: *The Biology of Crustacea,* Vol. 4 (D. C. Sandeman and H. L. Atwood, eds.), Academic Press, New York, pp. 241–292.

Wine, J. J., and Mistick, D. C., 1977, Temporal organization of crayfish escape behavior: Delayed recruitment of peripheral inhibition, *J. Neurophysiol.* **40:**904–925.

8

The Role of the Mauthner Cell in Fast-Starts Involving Escape in Teleost Fishes

ROBERT C. EATON and JOHN T. HACKETT

1. Introduction

The commonly observed "tailflip" startle response is one of the most characteristic behavior patterns of bony and cartilagenous fishes and amphibians. In the most familiar example, the behavior pattern is readily elicited in fish following a tap on the side of their aquarium. However, data from behavioral studies show that the response is an effective escape movement that enables the animal to avoid sudden attacks by predators. An example of this is shown in Figure 1 in which a small cyprinid fish uses a common startle response movement pattern to avoid a strike by a piscivorous snake.

Of the various fishes and amphibians having the startle pattern just described, teleosts are the only group in which the behavior has been studied thoroughly. Because of this, we concentrate here on the work done on teleosts. In the teleosts, considerable evidence shows that an action potential in one of a single pair of prominent neurons called the Mauthner cells (M cells) can initiate startle responses used in escape. This hypothesis was first proposed almost 70 years ago on the basis of neuroanatomical observations (Bartelmez, 1915). As shown in Figure 2, the M cell receives a conspicuous supply of primary afferents from the ear and the M-axon synapses on motor neurons in the spinal cord. Thus, a sudden

ROBERT C. EATON • Behavioral Biology Group, Department of Biology, E.P.O., University of Colorado, Boulder, Colorado 80309. *JOHN T. HACKETT* • Department of Physiology, School of Medicine, University of Virginia, Charlottesville, Virginia 22908.

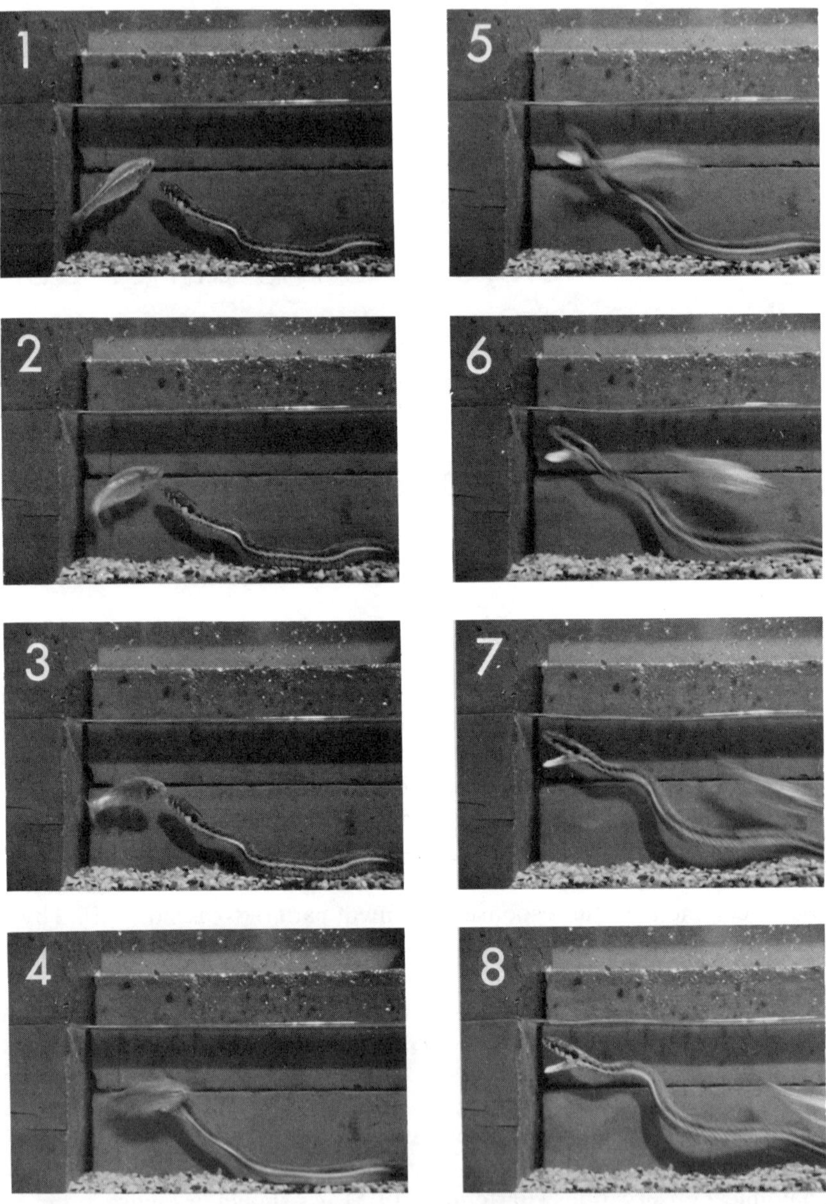

Figure 1. Montage of frames from a movie film showing use of a startle response by a small cyprinid fish in successfully avoiding a strike by a piscivorous snake. In this sequence the fish initiates a C-type fast-start moments before the snake begins its lunge. Although the fish moves only a small distance to the side, this displacement is sufficient to cause the snake to miss its target. The encounter took place in an aquarium and was filmed at 24 frames/sec (from the film, "The Predatory Behavior of Snakes," a film in the series "Aspects of Animal Behavior" produced by George A. Barthlomew and Robert G. Dickson, University of California, Los Angeles. ©Regents of the University of California).

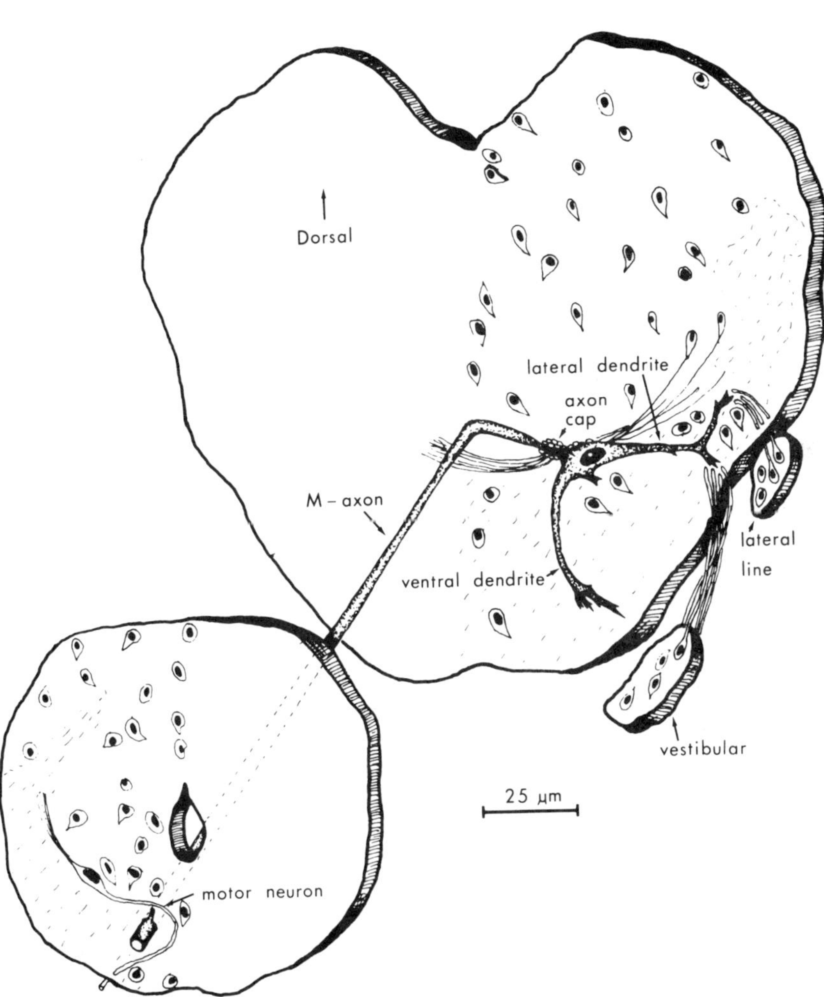

Figure 2. The basic M-cell circuit as seen in a zebrafish larva. The right M cell is shown in the transverse section through the hindbrain at the level of the eighth cranial nerve. Shown are excitatory afferents from the acoustico-vestibular and lateral line systems projecting to distant zones of branches on the lateral dendrite. The ventral dendrite branches among fibers of the tectobulbar tract from which it probably receives visual input. The M axon crosses the midline and descends into the spinal cord where excitatory connections are made with motoneurons innervating the trunk musculature (from Kimmel and Eaton, 1976).

acoustic stimulus can activate one M cell and cause a body contraction resulting in a rapid acceleration through the water.

However, from the simple neural circuit in Figure 2, one could not predict the richness of the apparent sensory-motor integration that takes place in this system when a fish is presented with a sudden stimulus that elicits a startle response. The behavior pattern itself corresponds to what some might call a fixed action pattern. To begin with, the threshold for the response appears to vary greatly, depending on external variables and on the state of the animal. This suggests that complex information processing may be taking place. However, the intensity and duration of the response seem to be independent of the stimulus strength. Thus, once triggered, the motor program goes to completion under most circumstances. This motor program consists of multiple mechanical phases, the first two of which displace the animal about one body length within 100 msec. This movement is accomplished through the participation of every major somatic muscle group—trunk, tail, fins, operculum, extraocular eye muscles, and jaw. The pattern is stereotypic when considered in terms of the timing of the various phases. However, although the configuration of the motor contractions is relatively fixed in the initial phase of the response, the contractions are quite variable in form, but not performance, during subsequent phases.

The significance of these observations is that the response is not a simple, graded, withdrawal to an aversive stimulus. Instead, it is a highly coordinated behavior pattern evidentally involving complex neuronal decision making and execution processes that involve major portions of the animal's sensory and motor systems. The purpose of this review is to describe what is currently known about the fast-start behavior and the neuronal processing that mediates this complex behavioral response. This processing is responsible for modulating the input to the M cell, for insuring that only one M cell fires to the stimulus, for activating a constellation of associated muscular responses such as fin contractions, and for initiating the various stages of the behavioral response.

2. Types of Startle Responses and Fast-Starts

Startle responses are thought of as short latency behavior patterns elicited by abrupt and unexpected stimuli. Numerous studies have been done to characterize the form and mechanical performance of these behavior patterns in fishes. The "tailflip" response mentioned above is only one of a rich variety of types of startle responses observed in fishes.

Movement patterns used for startle or escape can also occur in other contexts, such as predatory lunges at other fishes (Rand and Lauder, 1981; Webb and Skadsen, 1980; Vinyard, 1982) or for aggressive or other social displays (Wyman and Ward, 1973; Fernald, 1975). These behavior patterns may involve the same or parts of the same neural circuits. However, we concentrate here on those involving escape, as they are the only ones for which the neurobiology has been studied.

Experiments using high-speed motion picture cameras show that the two main variables of startle patterns are the configuration of the body during the response and the presence or absence of fin movements. For example, on one end of the spectrum are startle responses in which the animal erects its fins but does not contract the body and remains stationary in the water. In species with fin spines, this can have a clear defensive function (Eaton *et al.*, 1977*b*). Other startle responses consist of a high-performance turning maneuver involving a sudden acceleration. These are usually designated as "fast-starts" to distinguish them from other forms of startle behavior (Weihs, 1973; Webb, 1978*b*; Eaton and Bombardieri, 1978). Examples of fast-starts are shown for the trout, goldfish, catfish, and characid in Figure 3. Figure 4 illustrates the lateral body profiles of these fish. Fast-starts can also involve fin movements that in specialized animals such as the hatchetfish, *Gasteropelecus*, can cause the animal to jump from the water (Auerbach and Bennett, 1969*a*; Eaton *et al.*, 1977*b*). Some species with elongate bodies, such as the spiny eel (Figure 4F), can contract their bodies to cause a backward retraction of the head (Figure 3F).

Responses involving a sudden acceleration through the water, as for the upper four examples in Figure 3, are well studied. Although teleosts vary greatly in body morphology, many utilize one of the two forms of fast-start for escape, C starts and S starts. These are named according to the configuration of the body during the first contraction of the response. In one case, the animal's initial body contraction is on one side, so that the animal assumes the shape of a letter "C." Examples A–D in Figure 3 are typical C starts. C starts are most commonly observed in fish with relatively short body lengths. C starts have been shown to be initiated by the M cell in goldfish (Eaton *et al.*, 1981). S starts, on the other hand, are more often utilized by fish with longer, flexible bodies having a relatively large ratio of length to width. In this case the initial contraction is bilateral with major curves on opposite sides, anterior and posterior. Some fish, such as the kelp bass, *Paralabrax clathratus*, utilize both patterns (see Eaton and Bombardieri, 1978), but nothing is known about the neurophysiology of S starts.

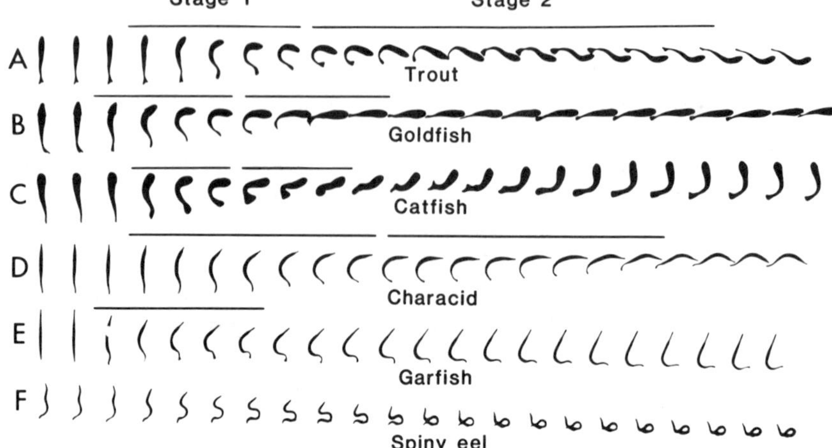

Figure 3. Dorsal view silhouettes of teleost fish during startle behavior in response to a sudden vibrational stimulus. Silhouettes were drawn from the projected image of a cine film at 5-msec intervals beginning 5 msec before the onset of a vibrational stimulus. For this comparison of form, fish lengths were standardized and successive silhouettes were drawn by moving the figure a constant distance to the left. Bars above the silhouettes indicate the two stages of the response. (A–D) The responses of the upper four fish are similar in form, and differ mainly in the duration of the stages. Stage duration varies linearly with fish length, so the larger fish take more time to get through the two stages. (E) The garfish did not give the second stage and the spiny eel simply retracted its head. Species names and lengths are given in caption to Figure 4, which illustrates the lateral profiles of these fish (from Eaton *et al.*, 1977*b*).

3. Stimulus Conditions for Eliciting Fast-Starts

Available data suggest that numerous types of stimuli can be utilized in eliciting fast-starts. These may activate the response through either a single modality such as vision (Eaton *et al.*, 1977*b*; Dill, 1974) or several simultaneously, as is probably the case for actual predatory encounters as in Figure 1. Effective stimuli include sound (Zottoli, 1977), or sound and mechanical vibration (Rodgers *et al.*, 1963; Eaton *et al.*, 1977*b*), body vibration (Eaton and Farley, 1975; Rock *et al.*, 1981), electrical field (Wilson, 1959; Webb, 1975), and visual stimuli (Rodgers *et al.*, 1963; Eaton *et al.*, 1977*b*).

Casual observations show that the predominant stimulus characteristic is similar to that for activating startle behavior in other animals. Stimuli chosen in laboratory situations to elicit the response are usually ones that have both a sudden onset and are "unexpected." The require-

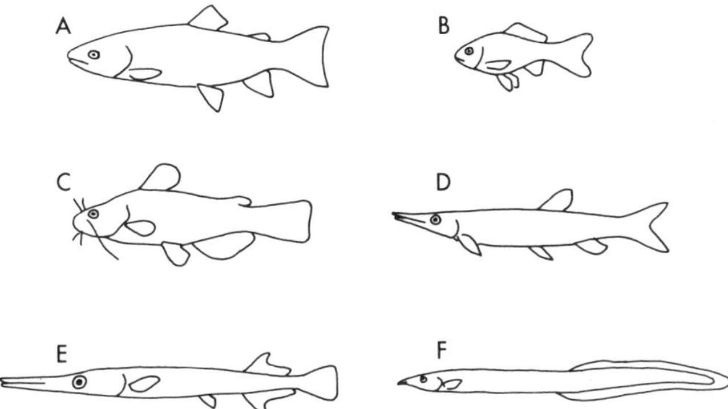

Figure 4. Lateral body and fin profiles of the six species of teleost fish used in the cine analysis of fast-starts in Figure 3. (A–F) Species names and standard lengths. (A) Rainbow trout, *Salmo gairdneri*, 13.0 cm. (B) Goldfish, *Carassius auratus*, 8.5 cm. (C) Catfish, *Ictalurus nebulosus*, 9.2 cm. (D) Characid, *Hydrocinos maculatus*, 18.4 cm. (E) Garfish, *Xenetodon cancila*, 12.0 cm. (F) Spiny eel, *Mastacembelus loennbergi*, 25.5 cm (from R. C. Eaton and R. A. Bombardieri, unpublished data).

ment that the stimulus is unexpected brings up the issue that the animal's experience with the stimulus will affect the outcome of any measurements of its sensitivity. As with other animals, prestimulus conditions (see Chapters 9 and 10, this volume) are thought to be critical, but are as yet unstudied in fish. In addition, the response shows habituation with repeated presentations of the stimulus (Thompson and Spencer, 1966) (see Section 11).

Only a few studies have provided quantitative information to characterize adequate stimuli for fast-starts. Such data are likely to vary with the species studied, experimental conditions, size of the fish, and many other variables. Therefore, values reported here are intended to provide only estimations for general stimulus configurations. Thresholds to vibrational or auditory stimuli have been provided by Eaton and colleagues for restrained zebrafish (*Brachydanio rerio*) larvae and by Blaxter and colleagues for free-swimming herring (Eaton and Farley, 1975; Eaton and Kimmel, 1980; Blaxter and Hoss, 1981; Blaxter *et al.*, 1981). For the zebrafish, a vibrating probe was placed against the side of the head next to the otic capsule. At frequencies of about 100 Hz, M-cell-initiated starts could be elicited at displacements of 10–35 μm.

Blaxter and colleagues examined thresholds to vibrational stimuli for herring, *Clupea harengus*, swimming in a tank. Threshold sound pressure values of 15 Pascals (1 Pa = 1 Newton/m^2) were recorded for a frequency of 100 Hz produced by a loudspeaker. Under the conditions of their experiment, these workers found that responses to single-cycle stimuli were equal in threshold to those for multiple cycles. But when the stimulus took longer to reach maximum amplitude, it had a weaker effect. Thus, the importance for sudden onset of the stimulus is confirmed by these experiments.

The threshold for escape to visual stimulation was first analyzed quantitatively by Dill (1974) for the adult zebrafish. The fish were presented with both real and model predators, whose size and velocity of approach were varied. An escape response is elicited when the "looming effect" reaches a certain threshold value as the image of the approaching predator expands on the retina of the prey. The threshold is measured as the rate of change of the angle, α, subtended by the approaching object measured at the prey's eye. Webb (1981) has named $d\alpha/dt$ at the start of the prey response as the apparent looming threshold (ALT). This is given by the following relation:

$$\frac{d\alpha}{dt} = \frac{4US_h}{4\,(D + d)^2 + S_h^2}$$

where, U = predator speed at time of prey response; S_h = predator shape (i.e., mean of maximum depth and width); D = distance between the predator's nose and the prey when the prey responded; and d = posterior distance of the predator's maximum depth and width from its nose (Webb, 1981). Webb determined ALTs for Northern anchovy, *Engraulis mordax*, attacked by clownfish, *Amphiprion percula*. Apparent looming thresholds varied inversely with the size of the prey and ranged from 32 rads/sec for 0.29 cm fish to 1.7 rad/sec for 1.2 cm fish. Dill's (1974) values for zebrafish were 0.43 rad/sec for 2.0 cm fish, but would be expected to be larger than Webb's values because of modifications in the formula.

4. Performance Measures of Fast-Starts

4.1. Response Latency

Response latency is defined as the interval from the stimulus onset to the first detectable movement of the animal in response to the stimulus. Latencies of fast-start responses in teleosts are comparable to the fastest reflex responses in both vertebrates and invertebrates. Latencies for various marine and fresh water species are listed in Table 1 along with stimulus and temperature parameters. In adult fishes, minimum response latencies vary among different species from about 6–40 msec and depend on a large number of external conditions (Table I).

Table I. Minimum Latencies of Fast-Start Responses in Teleost Fish under Various Conditions

Species	Latency (msec)	Stimulus	Temperature (°C)	Reference
Salmo gairdneri (rainbow trout)	9.2	Vibration	20	Eaton *et al.*, 1977*b*
Brachydanio rerio (adult zebrafish)	5.0	Vibration	26	Eaton *et al.*, 1977*b*
Carassius auratus (goldfish)	6.4	Vibration	24	Eaton *et al.*, 1977*b*
Clupea harengus	25	Vibration (above threshold)	10	Blaxter *et al.*, 1981
Clupea harengus	40	Vibration (at threshold)	10	Blaxter *et al.*, 1981
Salmo gairdneri (rainbow trout)	23	Electric shock	5	Webb, 1975
Salmo gairdneri	6	Electric shock	25	Webb, 1975
Perca flavescens (solitary, yellow perch)	24	Electric shock	15	Webb, 1980
Perca flavescens (schooling)	10	Electric shock	15	Webb, 1980

Stronger stimuli produce M-cell responses with shorter latencies (Eaton and Farley, 1975; Eaton and Kimmel, 1980). The corresponding result occurs in behavioral responses as well. For the herring, Blaxter and coworkers (1981) recorded decreases in latency by as much as 38% when the stimulus was changed from a threshold value to well above threshold (Table I). As might be expected, latency also depends strongly on temperature. The effect is predicted from the faster nerve condition time and the shorter muscle contraction time at higher temperature. Also shown in Table I are the findings of Webb (1980) that shorter response latencies were observed to occur in some fish, such as the perch, *Perca flavescens*, when schooling than when swimming as solitary individuals. So, social context may have a significant effect on this performance parameter. This intriguing finding would benefit from further study.

4.2. Mechanical Form

During the C start the animal is capable of accelerating its body up to 5 G in 20 msec (Webb, 1978*a*). One hundred milliseconds after the beginning, the animal is displaced about one body length from its initial position. Thus, it is not an exaggeration to describe this movement as an explosive acceleration. It is widely recognized that such C starts are multiphasic motor patterns with distinct mechanical stages that occur in a fixed sequence (Weihs, 1973; Kimmel *et al.*, 1974; Webb, 1978*b*; Eaton *et al.*, 1977*b*). This conclusion is based on studies employing high-speed motion picture cameras to record the response pattern. In the initial contraction, stage 1, the animal assumes a C-like body bend with head and tail bent to the same side (Figure 3A–E). During stage 2, the C-like bend proceeds down the body. How far the initial bend proceeds determines the escape trajectory. Typical escape trajectories may propel the animal either forward or to the side during the second stage (Figure 5B). Stage 3 varies from unpowered glides to the periodic lateral movements of the body and tail usually associated with steady swimming.

In some fishes, under certain conditions, stage 1 can occur without stage 2. One such example is shown for the garfish in Figure 3E. If given enough room, this fish is capable of performing a typical C start with both the first and second stages present. However, the fish in Figure 3E was tested in a small aquarium with little room to maneuver. The result was that following stage 1, the fish remained almost stationary except for a slight caudal recoil movement. This emphasizes the importance of the second stage in developing forward propulsion.

Early accounts, based on unaided visual observations, described the M-initiated patterns as a "tailflip." Perhaps a more descriptive terminology would be to describe the movement as a fast body bend, or simply "body bend," because the M-cell-initiated phase involves a contraction of the entire body: both the head and tail are bent to one side (e.g., Eaton *et al.*, 1977*b*). Only the second stage, which is not mediated by the M cell, might be compared with a tailflip.

Both the biomechanics and behavioral form of fast-starts have been extensively studied in teleosts (e.g., Eaton *et al.*, 1981; Webb, 1982). Results from behavioral analyses by Eaton and colleagues (1981, 1982) are shown in Figure 5. Figure 5A is an example of how the analysis was carried out on a single response that was filmed at 500 frames/sec. The x, y coordinates of the position of the rostrum were plotted every 2 msec as the goldfish executed a C-type fast-start. Figure 5B shows 17 superimposed response pathways. For the first 20 msec, stage 1, the movement followed a prescribed path, but after this it became variable with respect to the initial orientation. However, total distance covered remained fairly constant through stage 2 for these goldfish which were all close to the same length.

In Figure 5C, the angular velocity of the head of the fish is plotted as a function of time after the beginning of the response. The graph is the mean of 20 different responses. During stage 1 there was a sudden increase

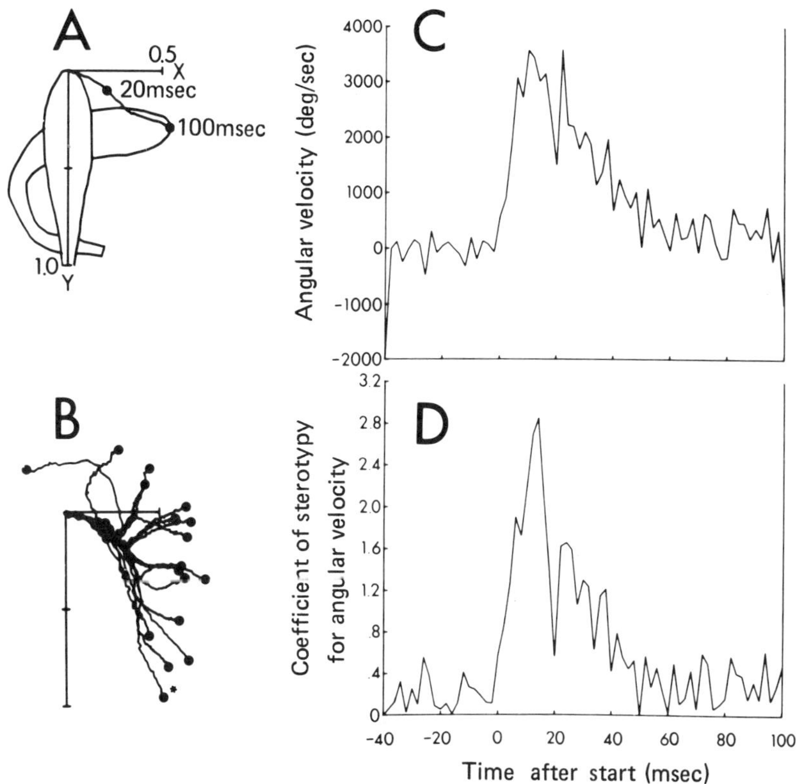

Figure 5. Quantitative performance analysis of C starts in goldfish. (A) Silhouettes are representative profiles of the fish at the beginning of the response, and at the end of stage 2 (100 msec). The *y* axis was the midline of the fish at 0 msec with the origin at the rostrum. (B) Seventeen superimposed response pathways of one fish. (C) Mean angular turning performance of 20 C starts in three goldfish. The 20 responses were synchronized so that 0 msec time was the frame just before the animal first moved in response to the stimulus. Angular velocity was determined from the angular rotation of the midline of the head at 2-msec intervals. (D) Coefficient of stereotypy for angular velocity of the responses in (C). Responses were obtained by dropping a ball into an aquarium holding the fish [(A, B) from Eaton *et al.*, 1982; (C, D) from Eaton *et al.*, 1981].

in the turning speed of the head of the animal, but this declined during the second stage that began after about 30 msec.

Analysis of the form and performance of the C start shows that the first stage has a high degree of stereotypy, whereas the second stage is more variable. This conclusion is based on actual measurements of the variability of the behavior pattern. An example of the analysis of one parameter, angular velocity, is shown in Figure 5D. Here the *y* axis is the coefficient of stereotypy for the angular velocity. This is a dimensionless value approximately equal to the inverse of the coefficient of variation (Barlow, 1968). The coefficient of stereotypy increased dramatically during stage 1 and then decreased to preresponse levels during stage 2. The high degree of stereotypy might be expected on the basis of the fact that, as we will show in Section 6.1, the initial stage of the response is invariably triggered by a single action potential from the M cell. Steering is a significant component of the pattern during the second stage. This tends to make the second stage less stereotyped than the first stage. Because of the speed of the movement, the ability to steer the response is probably important to the fish in avoiding obstacles or other fish. This has been recently highlighted by Blaxter and colleagues (1981), who studied fast-starts of herrings in schools. These investigators remarked on the ability of the herring to avoid collisions during the escape movement despite the presence of many nearby neighbors (see Figure 6A).

4.3. The Role of Fast-Starts in Predator and Object Avoidance

One of the main roles of the fast-start behavior is avoidance of predators, as indicated in the example of Figure 1. Another example of a successful C start used in predator avoidance has recently been illustrated by Lauder and Liem (1981) who recorded the interaction with a high-speed movie camera. The effectiveness of such responses has been demonstrated by Webb (1981) in a study of predatory attacks by clown fish on Northern anchovy larvae. In this laboratory experiment, the probability of escape was as high as 70% if the prey utilized a C start. C starts are also used in the escape from objects falling into the water from above the fish (Eaton *et al.*, 1981). In all these cases, speed and maneuverability are clearly important parameters.

Webb and Skadsen (1980) also analyzed attacks by tiger musky, *Esox* sp., on fathead minnows, *Pimephales promelas*. This study illustrates how common fast-start patterns can be used for attack and escape. Many predator–prey encounters were preceded by a stalk in which the musky closed to within striking distance with its body sometimes already retracted into the S posture before beginning the lunge. C starts were used by the minow to escape. Strike success varied widely from 14–100% depending on the strike pattern and experience of the predator. The strike trajectory was directed at or near the center of mass of the prey. This point is a logical target as it moves least during the C start and is also located near the point of maximum body profile (Webb, 1978*b*).

Escape in such a situation appeared to depend on successful detection of the predator and proper timing of the escape sequence relative to the attack. It frequently appeared that the escape was not initiated, as though the prey were unaware of the predator, or waited too long to respond (Webb and Skadsen, 1980). A similar relationship was also discovered for the Northern anchovy where 24–30% of larvae failed to escape because they began the response too late to avoid capture. Webb has posed the important question of why such a high proportion of the prey apparently leave escape initiation so late. The answer to this may relate to important features of the escape tactics. At present, escape tactics are not well understood and probably vary depending on the type of predator, attack speed and trajectory, availability of cover, and other environmental features. Lunging type encounters, as shown in Figure 1, have been modeled on theoretical grounds by Webb (1976). The details of Webb's model are quite complex but the main conclusion is that the outcome of a prey–predator interaction depends more on reaction latency and accurate timing than on performance differences between prey and predator. This suggests that proper timing for the firing of the M cell is crucial if escape is to occur. Thus, not only must the decision be made *whether* to fire the M cell, but also *when*. These considerations probably play a role in the design of the physiology of this cell, although as yet little is known about triggering of the M cell in complex stimulus situations approximating actual predatory attacks.

4.4. Directionality of the Escape Response

Since fast-starts are utilized in avoiding predators, it is reasonable to suppose that the escape would be directed out of the path of the attack. Both behavioral (Blaxter *et al.*, 1981; Eaton *et al.*, 1981) and electrophysiological experiments (Eaton and Kimmel, 1980) support this point of view. An example is shown in Figure 6A, which is a drawing made from video frames of a school of herring avoiding a vibrational stimulus to the side of the tank in which they were swimming. The experiment was designed so that the herring could not anticipate from which side the stimulus would occur. Within 20 msec after the stimulus (center panel), the majority of individuals can be seen to be in the first stage of C starts and turning away from the source of the vibration at the top of the frame. Eaton and colleagues (1981) found that goldfish fire the M cell on the side closest to a ball dropped into the water above the fish in 85% of the trials. This also resulted in turns with the initial stage oriented away from the threatening stimulus.

Eaton and Kimmel (1980) showed that zebrafish larvae were approximately five times more likely to activate the M cell on the same side of

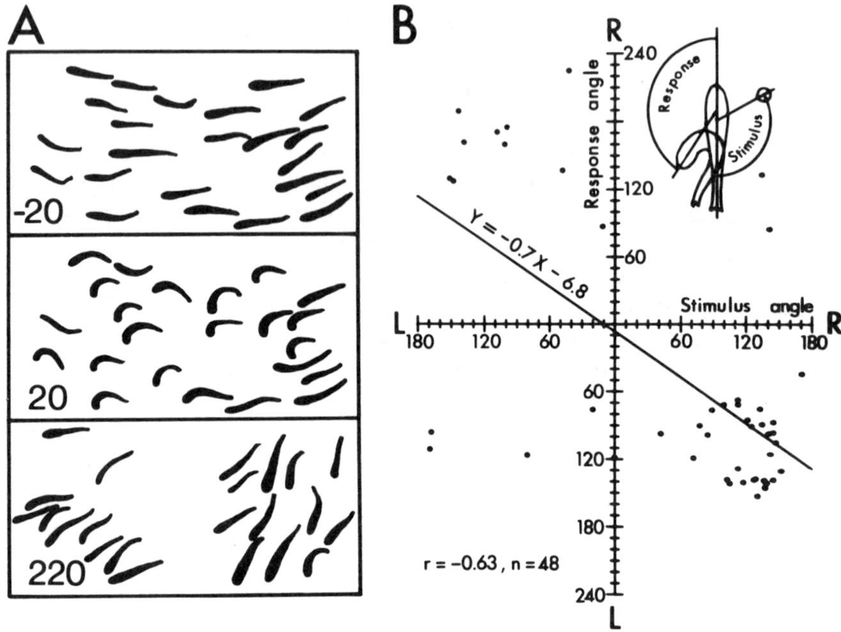

Figure 6. Directionality of C starts to vibrational and visual-vibrational stimuli. (A) Drawings made of three video frames at 20 msec before and 20 and 220 msec after a vibrational stimulus was delivered at the side of the tank in which a school of herring was swimming. The stimulus is at the top of the frame. Most fish oriented away from the stimulus within 20 msec. (B) Quantitative relationship between the stimulus angle and response angle for goldfish avoiding a ball dropped into an aquarium. These are all M-initiated responses as the fish were implanted with chronic microelectrodes to monitor M-cell activity. Each point was taken 100 msec after the response began and represents the angle of the fish relative to the initial orientation when the ball hit the water surface (see insert) after the response began. These data show a statistically significant orientation away from the ball during the second stage of the response. A similar directionality is also seen in the escape response of the cockroach (see Chapter 4, this volume) [(A) From Blaxter *et al.*, 1981; (B) from Eaton *et al.*, 1981].

a vibrational stimulus as the opposite M cell. Of course, because of the crossed axon of the M cell, this would result in C starts beginning with the initial turn on the side away from the stimulus. The second stage of the response also tends to be away from the side of the stimulus. An example is shown in Figure 6B in which response angle is plotted against stimulus angle 100 msec after a ball was dropped into the aquarium. As seen here, the orientation of the fish at the end of the second stage was highly related to the stimulus angle relative to the body axis, whereas the second stage orientation is quite variable relative to the initial orientation of the fish (Figure 5B). The mechanism for the initiation of directed responses at short latency to vibrational stimuli is unknown.

5. The Mauthner Cell

5.1. Occurrence of the M Cell in the Lower Vertebrates

M cells are found in representatives of all the lower aquatic classes of vertebrates: bony fishes (lungfish, gars, trout, goldfish, etcetera.), cartilaginous fishes (sharks and chimaeras), jawless fishes (lampreys), and amphibians (salamanders and frogs) (see Kimmel, 1982*a*). Altogether the morphology of the M cell has been described in over 200 species, and Zottoli (1978) has presented an extensive table showing the taxonomic relationships based on this literature. Much of the comparative morphological work was done by Stefanelli and colleagues who showed that the M cells are most prominent in actively swimming subcarangiform fish such as minnows. The M cell is absent or less well developed in fish with more sedentary life styles, or those that swim with eel-like movements (Stefanelli, 1951; Stefanelli, 1980). Examples where the M cells have not been identified include toadfishes, puffers, and moray eels. The M cells are also absent, among the amphibia, in toads and in adult terrestrial forms of species with aquatic larvae. In the latter case, the M cells may be present during the aquatic stages but regress when the legs develop and the animal emerges onto land (Fox and Moulton, 1968).

5.2. The Reticulospinal System and the M Cell

The M cells are located at the level of the eighth cranial nerve in the hindbrain (Figures 2, 7A). These neurons are members of the reticulospinal system that is a distributed nucleus situated on either side of the midline and extending from the midbrain to the caudal hindbrain (Figure 8). Neurons of the reticulospinal system receive sensory information from various modalities and also input from more rostral "integrative" areas of the brain. These brain-stem cells are thought to be major intermediaries in the initiation of motor commands (see Lawrence and Kuypers, 1968).

The M cell is distinguished from other reticulospinal neurons by four main criteria reviewed by Zottoli (1978).

1. The M cell has a conspicuously large soma that is located in the medulla oblongata at the level of the eighth cranial nerve. In many species the M cell soma diameter is as much as 100 μm.

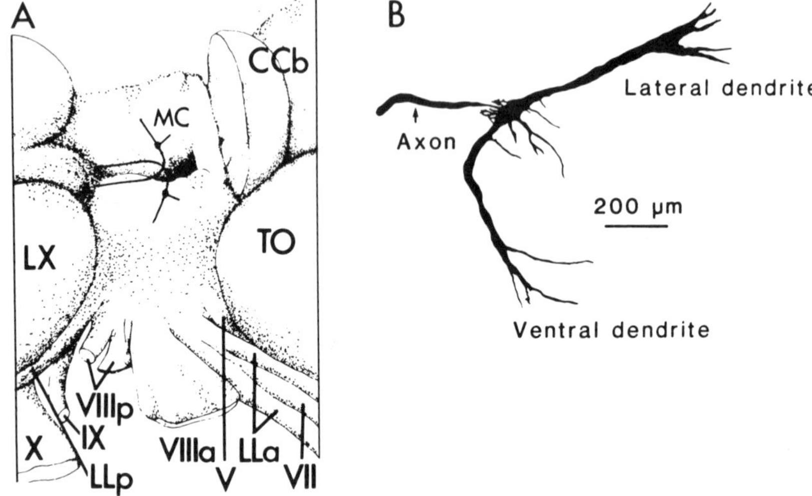

Figure 7. Configuration of the M cell and its relation to the cranial nerve roots of the goldfish. (A) Lateral view of the brain from the rostral portion of the vagal lobes (LX) to the caudal surface of the optic tectum (TO). Most of the corpus cerebellus (CCb) has been removed, allowing a diagrammatic projection of the M cells (MC) onto the medullary surface. The M cells lie about 1.5 mm below this surface and are near the roots of the eighth nerve as well as those of the fifth, sixth (not shown), seventh, ninth, tenth, and the posterior (LLp) and anterior (LLa) lateral line nerves. (B) Two-dimensional reconstruction of the M cell after cobalt injection (from serial 15 um sections) (from Zottoli, 1978, in: *Neurobiology of the Mauthner Cell,* (D. S. Faber and H. Korn, eds.), Raven Press, N.Y.).

2. There are two main M-cell dendrites (Figure 7B). A lateral dendrite extends nearly to the periphery of the brain and receives as its principal input, large primary afferents from the acoustico-vestibular system. One of these fibers is illustrated for a larval zebrafish in the micrograph of Figure 9A. A ventral dendrite in fish branches among fibers of the ventrolateral neuropil column. Principal fibers in this column are those from the tectobulbar tract that presumably gives synapses to the M-cell ventral dendrite that is extensively branched in this region.
3. The M cell possesses a specialized neuropil called the axon cap that surrounds the initial segment and axon hillock of the neuron. This neuropil is illustrated in Figure 9B for the larval zebrafish. The central core of the axon cap is densely packed with unmyelinated fibers, whereas the periphery of this neuropil is surrounded by a sheath of glial lamellae. This neuropil is very distinctive in both light and electron micrographs, and as described in Section 9.1, it is the site of an electrotonic inhibition of the M cell.

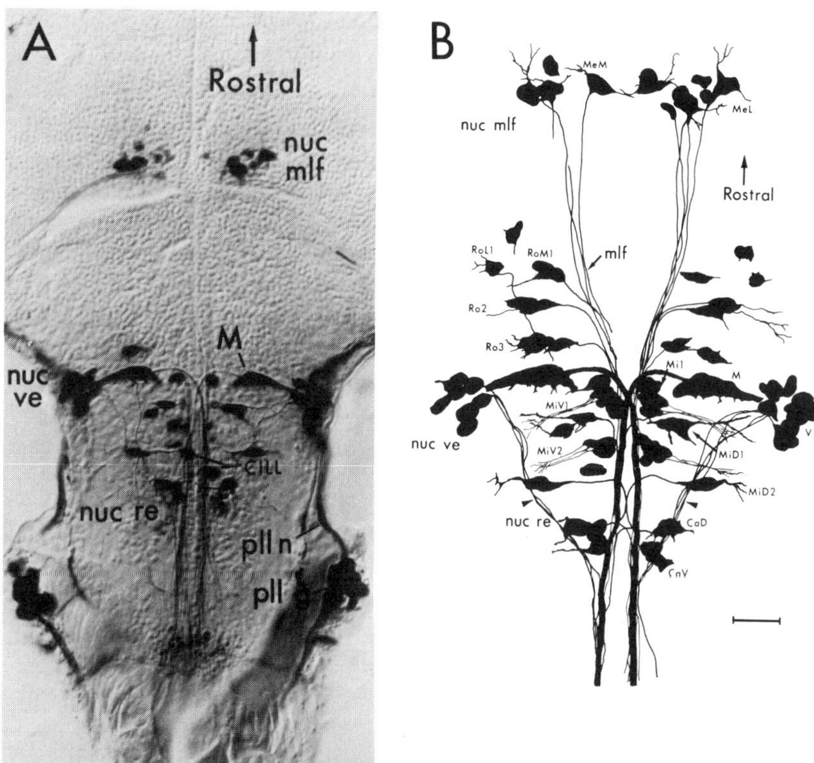

Figure 8. The M cell and other reticulospinal cells in the zebrafish larva six days after fertilization. (A) Micrograph of HRP-filled neurons following a complete transection of the trunk at the level of the cloaca. The sections include the reticular formation (re) and vestibular nucleus (nuc ve) of the hindbrain and the nucleus of the medial longitudinal fasciculus (nuc mlf). Also shown are the vestibular nucleus (nuc ve) and the posterior lateral line nerve (pll n) and ganglia (pll g), which were also damaged by the transection. Individually labeled cells include the M cell (M) and a pair of lateral line efferent neurons (CILL). (B) Dorsal view map of reticulospinal neurons reconstructed from micrographs prepared as in (A). The positions of three brain regions (nuc mlf, re, and nuc ve) are as in the micrograph. Individual cells are labeled with a code based on the positions of the cells in the rhombencephalon (R, rostral; M, middle; and C, caudal). Labeled cells on the left are found ventral to the M cells, labeled cells on the right are found at the level of the M cells. Scale bar: (A) 33 μm; (B) 25 μm (from Kimmel *et al.*, 1982*a*).

4. The M cell has a distinctively large axon that may be up to five times greater in diameter than adjacent fibers. In many fishes this axon and its myelin sheath is at least 0.1 mm in diameter. It descends into the spinal cord on the side of the body opposite the M-cell soma. The M axon courses within the dorsal bundle of the

medial longitudinal fasciculus (MLF) (Figure 8B). In mammals, the MLF also carries the axons that initiate startle responses (see Chapter 10, this volume).

Because of the large size of the Mauthner dendrites, it is thought that the cell is distinguished functionally because of the number and diversity of synaptic inputs it receives. Cochran *et al.*, (1980) have estimated that the anuran M cell has approximately 200,000 synaptic terminals on its surface. Many of these terminals have been categorized by Nakajima (1974) and Cochran *et al.*, (1980). The presence of such a large array of synaptic inputs suggests that the M cell plays an important role in integrating diverse signals to trigger the fast-start.

Recent studies have shown that the M cell is the most conspicuous member of a class of neurons with the general morphology described above (Kimmel *et al.*, 1982*a*). In the larval zebrafish there are at least three other pairs of reticulospinal, Mauthner-type cells that can be identified individually following anterograde transport and staining of horseradish peroxidase (HRP). One of these cells, MiD1, is compared with the M cell in Figure 10. These cells all have two principal dendrites and crossed-descending axons. It is not yet known if any of the other Mauthner-type cells have axon caps. A similar group has long been known in the lamprey, which has two pairs of Mauthner-type cells, as well as identified Mueller cells, which do not have crossed axons (Rovainen, 1979).

6. The Role of the M Cell in Triggering Fast-Starts

Although Bartelmez in 1915 proposed that startle responses in fishes are mediated by the M cell, early neurobiologists were more concerned with the incorrect idea that the M cells mediate the side-to-side movements

Figure 9. (A) Synaptic terminals on a distal portion of the M-cell lateral dendrite (ld) of the left M cell in a zebrafish larva six days after fertilization. A myelinated axon (a) can be seen forming an unmyelinated terminal (t) with large gap junctions (g) and nonsynaptic attachment points (p, puncta adherens). This axon was traced from the ipsilateral VIIIth nerve. A large synaptic bouton is present in the upper left of the field and forms a large chemical synapse (s) with the dorsal surface of the lateral dendrite. Another terminal from a myelinated axon is also present with a gap junction and puncta adherens. Scale: 1 μm. (B) The axon cap of the larval zebrafish M cell. The axon cap is a distinctive neuropil in which a strong electrotonic inhibition can block an action potential in the M-cell initial segment (is). This neuropil is surrounded by a glial sheath of lamallae that are believed to serve as a high-resistance barrier to extracellular current flow. The neuropil itself consists of a tightly woven network of fibers and terminals of various types. Some terminals synapse on each other (circles) and others synapse on the M-cell membrane by means of gap junctions (g) and chemical synapses (s). Scale: 1 μm (from Kimmel *et al.*, 1981).

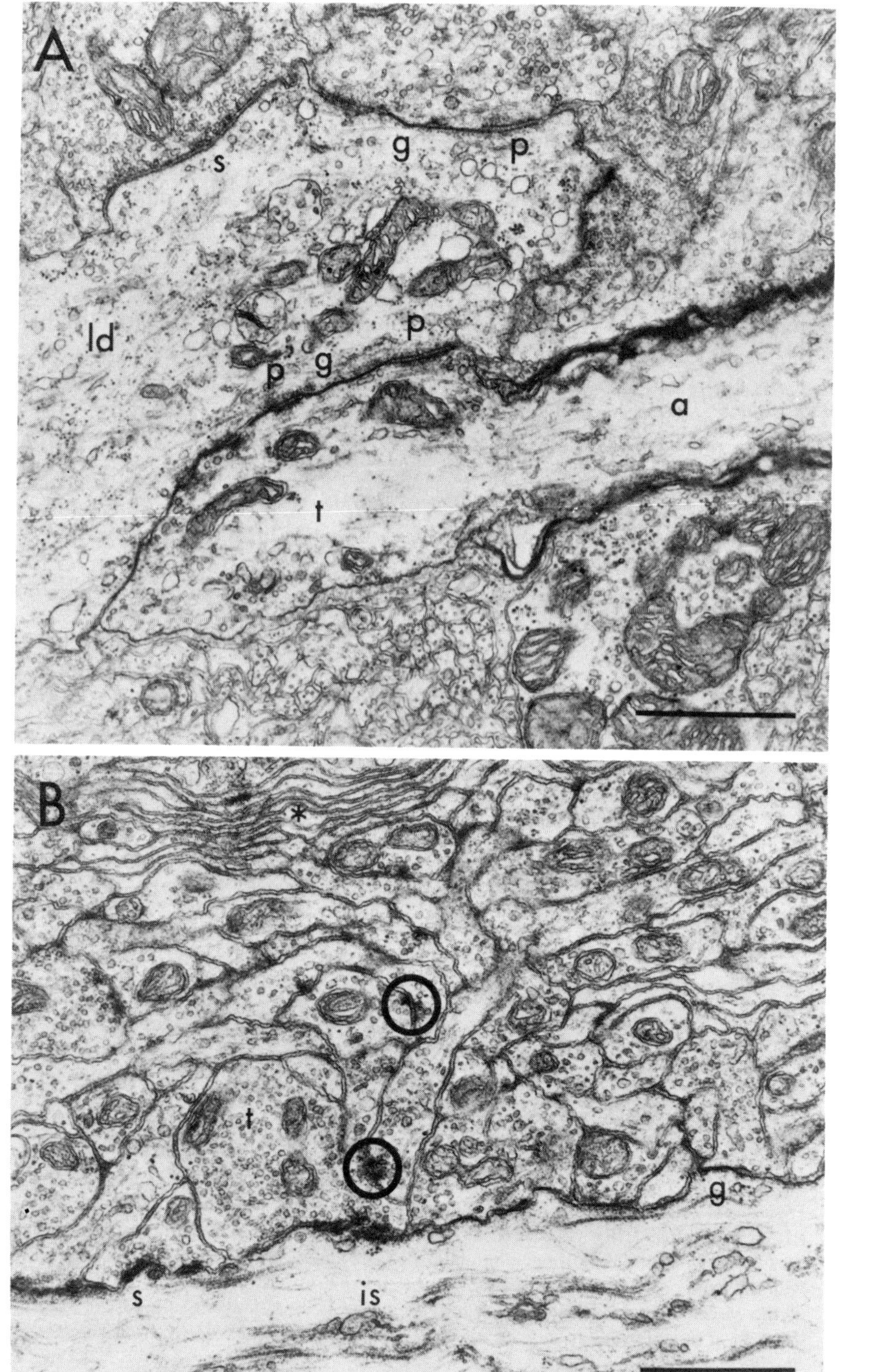
A
s
g
p
ld
p
p
g
a
t
B
*
t
g
s
is

of the tail during swimming: the startle response hypothesis of M-cell function did not receive any direct experimental tests until E. C. Berkowitz in T. H. Bullock's laboratory revived the issue in the early 1950s. Work since that time has been a steady improvement in the ability to show a direct relationship between the firing of the M cell and the onset of the behavioral response. Details of the early research are reviewed by Eaton and Bombardieri (1978).

Recent work on both fish and amphibian preparations demonstrates that the most common fast-start response pattern, the C start, is initiated by the M cells. This is one of the few cases in the vertebrate nervous system where it is possible to make a rigorous causal connection between the activity of a particular cell and a defined behavioral response. That is, it is possible to precisely correlate the firing of the M cell with the onset of a quantitatively defined response in free-swimming animals. In addition, intracellular stimulation of a single M cell results in an appropriate muscular activation believed to correspond to the first stage of the C start. Finally, hyperpolarization of the M axon prevents this muscular response following a stimulus that would drive the behavioral response.

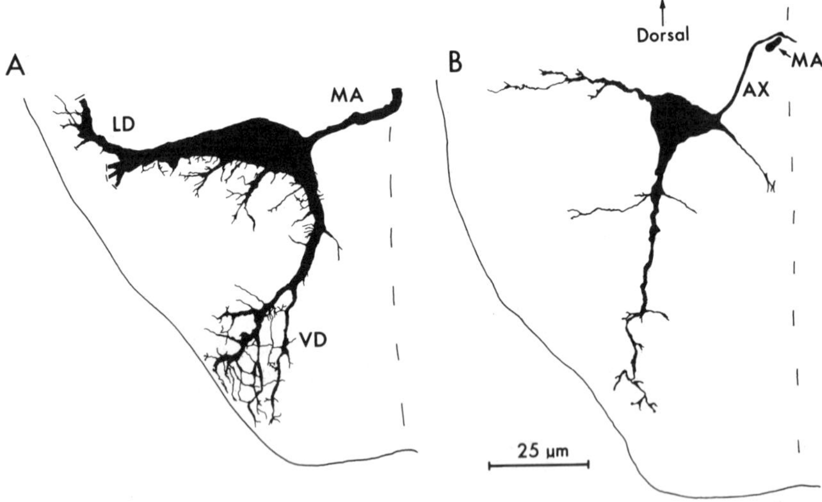

Figure 10. (A, B) Comparison of a reconstructed M-cell (A) and M-cell analogue, MiDl (B), prepared from HRP-filled neurons in the zebrafish larva. MiDl has an axon (AX) that crosses the M-axon (MA) of the contralateral M cell and then descends into the spinal cord with the axon of the ipsilateral M cell. The lateral dendrite of MiDl branches in the vicinity of the M-cell lateral dendrite. The ventral dendrite is also similar to the M-cell ventral dendrite in its extensive branching in the ventrolateral neuropil column (from Kimmel *et al.*, 1982*a*).

6.1. Evidence from Chronic Electrophysiological Recordings

In several studies it has been possible to record from the M neuron in freely swimming, or lightly restrained animals responding to auditory or vibrational stimuli (Eaton and Farley, 1975; Zottoli, 1977; Rock, 1980; Eaton *et al.,* 1981; Prugh *et al.,* 1983). Taken together, these studies provide a convincing connection between the firing of the M cell and the onset of a defined behavior pattern. Such studies are possible because the M-cell action potential (M spike) (Figure 11) is of very large amplitude in the vicinity of the axon initial segment. The explanation for this large amplitude is based on the large convergence resistance to current flow across the axon cap (Furshpan and Furukawa, 1962). This allows rapid identification of the spike and chronic recordings with implanted electrodes. In one of these studies, Zottoli (1977) recorded from one M cell of freely swimming goldfish and also recorded bilateral electromyographic (EMG) responses of the trunk musculature. He was able to show a nearly perfect correlation between the presence of the M spike and the occurrence of a large EMG of the contralateral trunk musculature. More recently, Prugh and colleagues (1983) were able to record the M spike with bipolar electrodes placed in a droplet of water in which a zebrafish larva was free to swim. These electrodes recorded M spikes on the order of 0.5 mV in amplitude outside the body of this small fish, which is only 3

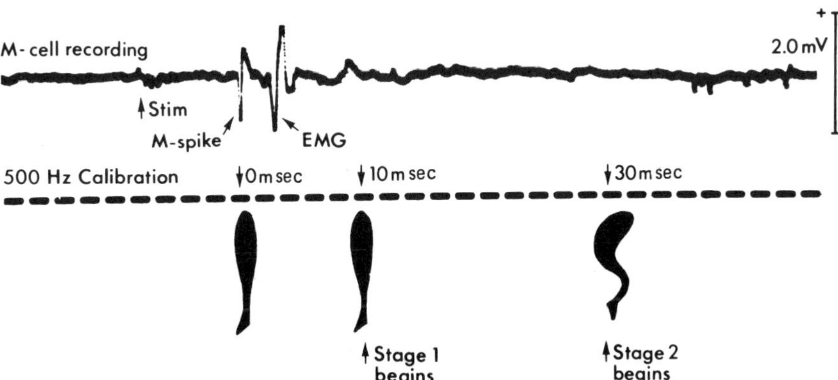

Figure 11. A simultaneous cine record and electrophysiological recording from a chronically implanted microelectrode near the left M cell in a goldfish responding to a ball dropped into the water above the fish. Upper trace, electrophysiological recording from the M-cell; lower trace, calibration signal; silhouettes, drawings made from dorsal view images from the cine record. Silhouettes were chosen from points on the film at the indicated times that correspond to when the M cell fired, at the beginning of stage 1, and at the beginning of stage 2. The M cell fired 8 msec after the ball struck the water. Following the M spike is a prominent EMG, volume conducted from the body musculature (from Eaton *et al.,* 1981).

mm in length. Behavioral responses were recorded by monitoring the output of a photocell when the fish moved in the path of a beam of light. As in previous studies, the M spike was closely correlated with the onset of the fast-start behavior.

Because of the speed of the fast-start response, it is difficult to know by direct visual observation, or transducers, what behavior pattern has actually happened, or when the M cell fired relative to the various mechanical stages of the response. To clear up these complexities, Eaton and colleagues (1981) chronically recorded from the M cell in goldfish while simultaneously filming fast-starts with a high-speed movie camera. Results from one such experiment are shown in Figure 11. In this figure are the electrophysiological recording from the M cell and silhouettes drawn from selected corresponding frames of the cine record. The stimulus was a ball dropped into the aquarium from above the fish. The impact of the ball hitting the water is indicated by the arrow ("stim") below the recording trace.

The basic finding was a perfect, time-locked correlation between the presence of an initial M spike and the onset of the behavioral response. The M spike was always followed by a response beginning with a contraction of the body musculature on the side opposite the monitored M cell. There were no fast-starts beginning on the side opposite the monitored M cell in which the M cell did not fire first.

The timing of the M cell's activity relative to the onset of the behavior is consistent with the hypothesis that this cell triggers the observed motor pattern. The interval from when the ball hit the surface of the water and when the M cell fired was about 7–8 msec. The M cell was followed by a prominent biphasic potential that represents a volume-conducted EMG of the musculature. As shown in Figure 12, the interval between the M spike and this EMG was 2–3 msec, a value that corresponds to the findings of Zottoli (1977) and to the more recent study, discussed below, of Hackett and Faber (1983*a*). The delay between the M spike and the EMG can be accounted for by the short conduction time of the M-axon spike in the spinal cord and by the delays for transmission between the M axon and motoneuron and the neuromuscular junction. However, it is now thought that the timing is not so limiting as to exclude the possibility of an additional relay neuron between the M axon and motoneuron (see Section 10.2). Finally, the movement of the fast-start began 6–10 msec after the EMG. This value corresponds to what is believed to be the activation time of the contractile apparatus itself (see Eaton *et al.*, 1981).

These results demonstrate that the M spike precedes the onset of the first stage of the C-type fast-start behavior. In cases when the initial contraction was on the same side of the body as the monitored M cell,

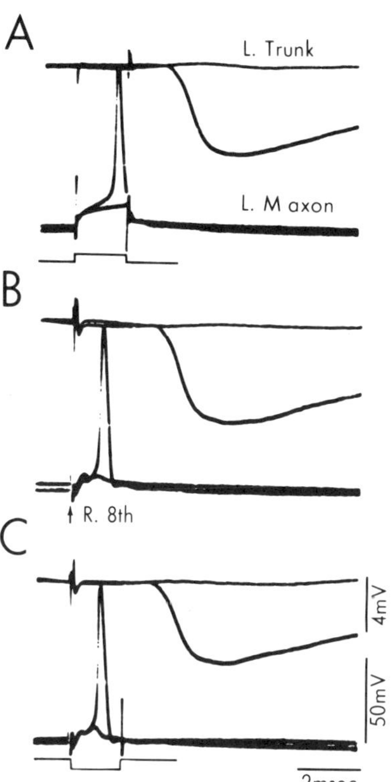

Figure 12. The M cell is sufficient and necessary to produce an EMG that corresponds temporally to the one seen in chronic recordings in Figure 11 when the M cell fired and the fish initiated a C start. A bipolar electrode recorded the EMG of the left trunk muscle, and a microelectrode penetrated the left M axon. (A) A depolarizing current pulse from the recording electrode produced an M spike, middle trace, and an all-or-nothing EMG that follows the M spike at an interval of 2.4 msec. Also shown is a superimposed below-threshold stimulus of the M axon. This failed to produce the EMG. (B) An above- and below-threshold stimulus to the right VIIIth nerve. This stimulus failed to produce any EMG when the M cell failed to fire. (C) Hyperpolarization of the M axon blocked an above-threshold stimulus to the eighth nerve and also failed to produce the characteristic EMG (from Hackett and Faber, 1983*a*; reprinted with permission of Pergamon Press, Ltd.).

no M spikes were ever recorded. Such responses were probably initiated by the unmonitored M cell on the opposite side of the brain. The M cell never fired after the behavioral response had already begun. This finding rules out the hypothesis that the two stages of the C-type fast-start are necessarily initiated by a sequential firing of the two opposite M cells. The M cell was also never observed to fire during swimming.

6.2. Evidence from Acute Electrophysiological Recordings

Results from acute recordings show that stimulation of the M cell with an intracellular microelectrode results in a characteristic EMG of the contralateral body musculature. The onset of this EMG is the same as that recorded in the chronic experiments just described. Thus, this EMG represents an easily measured electrical response of the muscles that are probably involved in the body contraction during the C start. Such an experiment has been done recently on the hatchetfish by Aljure *et al.* (1979) and on goldfish by Hackett and Faber (1983*a*). Similar results were also obtained in experiments on the bullfrog tadpole (Rock, 1980; Rock *et al.*, 1981).

Results from Hackett and Faber's (1983*a*) study on the goldfish M cell are shown in Figure 12. The goldfish was restrained and lightly anesthetized. The M axon was then penetrated with a microelectrode and an intracellular current pulse was used to activate this neuron. The result, shown in Figure 12A, was an EMG of the trunk musculature on the same side of the body as the M axon (or, on the side opposite the activated M-cell soma). The EMG followed the M spike by 2.4 msec, a value within the 2–3 msec range obtained by Eaton and colleagues (1981) from chronic recordings during M-initiated fast-starts in freely swimming animals (Figure 11). An electrical stimulus to the posterior eighth cranial nerve on the same side as the M-cell soma elicited an identical EMG. Regardless of how the M cell was activated, it was always followed by the EMG. Furthermore, the EMG occurred in an all-or-nothing manner. This finding demonstrates that the M-cell impulse is sufficient for producing a short latency muscle response that corresponds temporally with the fast-start behavior.

Hackett and Faber (1983*a*) have also shown that the M cell is necessary for the generation of the EMG. This is illustrated in Figure 12C. In this case, the M axon was hyperpolarized by an intracellular current pulse, before and after an EPSP evoked by an above-threshold stimulus of the posterior VIIIth nerve. The same stimulus that would normally fire the M cell failed to do so when the M cell was hyperpolarized. The corresponding EMG was also absent.

Thus, these studies, involving both chronic and acute recordings, provide a compelling case that the firing of the M cell is causally related to the onset of the C-type fast-start behavior. It is important to note, however, that there is evidence for alternative pathways that can initiate apparently identical motor patterns (see Section 12).

7. Sensory Inputs to the M Cell

Previous research has concentrated almost exclusively on the sensory input of the octavolateralis system to the M cell through its lateral dendrite. The octavolateralis system consists of a family of modalities including receptors for audition, gravity, angular acceleration, vibration, water flow, and electric fields (Bullock, 1981). Inputs to the M cell from auditory, equilibratory, and lateral line have now all been studied physiologically. In addition, initial physiological studies have been made to examine the visual, or tectal, input to the M cell. These inputs are bilateral and mediated through the ventral dendrite (A. R. Hordes and D. S. Faber, personal communication). In most cases, sensory afferents have not been studied with behaviorally relevant sensory signals. Instead, these analyses have utilized electrical stimulation of the afferents to gain a first-order understanding of the pathways involved and their synaptic actions on the M cell.

7.1. Eighth Nerve Afferents

Inputs from the eighth nerve mediate not only sensations of hearing but also vibration, equilibrium, and acceleration. The fast-start behavior in freely moving fishes can be readily activated by an auditory or vibrational stimulus, and the presumed pathway to the M cell is from hair cells in the otolithic organs, the sacculus, lagena, and utriculus to primary receptor cells in the eighth nerve to the lateral dendrite of the M cell. The posterior branch of the eighth nerve is the most thoroughly studied input to the M cell. In fishes, auditory responses recorded intracellularly from posterior eighth nerve fibers from the sacculus can respond with twice the frequency of impulses as the sound source (Furukawa and Ichii, 1967). Axons of this type probably contact the M cell as club endings.

Under normal circumstances, the auditory responses recorded from M cells in goldfish are subthreshold for spike generation in restrained animals (Diamond, 1971; Faber and Korn, 1978). However, it is possible to record sound-evoked EPSPs in the M cell (Diamond, 1971; Lin *et al.*, 1982). The sound-evoked EPSPs exhibit multiple, fast-rising components superimposed on an underlying slower depolarization. The amplitude of the fast components was maximal at distal dendritic recording sites where club endings make synaptic contact. The slower ones were more distributed, extending to the proximal region of the lateral dendrite.

Following posterior eighth nerve stimulation, EPSPs in the M cell have at least two components distinguishable by their peaks and by their depression in response to high-frequency stimulation. This is true for M cells in teleosts, lampreys, and the bullfrog tadpole (Furshpan, 1964; Rovainen, 1979; Hackett *et al.*, 1979). The two components are illustrated in the recordings of Figure 13. It is believed that the early component is produced by electrotonic synapses, whereas the late components are produced by chemical synapses. The first evidence that there was electrotonically mediated synaptic transmission in the vertebrate brain was based on the early component of this EPSP recorded from goldfish M cells (Furshpan, 1964).

That electrical stimulation of the posterior eighth nerve evokes the

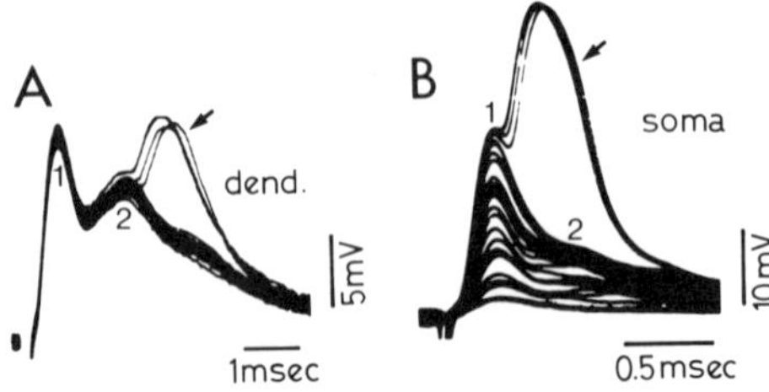

Figure 13. Multiple components of the eighth nerve evoked EPSP in the M cell lateral dendrite and soma. (A) Recording from the lateral dendrite, 325 μm from the axon hillock. The first component (1) is the electrotonic EPSP, whereas the second component (2) is chemically mediated. The stimulus was near threshold and an M spike (arrow) was elicited off the second EPSP. The electrotonic component was largest at this recording site, so the second EPSP must have been generated closer to the spike-initiating zone, the axon hillock, where it was above threshold and elicited the spike. (B) Recording from the soma in a different experiment at various stimulus intensities. In this case the spike (arrow) rose off the electrotonic component (1) and there was only a small chemically mediated EPSP (2) [(A) from Diamond (1968); (B) from Furshpan (1964); copyright 1964 by the American Association for the Advancement of Science].

M spike is consistent with the morphological observation that the large eighth nerve fibers contact the M cell as myelinated club endings that form gap junctions with the lateral dendrite (Figure 9A). Very short latency (0.10–0.15 msec) intracellular changes in potential were sometimes observed after an electrical stimulus to the ipsilateral posterior branch of the eighth nerve that innervates the sacculus and lagena in goldfish (Furshpan and Furukawa, 1962; Furshpan, 1964). These responses were recorded at different positions along the M-cell lateral dendrite; the largest were located in the region of the myelinated club ending contacts. Other units were identified as primary afferents by their short latency (0.1–0.4 msec) and by their fatigue-resistant action potentials evoked by stimulating the posterior eighth nerve. In some of these units a positive-going potential was recorded in response to antidromic stimulation of the M cell (Furshpan, 1964). This transmission of an impulse from M cell to posterior eighth nerve was without delay. Thus, it is likely that there was a passive spread of the impulse across low-resistance connections between the M cell and the posterior eighth nerve fibers. The fibers were not identified morphologically. Confirmation of these results with dye-marking experiments would exclude the possibility that efferent fibers or second-order neurons were examined.

Two mechanisms have been proposed to account fo the dual-component transmission of the eighth nerve to M cell. Furukawa (1966) has suggested that the second component of the eighth nerve-evoked EPSP is produced by the activation of small fibers from the sacculus that reach the M cell by polysynaptic pathways. It may be that more direct afferents that impinge on the M soma are activated by electrical stimulation of the posterior branch of the eighth nerve. On the other hand, part of the composite EPSP is possibly produced by a single synapse where both electrotonic and chemically mediated transmission coexist. The myelinated club endings make morphologically mixed junctions with the features of gap junctions and chemical synapses. Faber and colleagues (1980) found that cobalt ions injected into the M cell could uncouple the chemically mediated EPSP. This effect occurred at low iontophoretic current intensities and was seen only if the electrotonic component was intact. At high iontophoretic currents, the electrotonic component was quickly uncoupled. It was suggested that the injected cobalt ions cross the gap junctions and reduce transmitter release, but at high currents completely uncouple the gap junctions. The existence of dual modes of transmission is not resolved nor is the significance obvious. At many synapses where the morphological evidence indicates there may be dual mode of synaptic transmission, the physiological evidence reveals only electrotonic transmission (Bennett, 1977). In any case, the presence of electrotonic synapses undoubtedly contributes to the short latency of the behavioral response to auditory stimulation.

The M cell also receives afferents from the anterior branch of the eighth nerve. This nerve contains afferents from the utriculus and mediates both sound (Blaxter *et al.*, 1981) and equilibrium. In teleosts, the stimulation of the anterior branch of the eighth cranial nerve evokes EPSPs whose maximum amplitude is recorded from the proximal regions of the lateral dendrite (Zottoli and Faber, 1979). The input had an apparent monosynaptic latency of 0.56 msec, which was longer than the early component evoked by the posterior eighth nerve. The input, therefore, is likely to be chemically mediated. The maximum EPSP was subthreshold for generation of an action potential in the M cell, although its excitatory nature could be clearly shown by appropriate pairing with the posterior eighth nerve. Thus, it seems that this afference may have a modulatory role on M cell excitability (Zottoli and Faber, 1979).

7.2. Lateral Line Afferents

The lateral line system has hair cell receptors on both the head and trunk of the body. These transmit to the central nervous system from the head via branches of the seventh (facial) nerve and from the trunk via the posterior lateral line branch of the ninth nerve. The lateral line system is responsible for the detection of surface waves and local water movements or vibrations causing water displacements (Sand, 1981). It is the lateral line head receptors that appear to detect surface waves (Schwartz, 1967) and to give the fish the "distance-touch" perception responsible for close-range obstacle detection (Dijkgraff, 1963). According to Sand (1981), the trunk lateral line system is not well understood, although in some species it is essential for schooling (Partridge, 1981).

Only the lateral line input through the posterior lateral line nerve has been studied physiologically. It has not been possible to activate an M spike through lateral line stimulation either from an electrical shock to the nerve or from pulses of water directed at the fish (Korn *et al.*, 1974). Stimulation of both ipsilateral and contralateral posterior lateral line nerves evokes composite postsynaptic potentials in the M cell whose amplitudes do not reach threshold for action potential generation (Korn *et al.*, 1974). The postsynaptic potentials had latencies of 0.87–2.92 msec and could result from either monosynaptic or polysynaptic pathways. It is believed that these potentials were mixed EPSPs and IPSPs (inhibitory postsynaptic potentials) because the late components seemed to be due to a chloride conductance increase. Following blockage of the IPSPs by intramuscular injections of strychnine, it was possible with strong stimuli to bring the M cell to the firing level (Korn and Faber, 1975*b*; Faber and Korn, 1975). Mapping the response in the M cell after strychnine injections indicates that the distribution of the excitatory inputs is broader than without the strychnine and that inhibitory inputs contact mostly the soma region. The inhibition is mediated through a known class of neurons, the so-called passive hyperpolarizing potential (PHP) cells (see Section 9.3) (Faber and Korn, 1975).

In conclusion, although the M cell can be shown to receive primary or secondary afferents from several sensory modalities, there are few data yet available on how this information is integrated by the cell. Most data have been derived by electrical stimulation, a technique useful for deciding possible neural pathways. Particularly lacking, however, are studies such as that of Lin and associates (1982) in which the response of the cell to more behaviorally relevant stimuli are examined.

One reason for the lack of studies employing meaningful sensory stimuli is that in adult preparations it has not yet been possible to use such stimuli to bring the cell to threshold when the animal is restrained (Faber and Korn, 1978). This is true even for acoustic stimuli that are known to activate the cell in free-swimming adult animals (Zottoli, 1977; Eaton *et al.*, 1981). Only in larval or embryonic animals such as the zebrafish or bullfrog tadpole has it been possible to use physiological stimuli (vibration) to activate the cell (Eaton and Farley, 1975; Rock, 1980). It appears that in the adult, the M-cell system is inhibited when the animal is restrained. Consequently, the afferent circuitry remains as one of the most important areas to be studied in the M-cell system.

8. Initiation and Propagation of the M Spike

Mechanisms of signal generation and propagation are an important problem in understanding how the M cell produces the neural "decision" to fire and initiate the fast-start behavior. In general, for a neuron to perform its signaling function, an action potential is generated as a result of electrotonic spread of synaptic potentials to a low-threshold, spike-generating zone. In the M cell there is a high degree of coupling from the soma to the axon initial segment. This coupling leads to the condition where orthodromic spike initiation occurs nearly simultaneously in the axon initial segment and the first active site in the axon (Funch and Faber, 1982). In this cell there is preferential triggering of the initial segment–axon hillock action potential in the orthodromic direction. Spontaneous failures of the axon hillock spike occur during antidromic propagation. Further, EPSP attenuation from the soma to the M axon is much less than attenuation in the opposite direction of the axon spike (Faber and Funch, 1980). The EPSP caused by eighth nerve stimulation can be readily recorded in the M axon, distal to the initial segment. Thus, there is a high safety factor for orthodromic activation of the axon.

Funch and Faber (1982) have presented evidence for discrete active sites every 2.0–2.8 mm along the M axon. This is interesting because no typical nodes of Ranvier are found along the myelinated M axon. However, recently Yasargil and colleagues (1982) have studied the morphology of the M-axon collaterals and found staining properties similar to those described at nodes of Ranvier. Definitive evidence for the mechanism of M-axon propagation thus remains an open question.

The M cell does not usually fire repetitively in response to sustained depolarization, although it can fire at high frequency during repetitive phasic stimulation. The mechanisms for spike repolarization have been investigated by injecting agents that act to block voltage-dependent potassium channels (Kaars and Faber, 1981). It was found that these agents affected only the axon initial segment. It is unlikely, however, that repolarization is a limiting step in M-cell function given the abundance of synaptic mechanisms for inhibition and depression (Diamond, 1971).

Once the M spike has been triggered, its propagation velocity depends on the diameter of the M axon, which varies in different species of fish and in animals of the same species but different sizes. The conduction velocity is about 85 m/sec in a 10-cm goldfish at 19–23 °C (Funch *et al.*, 1981). In the newly hatched zebrafish larva, 3 mm long, the velocity is only about 2 m/sec (Eaton and Farley, 1975). In either of these cases, though, the spike would conduct to the end of the spinal cord in about 1 msec and would thus depolarize the entire axon almost simultaneously. This rapid conduction probably helps synchronize the massive C-like muscle contraction of the body during the first stage of the fast-start. A curious aspect in goldfish is that M-axon diameter, and hence conduction velocity, does not increase with size of the fish (Funch *et al.*, 1981). In a 10-cm goldfish, the M-axon diameter is about 65 μm, whereas in 14-cm fish it is about 45 μm. The corresponding decrease in conduction velocity is 35%.

9. Inhibitory Actions on the M Cell

From the pioneering work of Furukawa and colleagues (review, Faber and Korn, 1978; Bennett, 1977), there is evidence for three kinds of synaptic inhibition on the M cell: electrical, chemical, and possibly presynaptic. All known types of M-cell inhibition at the cranial level are probably mediated by the same collateral network that involves relay and inhibitory neurons (Hackett and Faber, 1983*b*). A summary diagram of the inhibitory network is provided in Figure 14. This shows that very similar to the crayfish lateral giants (see Wine and Krasne, 1982), the M cell is inhibited at many possible sites.

9.1. Electrical Inhibition within the Axon Cap

One of the most characteristic functional features of the M cell is the short latency electrical inhibition that occurs in both cells after either has fired an action potential. The functional significance of this electrical inhibition seems to be to prevent, at short latency, secondary activation of either cell following the first M spike. This would inhibit a conflicting command from being initiated during an inappropriate phase of the response.

This field-effect inhibition is produced by afferents within the axon cap of the M cell. It is distinguishable from IPSPs in the cell because it occurs earlier, and it is manifested as an extracellular positivity not readily detected intracellularly. Thus, the net transmembrane action was one of the hyperpolarization (hence the name, extrinsic hyperpolarizing potential, EHP) that could block action potential generation as shown in Figure 15A. The electrical inhibition is readily obtained for experimental purposes following antidromic activation of the M cell. The EHP has several characteristics (Furukawa and Furshpan, 1963).

1. The amplitude has a less steep spatial gradient than the antidromic impulse and is fairly uniform throughout the axon cap
2. The response fatigues at 2-to 5-Hz stimulation frequency

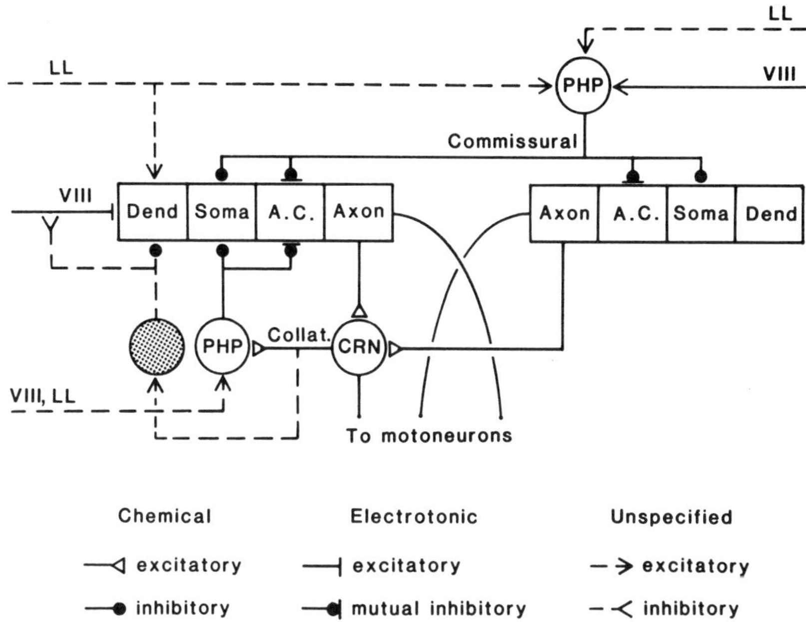

Figure 14. Summary diagram of the inhibitory network of the M-cell system at the cranial level. For simplicity, some of the known excitatory connections are not shown. The inhibitory regions of the M cells are shown as four-compartment boxes dividing them into dendrite (Dend), Soma, Axon Cap (A.C.), and Axon. Identified populations of neurons include commissural PHP cells; cranial relay neurons (CRN); and collateral PHP cells activated by CRNs. An unidentified population of inhibitory interneurons is shown as a strippled circle. (——) Monosynaptic pathways, (–––) established but not necessarily monosynaptic.

3. The EHP can be elicited in the absence of an antidromic response in the M cell
4. The EHP can block antidromic impulses from invading the axon hillock of the M cell (Figure 15A3). This action can also be mimicked by an anodic current pulse through a microelectrode positioned in the axon cap (Furukawa and Furshpan, 1963).

Furukawa and Furshpan's (1963) hypothesis for the generation of the EHP involves neurons, now called PHP cells, which were subsequently discovered by Faber and Korn (1973). The PHP cells are activated by the recurrent collateral network. These neurons generate a positive extracellular field potential by the failure of action potentials to invade their terminals within the axon cap region. In fact, a single impulse in a PHP cell causes a miniature EHP recorded in the cap (Faber and Korn, 1978).

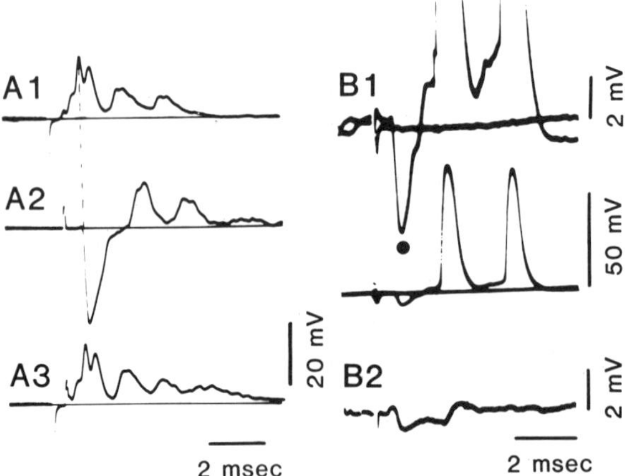

Figure 15. (A) Inhibitory effect of the EHP on the antidromic M spike. Traces are extracellular recordings made within the axon cap. (A1) EHP evoked by an electrical stimulus to the contralateral eighth nerve. (A2) Antidromic M spike evoked by an electrical stimulus to the spinal cord. (A3) When the antidromic M spike and peak of the EHP were timed to coincide, the antidromic M spike was blocked. (B) PHP evoked by spinal stimulation. (B1) High gain (upper trace) and low gain (lower trace) of an intracellular recording from a PHP neuron showing an all-or-nothing PHP (dot), followed by an EPSP and two spikes. (B2) Extracellular recording from the same PHP neuron showing a small negative field potential in this immediate vicinity. Subtraction of the intracellular and extracellular potentials yields a transmembrane hyperpolarization of 4.6 mV during the PHP. As the difference in timing between the M spike and PHP was never larger than 120 μsec, this interaction is not chemically mediated [(A) from Furukawa and Furshpan, 1963; (B) from Korn and Faber, 1975].

This positive-current flow does not return entirely through the extracellular spaces of the axon cap because the surrounding glia provide a high-resistance barrier to extracellular current. Instead, a significant fraction of the current flows across the M-axon initial segment and out through the M-cell soma membrane. The extracellular positivity results in a net transmembrane hyperpolarization of the M-cell membrane at the most critical site for M-cell action potential generation, the axon initial segment. No M-cell conductance increase is required for its effect. As the inhibitory mechanism does not involve low-resistance junctions, it differs from electrotonic synaptic transmission. A similar field-effect inhibition has been shown in the rat cerebellum (Korn and Axelrad, 1980), so this is not a mechanism unique to the M cell.

9.2. Chemically Mediated Collateral Inhibition of the M-Cell Soma

The collateral IPSP begins about 1 msec later than the beginning of the EHP and slightly overlaps with it. An experiment illustrating the collateral IPSP is shown in Figure 16. The IPSP is due to a chloride conductance increase (Furukawa and Furshpan, 1963). In the case of Figure 16, the IPSP is revealed as a depolarization because the M cell was pressure injected with KCl that caused a positive shift in the chloride equilibrium potential. As shown here, the associated conductance change lasts more than 45 msec, and probably serves to prevent secondary firing of the M cell after the initial M spike.

The IPSPs are susceptible to fatigue, as is the EHP, a factor that led Furukawa and Furshpan (1963) to suggest that both synaptic mechanisms were mediated by common neurons of the collateral network. As the IPSPs are blocked by strychnine, glycine is the putative neurotransmitter. However, this is still debated (Faber and Korn, 1978), because the IPSPs

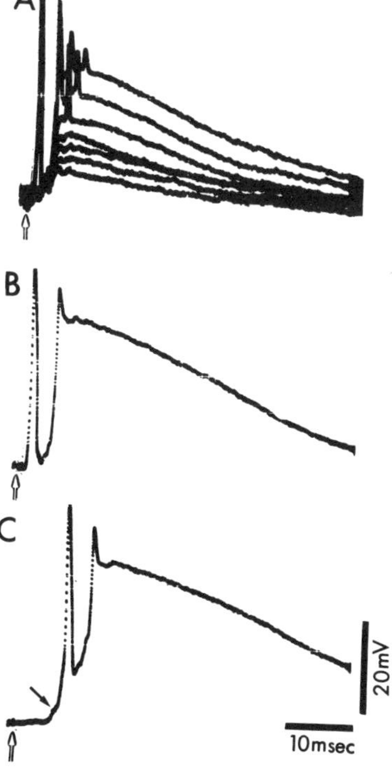

Figure 16. Mauthner cell inhibitory postsynaptic potentials. (A) Seven superimposed traces separated by 10 sec following pressure injection (250 msec, 10 psi) of KCl from the recording electrode that was located in the soma. The first spike was evoked by stimulation of the M axon in the spinal cord (hollow arrow indicates the beginning of the stimulus artifact). After the antidromic spike, the slow response is the collateral IPSP, which is inverted by the KCl injection. The largest response occurred 10 sec after the KCl injection. The depolarizing IPSP in the three largest responses crossed threshold and produced action potentials. The amplitudes of these spikes were smaller than the antidromic response because of the large conductance increase during the IPSP. (B) A response evoked by a single antidromic stimulus during another trial similar to (A), but with a steady leak of KCl from the pipette. (C) Click-evoked response. Note the similarity to (B), but the first spike arose from a postsynaptic potential (solid arrow). Before the KCl injection the click was not effective in eliciting a M-spike (from Hackett, unpublished).

can be blocked by injected strychnine without affecting the response to iontophoretically applied glycine (Diamond *et al.*, 1973). This difference in sensitivity to strychnine may be due to a presynaptic action that results in a decrease in glycine release from the inhibitory neuron terminals, and it may also involve a regional difference in the postsynaptic sensitivity to applied strychnine.

9.3. Neurons Responsible for Collateral Inhibition

The electrical and chemical inhibitions acting on the goldfish M cell are produced by a pathway that is probably disynaptic: The M cell first excites a pool of relay neurons that in turn activate inhibitory neurons, the PHP cells, with a mean delay of 0.8 msec. The relay neurons also excite cranial motoneurons, discussed in Section 10.1. The experiment in Figure 17 illustrates the role of the relay neurons (CRN) in the collateral inhibition of the M cell. An action potential in only one relay neuron can elicit a sizable EHP recorded in the axon cap of the M cell (Figure 17B). In addition, a single relay neuron impulse is capable of eliciting about one third of the inhibition caused by the entire collateral inhibitory network of PHP cells activated by the M axon (Figure 17C). This suggests that only a few relay neurons are required to inhibit the M cell.

The PHP neurons innervate the M-cell soma and the cap dendrites that extend from near the axon hillock and penetrate the axon cap (Figure 7B) (Korn and Faber, 1975*a*; Faber and Korn, 1978). An impulse in a single PHP neuron produces both electrical and chemical inhibition (Korn and Faber, 1976). A distinguishing feature of these cells is that they are electrically inhibited by the M cell. This is illustrated in Figure 15B. M-cell activation causes a PHP recorded from their axons (Faber and Korn, 1973). The PHP is a net transmembrane hyperpolarization with a latency and time course that are the same as the antidromic field potential of the M spike recorded in the axon cap. Furthermore, the M-cell antidromic field potential and the PHP have the same threshold and all-or-nothing character. Thus, the PHP is produced by M-cell action currents. Finally, the PHP was found to block impulse initiation in the PHP cell and therefore was inhibitory in function. The PHP may be important in synchronizing this pool of interneurons (Faber and Korn, 1978).

The PHP neurons have been marked in numerous studies with intracellular dyes for morphological identification (Korn *et al.*, 1978; Triller and Korn, 1978, 1982*a,b*; Zottoli and Faber, 1980; Korn and Faber, 1975*a*). The somas of these neurons are bipolar, round, or ovoid-shaped and clustered within an apparent nucleus below the M-cell soma, on each side of its ventral dendrite, at about half the distance between the fourth

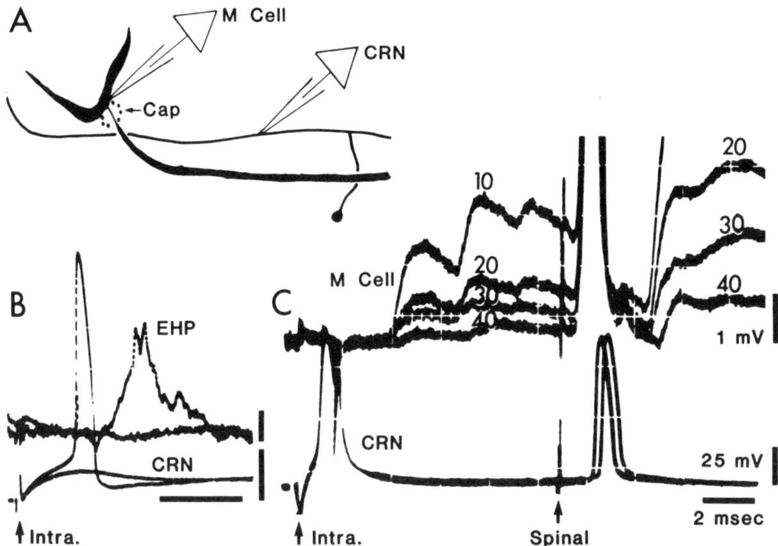

Figure 17. Inhibition evoked by the cranial relay neurons. (A) Schematic diagram of the circuit and location of the electrodes. One microelectrode penetrated the longitudinal axon of a CRN and was used for recording and stimulating. A second microelectrode either recorded from the axon cap (Cap) or the M-cell body. (B) Extracellular recording from the Mauthner axon cap. Activation of the CRN evoked, at a latency of 0.7 msec, a positive extracellular response in the axon cap. This was comparable in every way to the electrically mediated inhibition (EHP) reported by others: It is of shorter latency than the chemical collateral inhibition; only a small fraction of the response is recorded intracellularly; and the effect could be mimicked by anodal current (not shown) injected into the axon cap. (C) CRN-evoked inhibition compared with that evoked by the full collateral network following an antidromic M-cell activation by a spinal stimulus. Intracellular stimulation (Intra.) evoked a CRN spike (bottom traces) and in turn produced an M-cell IPSP (upper traces). The M cell was pressure injected with KCl from the recording microelectrode 10 sec before the first response was evoked and three sweeps were taken at subsequent 10 sec intervals. The amplitude of the CRN IPSPs is about one third that evoked by the entire collateral network when the M cell was activated by the spinal stimulus (Spinal). This can be seen by comparing the IPSPs with the same sweep number, before and after the spinal stimulus. Thus, the CRN and collateral IPSPs are equally dependent on KCl injections and evidently have the same inhibitory mechanism. The multiple peaked IPSPs in the M cell recording are most likely due to repetitive firing of the inhibitiory neurons postsynaptic to the CRN (from Hackett and Faber, 1983*b*).

ventricle and the ventral surface of the brainstem. There are two terminal fields: one within the M-cell axon cap and another outside this region. The dendrites extend ventrally as far as the border of the brainstem to the vicinity of the abducens nucleus.

Of particular interest in these studies has been the exploration for synaptic contacts correlated with the multiple actions of the neurons. The stained unmyelinated club endings have been traced to Gray type 2 synapses that contacted the soma and axon hillock within the axon cap region. As these endings in the cap region produced electrical inhibition and also end with chemical synapses, they have a dual function. Those endings that passed the axon cap impinged on the soma and initial portion of its main dendrites and were classified as small vesicle boutons. These endings presumably act to inhibit the M cell by chemical synaptic transmission.

In addition to inhibition produced by the collateral network, studies of the PHP cells have disclosed that there is at least a second type of these neurons. The terminal contacts established by these neurons appeared to be no different from those of the collateral cells. However, their somas were located medial and slightly dorsal to the lateral vestibular nucleus (Zottoli and Faber, 1980). These PHP cells project to both M cells as well as to vestibular and reticular complexes where they may also be inhibitory. Since they can be activated by eighth nerve volleys and assuming all of their terminals are inhibitory, they probably mediate commissural inhibition of the M cell (Furukawa and Furshpan, 1963; Faber and Korn, 1978). Triller and Korn (1982) have suggested that the commissural PHP cells prevent the firing of other neurons projecting down the spinal cord. As eighth nerve volleys may first activate PHP cells, a means is provided for modulating afferent input and subsequent firing of the ipsilateral M cell (Faber and Korn, 1978).

9.4. Dendritic Location of Presynaptic and Postsynaptic Inhibition

Besides the electrical and the chemical postsynaptic inhibition of the M axon and soma, respectively, there is evidence for a third type of inhibition (Furukawa *et al.*, 1963) that is mediated either presynaptically or by remote dendritic inputs. However, it may be quite difficult to distinguish presynaptic inhibition from postsynaptic inhibition occurring at remote dendritic sites. Diamond (1968) has shown that the response to iontophoretically applied putative inhibitory transmitter substances to the dendrites of the M cell may not be recorded at the soma because of decrement of the responses. Diamond thereby established the conceptual basis of dendritic remote inhibition as it is presently understood for this and other central neurons.

10. Output Circuitry of the M Cell

Following M cell activation, the motor response is widespread. Simultaneous with the unilateral bend of the body during the C start, there is bilateral and symmetrical contraction of the extraocular muscles, operculular, and mandibular muscles. Presumably, these cranial components of the response help streamline the head of the fish by closing the mouth and retracting the eyes. Also, with the rapid reduction in the mouth cavity, water is expelled out the gill slits and may have a propulsive or stabilizing action. In addition to these cranial components, the fins may be adducted at the same latency as the body bend. Figure 18 is a diagram of the circuitry as it is presently understood.

10.1. Cranial and Pectoral Fin Components

The cranial and pectoral fin components of the fast-start have been found to be mediated through the relay neurons interposed between the

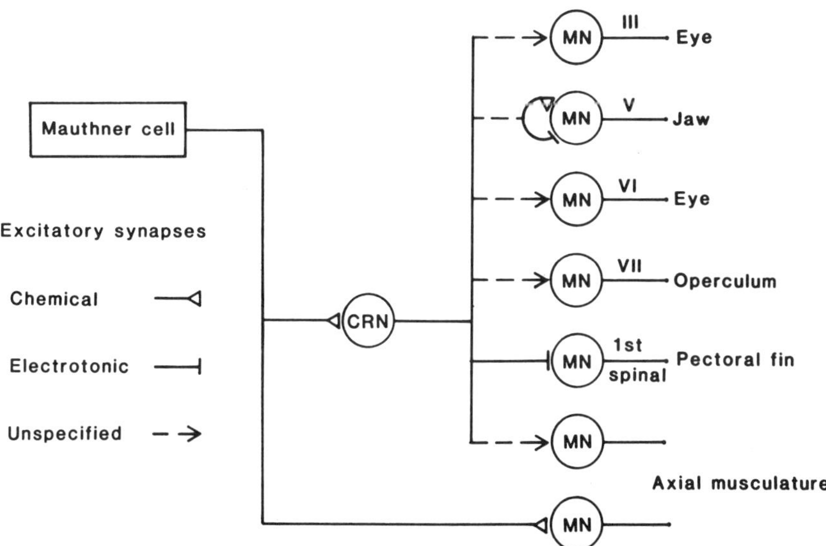

Figure 18. Summary diagram of the output network of the M-cell system derived from the goldfish and hatchetfish. Conventions as per Figure 14. The M cell (MC) is shown activating both a relay neuron (CRN) and motoneurons (MN). Relay neurons are responsible for bilateral activation of the motoneurons causing some muscle contractions associated with the first stage of the C start. These include the operculum and extraocular eye muscles that are innervated by the fifth, seventh, and third and sixth cranial motoneurons, respectively. Only the relay neuron output to the pectoral fins in the hatchetfish has been studied in detail (Auerbach and Bennett, 1969*a,b*). In a recent unpublished study, Hackett showed that at the connection between the relay neuron and fifth motoneurons, chemical and electrotonic transmissions coexist. The output of the M cell to the axial musculature is unilateral and to the side opposite the M-cell soma.

M cell and motoneurons. A similar arrangement has also been found in the crayfish lateral giant fiber output and the *Drosophila* giant fiber system (see Chapters 5 and 7, this volume). In fish, individual relay neurons are postsynaptic to both M axons, but their single impulses evoke activity in the muscles of only one side. An impulse in either M axon is sufficient for activation of the relay neuron. As shown in Figure 19A, hyperpolarization of a relay neuron sufficient to block spike initiation generally did not influence the muscular response, even though a single impulse in the neuron was sufficient to generate a maximum mandibular electromyogram (Figure 19B). Repetitive stimulation at increasingly higher frequencies revealed at least two components of the mandibular response, an indication of different motoneuron pools. A few neurons were found that innervated either opercular or extraocular muscles. Thus, they probably innervated groups of cranial muscles in a segmental manner.

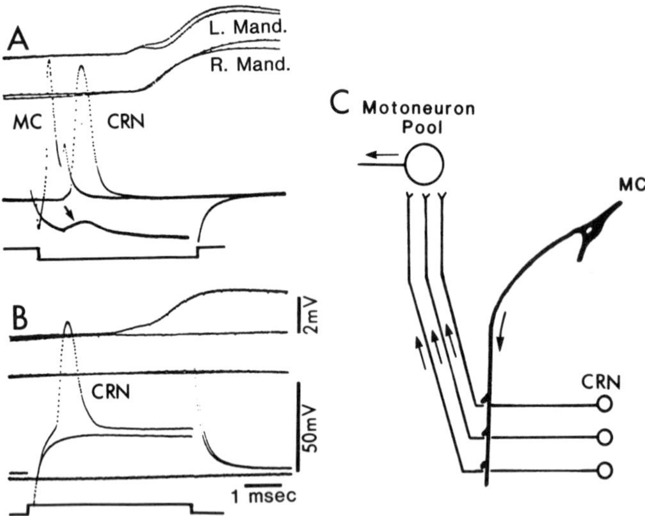

Figure 19. The electrophysiological characteristics of cranial relay neurons (CRN). CRNs were routinely penetrated within 50 μm of the M axon. (A) Intracellular activation of the Mauthner cell (MC) axon impulse evoked an action potential in CRN and produced EMGs in the left and right mandibular muscles (upper traces). Hyperpolarization of the CRN revealed the underlying EPSP (arrow). Note that the EMGs are only slightly altered, if at all, in the absence of the CRN spike. Thus, this CRN impulse is not necessary for the muscular response. (B) A CRN spike is sufficient for eliciting the EMG in the left mandibular muscle, only. Therefore, several CRNs are connected in parallel each with an output that is unilateral. (C) Diagram showing the divergence from one M axon to several CRNs that in turn innervate a motoneuron pool (from Hackett and Faber, 1983*a*; reprinted with permission from Pergamon Press, Ltd.).

Horseradish peroxidase (HRP) staining techniques have been used to determine the location of the relay and motoneurons that mediate the cranial components of the fast-start. Retrograde transport of HRP was used to determine the location of the motoneurons. The M axon and its postsynaptic neurons were injected with HRP from intracellular micropipettes. The relay neuron cell bodies are contralateral to the motoneurons they innervate. Their axons appear to make contact with the ipsilateral M axon. They then cross the midline to reach the opposite M axon, with which they also apparently establish contact. The crossed process bifurcates and runs in both directions. The terminals branching from this rostral process innervate cranial motor nuclei. Both the spinal and cranial relay cells have dendritic processes so they can probably be activated independently of the M cell (Model *et al.*, 1972; Hackett and Faber, 1983*a*). In the hatchetfish such relay neurons innervate the pectoral fin motoneurons (Auerbach and Bennett, 1969*a*). There is evidence for electrotonic synaptic transmission between these relay neurons and the motoneurons (Auerbach and Bennett, 1969*b*). Kimmel and colleagues (1982*b*) have identified six pairs of these neurons in the zebrafish larva.

10.2. Axial Musculature

There are several lines of evidence that suggest that the M axons do not synapse directly on all the motoneurons involved in the initial body bend of the C start. Diamond's (1971) study of the spinal components of the circuit is still the most extensive, but was done before the advent of intracellular micromarking. Therefore, the units he recorded were unidentified. Diamond recorded from three types of units postsynaptic to the M axon in the spinal cord. Two of these were probably motoneurons, whereas the third type may have been either an electrotonically isolated part of a motoneuron, or a relay cell. This latter possibility is supported by a fortuitous recording of Hackett and Faber (1983*a*) in which they penetrated such a relay cell. A directly activated spike in this neuron elicited a trunk EMG, but as with the cranial relay neurons, a hyperpolarizing pulse did not completely block the contraction following intracellular M-cell stimulation. This indicates that there are several relay neurons in the pathway. Furthermore, the latency was the same from an action potential in the M axon to the mandibular muscle EMG or to the trunk EMG. These results argue that relay neurons may play a major role in activation of the trunk musculature. Morphological studies also support this contention. Light and electron microscopic studies showed that M axons have connections not only with motoneurons, but also with intercalary cells (Celio *et al.*, 1979).

10.3. Crossed Spinal Inhibition

An impulse in one M axon will completely inhibit motoneuron activation evoked by the other M axon for a period of about 15 msec. Again, the effect of this inhibition is presumably to prevent an inappropriate secondary activation of the motoneurons during the first stage of the behavioral response. The inhibition recorded intracellularly from the axial motoneurons begins slightly later or at the same time as the monosynaptic excitation of the same units. Following this period of absolute suppression, there is a period of partial inhibition of the motor response that may last as long as 50 msec. Yasargil (see Faber and Korn, 1978) has hypothesized that the M axon makes electrotonic synapses with interneurons that cross the midline to chemically inhibit motoneurons on the opposite side of the body. Electron micrographs (Celio *et al.*, 1979) have revealed gap junctions between the M-axon collaterals and neurons that presumably mediate electrotonic interactions and the crossed spinal inhibition.

10.4. Inhibitory Control of Afferents

There is evidence that the M cell, as with the crayfish lateral giants, controls its own afferent pathways (see Chapter 7, this volume). This must be an important feature of the system to prevent reafference due to the rapid acceleration during the C start. Furukawa (1966) showed that excitation of the M axon elicited inhibition of sound-evoked action potentials recorded intracellularly from the eighth nerve. This was interpreted as a suppression at the level of the first-order sensory receptors mediated by efferent fibers terminating on the hair cells. Similarly, compound efferent potentials recorded from the posterior lateral line nerve were evoked by antidromic stimulation or intracellular stimulation of the M cell (Russell, 1976). The responses recorded in the lateral line nerves were always elicited by vibrations caused by tapping on the recording bench, but they did not invariably respond to direct stimulation of the M cell. And when they occurred, the latencies of their responses were variable. Obviously, many neurons other than M cells are effective in eliciting the response of the lateral line nerve. Figure 20 illustrates three types of tentative efferents identified in HRP-stained brains by Metcalfe and Kimmel (1982) in zebrafish. The somata of two of these cells (ROLE and RELL) are located just dorsal to the M axon in the medulla and send axons out the eighth nerve and posterior lateral line nerve, respectively. The close position of the M axon (not shown in Figure 20) to these neurons is suggestive of a direct synaptic connection.

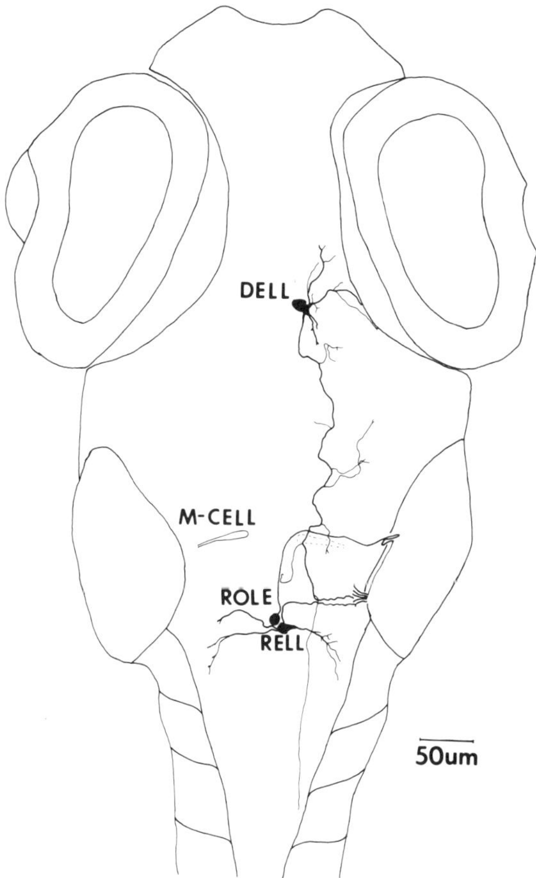

Figure 20. Diagram of three identified efferent neurons in the larval zebrafish brain. DELL and RELL are diencephalic and rhombencephalic efferents of the lateral line, respectively, and ROLE is a rhombencephalic octavolateral efferent neuron. These or similar cells may be responsible for command-derived inhibition that prevents reafference following the sudden movement caused by firing of the M cell. From Metcalfe, in preparation (see Metcalfe and Kimmel, 1982).

11. Fatigue in the M-Cell System and Response Changes Possibly Underlying Habituation

In general, synaptic connections in the M-cell system do not follow at high rates of stimulation. As Diamond (1971) pointed out, the M-cell system is not adapted for repetitive activation, but instead is intended for occasional use. Thus, the system has a high safety factor for one-time activation and a low safety factor for multiple activation within a short interval of time. Excitation of the spinal and cranial motoneurons, the collateral inhibition, and to a lesser extent the crossed spinal inhibition are all depressed with repeated stimulation at about 1/sec (Diamond, 1971). However, the PHP cell–M cell inhibitory synapses are depressed only 50% at stimulus rates of about 15–20/sec (Faber and Korn, 1982). Thus, the inhibitory synapses are fairly resistant to fatigue that rather occurs at an earlier stage of the recurrent pathway. This hypothesis was born out in experiments that show that 2/sec stimulation of the M axon to relay neuron synapse produces a level of fatigue that requires 20/sec stimulation at the relay neuron–PHP cell synapse (Hackett and Faber, 1983*b*).

Depression at the M cell–relay neuron synapse has been the subject of detailed studies by Bennett and colleagues (Auerbach and Bennett, 1969*a*; Highstein and Bennett, 1975). The depression was recorded as a decrease in the size of the relay neuron EPSPs. At high-frequency activation the EPSPs decreased in amplitude below that of spontaneously occurring miniature EPSPs. The suggested explanation for the depression was a presynaptic mechanism by which quantal size decreased locally at the site where there was evoked release of transmitter.

Recent evidence of Aljure and colleagues (1980) suggests that some of the "fatigue" of the M-cell network may be due to active processes possibly also involved in mediating habituation of the fast-start, previously demonstrated by Rodgers and colleagues (1963). Relay neuron responses readily follow M-cell stimulation at 1/sec, although the resulting pectoral muscle EMGs are progressively blocked at this rate. However, the EMGs are not blocked by direct stimulation of the motor nerve itself at 10/sec. Therefore, the depression of the muscle activity probably occurs at the motoneurons.

To test whether the motoneuron depression was due to synaptic fatigue or to an active process, Bennett and Day (1981) severed the anterior spinal cord. This, of course, separated the M-cell soma from its axon. Despite this, the M axon recovers and can conduct spikes (see also Eaton *et al.*, 1982). Following M-axon recovery, Bennett and Day found that the motoneuron depression was now absent. This confirms that the depression is due to descending, presumably inhibitory, control. It was suggested that the effect of this control would be to inhibit the fast-start at a level after the M cell already initiated its spike.

Experiments on zebrafish larvae suggest that the system can also be controlled at a level before the initiation of the M spike (Eaton *et al.*, 1977*a*). As shown in Figure 21, there appeared to be progressive failure

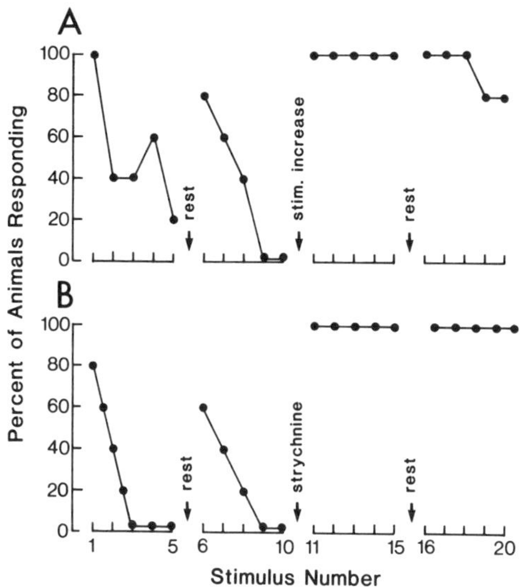

Figure 21. Course of habituation-like response decrements in the larval zebrafish responding to repetitive vibrational stimulation. Animals were given sets of five stimuli at 5-sec intervals, followed by a 6-min rest period. (A) The probability of obtaining a large, "all-or-nothing action potential" believed to be the M spike decreased with each succeeding stimulus in the series. During these experiments, there was an exact correspondence between the large spike and the presence of the startle response. (B) Response failure could be overcome by increasing stimulus intensity or by injecting the animal with strychnine. Control data not shown (from Eaton *et al.*, 1977*a*).

of M-cell activation to repeated presentation of a vibrational stimulus. As with habituation, responsiveness returned spontaneously when the preparation was allowed to rest, or if the stimulus intensity was increased. In these experiments the decrease in probability for M-cell activation was accompanied by a corresponding failure of the behavioral response to occur. There were no cases where the behavior was absent when the M cell fired. Thus, this response decrement must be occurring in the sensory pathway or in the M cell itself. The eighth nerve afferents fire without decrement to repeated vibrational stimulation in this preparation and are probably electrotonically coupled to the lateral dendrite of the M cell (Eaton and Farley, 1975; Eaton *et al.*, 1977*a*). Further, the M cell itself fires to repeated antidromic electrical stimulation of its axon. Thus, the most likely site underlying this response decrement is the M-cell soma. Also consistent with these findings in zebrafish is the fact that in goldfish, when the behavior fails to occur to a stimulus, the M spike is also absent

(Zottoli, 1977; Eaton *et al.*, 1981, 1982). In zebrafish, the decrement could be overcome by injections of strychnine, but not saline, thus suggesting an inhibitory mechanism. This possibility is in contrast to the monosynaptic mechanism demonstrated for habituation of the well-known gill withdrawal reflex in the mollusk, *Aplysia* (Kandel, 1976).

12. Non-Mauthner Fast-Start Circuits

Not all C starts require the M cells. The existence of non-Mauthner fast starts was first suggested by Diamond (1971) and later observed in physiological experiments on restrained zebrafish embryos by Eaton and colleagues (1977*a*). In the experiments on zebrafish, the animals were restrained and recordings were done with metal microelectrodes from the brain. Fast-starts were elicited by a vibrating probe at the body surface and were monitored when the animal's movements interrupted a beam of light impinging on a photocell. Some fast-starts that did not have an initial large spike were found to be identical in form to those that were accompanied by an initial spike thought to be from the M cell. The large spike correlates with the physical presence of the M cell in these animals (Eaton and Kimmel, 1980).

Two more recent studies have confirmed the existence of the non-Mauthner fast-starts. Kimmel and colleagues (1980) used ionizing radiation to cause developmental deletions of individual M cells in embryonic zebrafish. Eaton and associates (1982) used small electrolytic lesions to delete M cells from adult goldfish brains. In both studies, high-speed motion pictures were used to analyze the behavior. The results were that presence or absence of the M cells made no difference in the probability of whether the fast-starts could be elicited after either vibrational or visual-vibrational stimulation. Figure 22 gives a comparison between M-cell-initiated and non-Mauthner responses in the goldfish. This figure shows that the non-Mauthner responses (A, upper series of silhouettes) are sometimes indistinguishable from the M-cell-initiated responses (B), as both are well within the range of variability of known M-cell-initiated responses.

There were, however, some statistically significant differences in the behavioral response when the M cells were missing. But, these differences depended on the experiment. In the zebrafish, the mechanical performance was on the average 32% weaker when the animal gave a response opposite the side of the deleted M cell. There was no significant difference in latency between the Mauthner and non-Mauthner responses. In the goldfish, the mechanical performance, as shown in Figure 22, was remarkably similar whether or not the M cell was present on the side op-

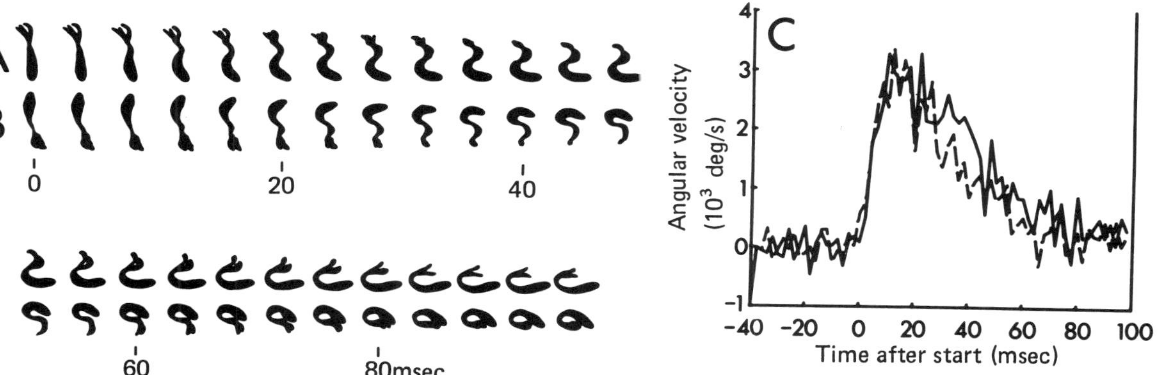

Figure 22. Comparison of M-cell-initiated and non-Mauthner fast-starts in goldfish in which one M cell was lesioned by an electrolytic current. The behavior was filmed at 500 frames/sec. (A, B) The upper response (A) is a non-Mauthner response, whereas the lower one (B) is a putative M-cell-initiated response. The first silhouette of each series of 2 msec before the movement began. Each subsequent image is at 4-msec intervals and was made in the same way as Figure 2. The latter response is called putative because the animal was not implanted with a chronic microelectrode near the intact M cell, and it cannot be proved that the response was necessarily preceded by an M spike. But, both these responses are well within the range of variability seen for known M-initiated responses. (C) Quantitative comparison of the angular velocity of M-cell-initiated (——) and non-Mauthner responses (–––). The curves were formed by calculating the mean angular velocity every 2 msec after the beginning of 13 M-cell-initiated and 16 non-Mauthner responses. Responses were synchronized as for the analysis in Figure 6C. A recording electrode was implanted near the healthy M cell to show that the responses were indeed accompanied by an initial M spike. Responses in this figure were all elicited by dropping a ball into the aquarium (from Eaton *et al.*, 1982).

posite the response. To determine this, performance parameters were studied at 2-msec intervals. Included in this analysis were comparisons of the peak angular velocity, peak displacement speed, time to peak angular velocity, position of the fish at various intervals after the beginning of the response, and so on. Figure 22C shows a comparison of the mean angular velocity at 2-msec intervals of 13 M-cell-initiated and 16 non-Mauthner responses. No movement parameters were found that were significantly different when the M cell was lesioned. However, in contrast to the zebrafish, in the goldfish the non-Mauthner responses were significantly longer in latency.

The difference in results between the studies on zebrafish and goldfish are probably due to differences both in the preparation itself and the techniques to delete the cell. In any case, to make the deletions in these studies, damage to other neurons was unavoidable. Thus, it is difficult to make firm statements about the meaning of the difference in performance between M-cell-initiated and non-Mauthner responses in these studies. Such differences could be either artifactual or represent real differences in the neurophysiology of the circuits initiating the behavior.

The most important point is that in both goldfish and zebrafish some non-Mauthner responses were identical in both latency and performance to those initiated by the M cell. It appears that the M cell predominates most of the time in the presence of an adequate stimulus for startle. But other circuits can also perform this function and, at least in the zebrafish embryo, do it some of the time under normal conditions in a manner identical to the M cell. The M-cell analogues already found by Kimmel and colleagues (1982*a*) in the zebrafish (Figure 10B) are good candidates for this function.

This new evidence on non-Mauthner fast-start circuits directly relates to the problem of deciding whether the M cell is a vertebrate command neuron (Eaton, 1983). Kupfermann and Weiss (1978) in their often-cited review, chose the M cell as being a putative command neuron. Such neurons are thought to be the critical decision point in the triggering of complex, stereotypic behavior patterns such as the teleost C start. From recording experiments such as those of Hackett and Faber (1983*a*), it might be decided that the M cell is a command neuron because it is necessary and sufficient for a defined part of the behavioral response. However, the existence of alternative fast-start circuits suggests that the M cell may not be a critical decision-making point for the initiation of a response elicited by a naturalistic stimulus. How the alternative pathways functionally relate to the M cell is obviously an important question in deciding whether the M cell acts as a single command element or is part of a system that makes a collective decision.

13. Discussion and Conclusions

The neural basis of fast-start behavior in fish and the role of the M cell were last reviewed by Eaton and Bombardieri in 1978. Since that time there have been considerable advances in understanding both the behavior itself and its neural basis. One reason for this continuing progress is that it is possible to use intracellular microelectrodes to simultaneously record not only from the M cell, but also other presynaptic or postsynaptic cells in the network. This enables a precise determination of the nature of the associated neuronal relationships. Furthermore, it is possible to record from the M cell in free-swimming animals. Thus, even though the M cell is a vertebrate neuron, it is as accessible as some giant fiber systems in invertebrates.

For the first time, the activity of the M cell can be shown to be causally related to the onset of the first stage of the C-type fast-start. Experiments using high-speed motion picture cameras have been coupled with simultaneous chronic recording from the M cell. These studies show that the M cell fires just before the onset of the first stage of the C-start pattern and that the M cell does not directly initiate the subsequent contractions during this behavior pattern (Eaton *et al.*, 1981).

In addition to these neuroethological studies, recordings from acute preparations show that the firing of the M cell is both necessary and sufficient for a large EMG occurring in the body musculature on the side opposite the activated M-cell soma (Hackett and Faber, 1983*a*). This EMG occurs at the same interval following the M spike as large EMGs recorded during fast-starts in free-swimming animals. Thus, a strong conclusion can be made that the firing of the M cell is normally the causal initiating event in the C-type fast-start. Depending on one's definition, the M cell could be described as a vertebrate command neuron (Eaton, 1983; Rock *et al.*, 1981).

Other important advances include physiological and behavioral studies on amphibians where it can also be shown that the firing of the M cell is causally related to the C-type fast-start (Rock *et al.*, 1981). In addition, the first intracellular recordings have now been done from the M cell in a noncyprinid teleost fish, the flounder (Zottoli, 1981). Such studies will probably be only the beginning for fruitful comparative analyses of M-cell function.

There is now a much fuller understanding of the M-cell network responsible for initiating the C start. But we are just beginning to appreciate how a single-action potential in one M cell results in a behavior pattern with a duration of 100 times longer than the initiating neuronal event. The PHP cells have been described in considerable detail (e.g.,

Triller and Korn, 1982*a,b*). Auerbach and Bennett (1969*a*) were the first to describe relay neurons between the M-cell axon and its motor neurons, specifically those mediating pectoral fin movements in hatchetfish. Recent work by Hackett and Faber (1983*a*) demonstrates a similar set of neurons that activate cranial and possibly trunk components of the C start of the goldfish. The cranial relay neurons were also found to be presynaptic to PHP cells mediating collateral and commissural inhibition of the M cell (Hackett and Faber, 1983*b*). Such relay cells might be a general feature in the outputs of not only the M-cell system, but also of functionally analogous systems in invertebrates (see Chapters 5 and 7, this volume).

Behavioral and physiological work demonstrates the existence of alternative neural circuits that can activate C-type fast-starts in the absence of the M cell (Kimmel *et al.*, 1980; Eaton *et al.*, 1982). In the case of the goldfish, the non-Mauthner and M-cell-initiated responses are remarkably similar in motor performance. However, the M-cell-initiated responses are significantly shorter in latency, although the meaning of this difference is not yet clear. Neuroanatomical work by Kimmel and associates (1982) shows the existence of several pairs of M-cell analogues that might be responsible for mediating fast-starts in the absence of the M cell.

The existence of the non-Mauthner circuits mediating fast-starts somewhat clouds the picture regarding whether a given behavior pattern can be described as M-cell-initiated in the absence of concomitant electrophysiological recordings (see Diamond, 1971; Eaton *et al.*, 1977*b*). It seems clear that caution is advised until we gain a better understanding of the role of the alternative circuits. Nevertheless, it seems quite likely that most short latency C-type fast-starts are probably initiated by the cell. The available evidence connecting the M cell to this behavior is quite strong, as described above. Added to this is the observation that in normal animals non-Mauthner responses have been seen only in embryonic zebrafish, and there only rarely (see Eaton and Bombardieri, 1978). Such responses have never been observed in adult goldfish: when the M cell is present it appears to initiate the response. However, the absence of non-Mauthner fast-starts in previous experiments may be due to the type of stimuli used or other unknown variables. Therefore, it seems most justified to describe these behavior patterns as simply C starts, or at best "Mauthner-type" responses.

In addition to these new findings on fast-start behavior and the M cell, the M cell has continued to serve as an important model for studies in neurodevelopment (Kimmel, 1982*b*) and synaptology (Faber and Korn, 1982). Eaton and Nissanov (1984) have recently reviewed the behavioral role of the M cell during embryonic and larval stages.

What are some of the important new advances that can now be seen on the horizon of research on fast-start behavior and its neural mechanisms in fish? Work already underway will extend our understanding of the behavioral role of the C-type fast-start in predator avoidance (Weihs and Webb, 1984; Webb, 1984). New knowledge of feeding mechanisms of fish predators should also help us better understand the necessary conditions for escape (Lauder and Liem, 1981). Such research could help considerably in comprehending the functional design of the M-cell system, particularly with regard to how the firing of the M cell is timed relative to a predatory attack.

Technological improvements in both behavioral analysis and recording can be expected to play an important part in the future. For example, recently Prugh and associates (1983) have devised a method for noninvasive recording of what are thought to be M spikes *outside* the body of the larval zebrafish, much like what can be done to record giant fiber responses in earthworms (see Chapter 3, this volume). In addition, Bennett and Day (1981) have succeeded in recording M spikes from electrodes placed in the caudal peduncle of the adult hatchetfish. Simplification of recording methods such as these would greatly extend the ability to perform meaningful neuroethological experiments on the M-cell system. Second, direct computerized analysis of fish movement patterns has already been achieved with solid-state "matrix" cameras (Wieland and Eaton, 1983). These cameras are based on arrays of photo detectors and currently can be operated at hundreds of frames per second. The image of the fish is automatically digitized by a computer that calculates performance parameters such as response latency, acceleration, trajectory taken, and so on. These devices overcome many of the difficulties when using movie films to achieve fine-time resolution of movement patterns.

Certainly, as stated earlier, more effort needs to be put into understanding sensory processing by the M cell in response to naturalistic stimuli that would normally activate the behavioral response. This information would be of considerable benefit in developing models of how the M-cell network processes information to "decide" to initiate a behavioral response. Many of the sensory inputs to the M cell are now simply described as "modulatory" because electrical stimulation of the input fails to bring the M cell to threshold in a restrained animal. Such negative results are problematic to interpret given that an electrical stimulus is likely to simultaneously activate afferents, efferents, excitatory, and inhibitory fibers in a way quite unlike a natural stimulus. Better understanding of the sensory inputs, coupled with an increased knowledge of the origins of the M-cell afferents, would greatly improve our ability to understand the modulation of the M-cell responsiveness. Carefully conducted tracer studies, such as with HRP, are desirable, since most previous work has been done with normal, silver-stained material (Zottoli, 1978).

ACKNOWLEDGMENTS. We would like to thank Drs. D. S. Faber, J. H. S. Blaxter, C. B. Kimmel, P. W. Webb, and C. M. Wieland for their comments on the manuscript. Support was provided by grants from the National Science Foundation (BNS81–12423) to R.C.E. and J.T.H. (BNS81–12742).

14. References

Aljure, E., Day, J. W., and Bennett, M. V. L., 1980, Postsynaptic depression of Mauthner cell-mediated startle reflex, a possible contributor to habituation, *Brain Res.* **188:**261–268.

Auerbach, A. A., and Bennett, M. V. L., 1969*a*, Chemically mediated transmission at a giant synapse in the central nervous system of a vertebrate, *J. Gen. Physiol.* **53:**183–210.

Auerbach, A. A., and Bennett, M. V. L., 1969*b*, A rectifying electrotonic synapse in the central nervous system of a vertebrate, *J. Gen Physiol.* **53:**211–237.

Barlow, G., 1968, Modal action patterns, in: *How Animals Communicate* (T. S. Seboek, ed.), Indiana University Press, Bloomington, pp. 98–134.

Bartelmez, G. M., 1915, Mauthner's cell and the nucleus motorius tegmenti, *J. Comp. Neurol.* **25:**87–128.

Bennett, M. V. L., 1977, Electrical transmission: A functional analysis and comparison to chemical transmission, in: *Handbook of Physiology,* Sec. 1, *The Nervous System,* (E. R. Kandel, ed.), American Physiological Society, Bethesda, Maryland, pp. 357–416.

Bennett, M. V. L., and Day, J. W., 1981, Mauthner fiber reflex: Descending activity mediates "habituation" of the response to Mauthner fiber stimulation, *Soc. Neurosci. Abstr.* **7:**843.

Blaxter, J. H. S., and Hoss, D. E., 1981, Startle response in herring: The effect of sound stimulus frequency, size of fish and selective interference with the acoustico-lateralis system, *J. Mar. Biol. Assoc. U. K.* **61:**871–879.

Blaxter, J. H. S., Gray, J. A. B., and Denton, E. J., 1981, Sound and startle responses in herring shoals, *J. Mar. Biol. Assoc. U. K.* **61:**851–869.

Bullock, T. H., 1981, Comparisons of the electric and acoustic sense and their central processing, in: *Hearing and Sound Communication in Fishes* (W. N. Tavolga, A. N. Popper, and R. R. Fay, eds.), Springer-Verlag, New York, pp. 525–570.

Celio, M. R., Gray, E. B., Yasargil, G. M., 1979, Ultrastructure of the Mauthner axon collateral and its synapses in the goldfish spinal cord, *J. Neurocytol.* **8:**19–30.

Cochran, S. L., Hackett, J. T., and Brown, D. L., 1980, The anuran Mauthner cell and its synaptic bed, *Neuroscience* **5:**1629–1646.

Diamond, J., 1968, The activation and distribution of GABA and l-glutamate receptors on goldfish Mauthner neurons; an analysis of dendritic remote inhibition, *J. Physiol. (London)* **194:**669–723.

Diamond, J., 1971, The Mauthner cell, in: *Fish Physiology,* Vol. 5 (W. S. Hoar and D. J. Randall, eds.), Academic Press, New York, pp. 265–346.

Diamond, J., Roper, S., Yasargil, G. M., 1973, The membrane effects, and sensitivity to strychnine of neural inhibition of the Mauthner cell, and its inhibition by glycine and GABA, *J. Physiol. (London)* **232:**87–111.

Dijkgraff, S., 1963, The functioning and significance of the lateral-line organs, *Biol. Rev.* **38:**51–105.

Dill, L. M., 1974, The escape response of the zebra danio (Brachydanio rerio). I. The stimulus for escape, *Anim. Behav.* **22:**711–722.

Eaton, R. C., 1983, Is the Mauthner cell a vertebrate command neuron? A neuroethological perspective on an evolving concept, in: *Advances in Vertebrate Neuroethology* (J.-P. Ewert, R. R. Capranica, and D. Ingle, eds.), Plenum Press, New York, pp. 629–636.

Eaton, R. C., and Bombardieri, R. A., 1978, Behavioral functions of the Mauthner neuron, in: *Neurobiology of the Mauthner Cell* (D. S. Faber and H. Korn, eds.), Raven Press, New York, pp. 221–244.

Eaton, R. C., and Farley, R. D., 1975, Mauthner neuron field potential in newly hatched larvae of the zebrafish, *J. Neurophysiol.* **38:**502–512.

Eaton, R. C., and Kimmel, C. B., 1980, Directional sensitivity of the Mauthner cell system to vibrational stimulation in zebrafish larvae, *J. Comp. Physiol.* **140:**337–342.

Eaton, R. C., and Nissanov, J., 1984, A review of Mauthner-initiated escape behavior and its possible role in hatching in the developing zebrafish, *Brachydanio rerio, Environ. Biol. Fish.* (in press).

Eaton, R. C., Farley, R. D., Kimmel, C. B., and Schabtach, E., 1977*a*, Functional development in the Mauthner cell system of embryos and larvae of the zebra fish, *J. Neurobiol.* **8:**151–172.

Eaton, R. C., Bombardieri, R. A., and Meyer, D., 1977*b*, The Mauthner initiated startle response in teleost fish, *J. Exp. Biol.* **66:**65–81.

Eaton, R. C., Lavender, W. A., and Wieland, C. M., 1981, Identification of Mauthner-initiated response patterns in goldfish: Evidence from simultaneous cinematography and electrophysiology, *J. Comp. Physiol.* **144:**521–531.

Eaton, R. C., Lavender, W. A., and Wieland, C. M., 1982, Alternative neural pathways initiate fast-start responses following lesions of the Mauthner neuron in goldfish, *J. Comp. Physiol.* **145:**485–496.

Faber, D. S., and Funch, P. G., 1980, Differential properties of orthodromic and antidromic impulse propagation across the Mauthner cell initial segment, *Brain Res.* **190:**255–260.

Faber, D. S., and Korn, H., 1973, A neuronal inhibition mediated electrically, *Science* **179:**577–578.

Faber, D. S., and Korn, H., 1975, Inputs from the posterior lateral line nerves upon the goldfish Mauthner cell. II. Evidence that the inhibitory components are mediated by interneurons of the recurrent collateral network, *Brain Res.* **96:**349–356.

Faber, D. S., and Korn, H., 1978, Electrophysiology of the Mauthner cell: Basic properties, synaptic mechanisms, and associated networks, in: *Neurobiology of the Mauthner cell* (D. S. Faber and H. Korn, eds.), Raven Press, New York, pp. 47–131.

Faber, D. S., and Korn, H., 1982, Transmission at a central inhibitory synapse. I. Magnitude of unitary postsynaptic conductance change and kinetics of channel activation, *J. Neurophysiol.* **48:**654–678.

Faber, D. S., Kaars, C., and Zottoli, S. J., 1980, Dual transmission at morphologically mixed synapses: Evidence from postsynaptic cobalt injections, *Neuroscience* **5:**433–440.

Fernald, R. D., 1975, Fast body turns in a cichlid fish, *Nature* **258:**228–229.

Fox, H., and Moulton, J. M., 1968, Mauthner cells and the thyroid hormonal level in larvae of *Rana temporaria, Arch. Anat. Microsc. Morphol. Exp.* **57:**107–119.

Funch, P. G., and Faber, D. S., 1982, Action-potential propagation and orthodromic impulse initiation in the Mauthner axon, *J. Neurophysiol.* **47:**1214–1231.

Funch, P. G., Kinsman, S. L., Faber, D. S., Koenig, E., and Zottoli, S. J., 1981, Mauthner axon diameter and impulse conduction velocity decreases with growth of goldfish, *Neurosci. Lett.* **27:**159–164.

Furshpan, E. J., 1964, Electrical transmission at an excitatory synapse in a vertebrate brain, *Science* **144:**878–880.
Furshpan, E. J., and Furukawa, T., 1962, Intracellular and extracellular responses of the several regions of the Mauthner cell of the goldfish, *J. Neurophysiol.* **25:**732–771.
Furukawa, T., 1966, Synaptic interaction at the Mauthner cell of the goldfish, *Prog. Brain Res.* **21A:**44–70.
Furukawa, T., and Furshpan, E. J., 1963, Two inhibitory mechanisms in the Mauthner neurons of goldfish, *J. Neurophysiol.* **26:**140–176.
Furukawa, T., and Ichii, Y., 1967, Neurophysiological studies on hearing in goldfish, *J. Neurophysiol.* **30:**1377–1403.
Furukawa, T., Fukami, Y., and Asada, Y., 1963, A third type of inhibition in the Mauthner cell of the goldfish, *Jpn. J. Physiol.* **14:**386–399.
Hackett, J. T., and Faber, D. S., 1983*a*, Mauthner axon networks mediating supraspinal components of the startle response, *Neuroscience* **8:**317–331.
Hackett, J. T., and Faber, D. S., 1983*b*, Relay neurons mediate collateral inhibition of the goldfish Mauthner cell, *Brain Res.* **264:**302–306.
Hackett, J. T., Cochran, S. L., and Brown, D. L., 1979, Functional properties of afferents which synapse on the Mauthner neuron in the amphibian tadpole, *Brain Res.* **176:**148–152.
Highstein, S. M., and Bennett, M. V. L., 1975, Fatigue and recovery of transmission at the Mauthner fiber and giant fiber synapse of the hatchetfish, *Brain Res.* **98:**229–242.
Kaars, C., and Faber, D. S., 1981, Myelenated central vertebrate axon lacks voltage-sensitive potassium conductance, *Science* **212:**1063–1065.
Kandel, E. R., 1976, *Cellular Basis of Behavior,* W. H. Freeman and Co., San Francisco.
Kimmel, C. B., 1982*a*, Reticulospinal and vestibulospinal neurons in the young larva of a teleost fish, *Brachydanio rerio, Prog. Brain Res.* **57:**1–24.
Kimmel, C. B., 1982*b*, Development of synapses on the Mauthner neuron, *Trends Neurosci.* **5:**47–50.
Kimmel, C. B., and Eaton, R. C., 1976, Development of the Mauthner cell, in: *Simpler Networks and Behavior* (J. C. Fentress, ed.), Sinauer Associates Publishers, Sunderland, Massachusetts, pp. 186–202.
Kimmel, C. B., Patterson, J., and Kimmel, R. O., 1974, The development and behavioral characteristics of the startle response in the zebrafish, *Dev. Psychobiol.* **7:**47–60.
Kimmel, C. B., Eaton, R. C., Powell, S. L., 1980, Decreased fast-start performance of zebrafish lacking Mauthner neurons, *J. Comp. Physiol.* **140:**343–350.
Kimmel, C. B., Sessions, S. K., and Kimmel, R. J., 1981, Morphogenesis and synaptogenesis of the zebrafish Mauthner neuron, *J. Comp. Neurol.* **198:**101–120.
Kimmel, C. B., Powell, S. L., and Metcalfe, W. K., 1982*a*, Brain neurons which project to the spinal cord in young larvae of the zebrafish, *J. Comp. Neurol.* **205:**112–127.
Kimmel, C. B., Metcalfe, W. K., Schabtach, E., 1982*b*, Reticular neurons with T-shaped axons in embryos of the zebrafish, *Soc. Neurosci. Abstr.* **8:**764.
Korn, H., and Axelrad, H., 1980, Electrical inhibition of Purkinje cells in the cerebellum of the rat, *Proc. Natl. Acad. Sci. U.S.A.* **77:**6244–6247.
Korn, H., and Faber, D. S., 1975*a*, An electrically mediated inhibition in goldfish medulla, *J. Neurophysiol.* **38:**452–471.
Korn, H., and Faber, D. S., 1975*b*, Inputs from the posterior lateral line nerves upon the goldfish Mauthner cell. I. Properties and synaptic localization of the excitatory component, *Brain Res.* **96:**342–348.
Korn, H., and Faber, D. S., 1976, Vertebrate central nervous system: Same neurons mediate both electrical and chemical inhibitions, *Science* **194:**1166–1169.
Korn, H., Faber, D. S., and Mariani, J., 1974, Existence de projections des nerfs posterieurs

de la ligne laterale sur la cellule de Mauthner; leur effet antagoniste sur l'activation de ce neurone par les afferences vestibulares, *C. R. Seances Acad. Sci. (D)* **279:**413–416.

Korn, H., Triller, A., and Faber, D. S., 1978, Structural correlates of recurrent collateral interneurons producing both electrical and chemical inhibitions of the Mauthner cell, *Proc. R. Soc. London, Ser. B. Biol. Sci.* **202:**533–539.

Kupfermann, I., and Weiss, K. R., 1978, The command neuron concept, *Behav. Brain Sci.* **1:**3–39.

Lauder, G. V., and Liem, K. F., 1981, Prey capture by *Luciocephalus pulcher:* Implications for models of jaw protrusion in teleost fishes, *Environ. Biol. Fish.* **6:**257–268.

Lawrence, D. G., and Kuypers, H. G. J. M., 1968, The functional organization of the motor system in the monkey. II. The effects of lesions of the descending brain-stem pathways, *Brain* **91:**1–36.

Lin, J.-W., Wood, M. R., and Faber, D. S., 1982, Saccular nerve input to the lateral dendrite of the goldfish Mauthner cell: A combined electrophysiological and morphological study, *Soc. Neurosci. Abstr.* **8:**764.

Metcalfe, W. K., and Kimmel, C. B., 1982, Three types of posterior lateral line efferent neurons in larval zebrafish, *Soc. Neurosci. Abstr.* **8:**763.

Model, P. G., Spira, M. E., and Bennett, M. V. L., 1972, Synaptic inputs to the cell bodies of the giant fibers of the hatchet fish, *Brain Res.* **45:**288–295.

Nakajima, Y., 1974, Fine structure of the synaptic endings on the Mauthner cell of the goldfish, *J. Comp. Neurol.* **156:**375–402.

Partridge, B. L., 1981, Lateral line function and the internal dynamics of fish schools, in: *Hearing and Sound Communication in Fishes* (W. N. Tavolga, A. N. Popper, and R. R. Fay, eds.), Springer-Verlag, New York, pp. 515–522.

Prugh, J. I. P., Kimmel, C. B., and Metcalfe, W. K., 1983, Noninvasive recording of the Mauthner neuron action potential in larval zebrafish, *J. Exp. Biol.* **101:**83–92.

Rand, D. M., and Lauder, G. V., 1981, Prey capture in the chain pickerel, *Esox niger:* Correlations between feeding and locomotor behavior, *Can. J. Zool.* **59:**1072–1078.

Rock, M. K., 1980, Functional properties of the Mauthner cell in the tadpole *Rana catesbeiana, J. Neurophysiol.* **44:**135–150.

Rock, M. K., Hackett, J. T., and Brown, D. L., 1981, Does the Mauthner cell conform to the criteria of the command neuron concept? *Brain Res.* **204:**21–27.

Rodgers, W. L., Melzack, R., and Segal, J. R., 1963, "Tail-flip" response in goldfish, *J. Comp. Physiol. Psychol.* **56:**917–923.

Rovainen, C. M., 1979, Neurobiology of lampreys, *Physiol. Rev.* **59:**1007–1077.

Russell, I. J., 1976, Central inhibition of lateral line input in the medulla of the goldfish by neurons which control active body movements, *J. Comp. Physiol.* **111:**335–358.

Sand, O., 1981, The lateral line and sound reception, in: *Hearing and Sound Communication in Fishes* (W. N. Tavolga, A. N. Popper, and R. R. Fay, eds.), Springer-Verlag, New York, pp. 459–480.

Schwartz, E., 1967, Analysis of surface-wave perception in some teleosts, in: *Lateral Line Detectors* (P. H. Cahn, ed.), Indiana University Press, Bloomington, pp. 123–134.

Stefanelli, A., 1951, The Mauthnerian apparatus in the Ichthyopsida: Its nature and function and correlated problems of neurohistogenesis, *Q. Rev. Biol.* **26:**17–34.

Stefanelli, A., 1980, I neuroni di Mauthner degli Ittiopside. Valutizioni comparative morfologiche e funzionali, *Atti Accad. Naz. Lincei Mem. Cl. Sci. Fis. Mat. Nat. Sez. III,* Series 8, **16:**1–45.

Thompson, R. F., and Spencer, W. A., 1966, Habituation: A model phenomenon for the study of neuronal substrates of behavior, *Psychol. Rev.* **73:**16–43.

Triller, A., and Korn, H., 1978, Mise en evidence electrophysiologique et anatomique de neurones vestibulaires inhibiteurs commissuraux chez la Tanche (*Tinca tinca*), *C. R. Seances Acad. Sci. Ser. D.* **286:**89–92.

Triller, A., and Korn, H., 1982*a*, Morphologically distinct classes of inhibitory synapses arise from the same neurons: Ultrastructural identification from crossed vestibular interneurons intracellularly stained with HRP, *J. Comp. Neurol.* **203:**131–155.

Triller, A., and Korn, H., 1982*b*, Transmission at a central inhibitory synapse. III. Ultrastructure of physiologically identified and stained terminals, *J. Neurophysiol.* **48:**708–736.

Vinyard, G. L., 1982, Variable kinematics of Sacramento perch (*Archoplites interruptus*) capturing evasing and nonevasive prey, *Can. J. Fish. Aquat. Sci.* **39:**209–211.

Webb, P. W., 1975, Acceleration performance of rainbow trout, *Salmo gairdneri,* and green sunfish, *Lepomis cyanellus, J. Exp. Biol.* **63:**451–465.

Webb, P. W., 1976, The effect of size on the fast-start performance of rainbow trout *Salmo gairdneri,* and a consideration of piscivorous predator-prey interactions, *J. Exp. Biol.* **65:**157–177.

Webb, P. W., 1978*a*, Temperature effects on acceleration of rainbow trout, *Salmo gairdneri, J. Fish. Res. Board Can.* **35:**1417–1422.

Webb, P. W., 1978*b*, Fast-start performance and body form in seven species of teleost fish, *J. Exp. Biol.* **74:**211–226.

Webb, P. W., 1980, Does schooling reduce fast-start response latencies in teleosts? *Comp. Biochem. Physiol. A. Comp. Physiol.* **65:**321–324.

Webb, P. W., 1981, Responses of northern anchovy, *Engraulis mordax,* larvae to predation by a biting planktivore, *Amphiprion percula, Fish. Bull.* **79:**727–735.

Webb, P. W., 1982, Fast-start resistance of trout, *J. Exp. Biol.* **96:**93–106.

Webb, P. W., and Skadsen, J. M., 1980, Strike tactics of *Esox, Can. J. Zool.* **58:**1462–1469.

Weihs, D., 1973, The mechanism of rapid starting of slender fish, *Biorheology* **10:**343–350.

Weihs, D., and Webb, P. W., 1984, Optimal avoidance and evasion tactics in predatory-prey interactions, *J. Theor. Biol.* (in press).

Wieland, C. M., and Eaton, R. C., 1983, An electronic cine camera system for the automatic collection and analysis of high-speed movement of unrestrained animals, *Behav. Res. Methods Instr.* **15:**437–440.

Wilson, D. M., 1959, Function of giant Mauthner's neurons in the lungfish, *Science* **29:**841–842.

Wine, J. J., and Krasne, F. B., 1982, The cellular organization of crayfish escape behavior, in: *The Biology of Crustacea,* Vol. 4 (D. C. Sandeman and H. L. Atwood, eds.), Academic Press, New York, pp. 241–292.

Wyman, R. L., and Ward, J. A., 1973, The development of behavior in the cichlid fish *Etroplus maculatus* (Bloch), *Z. Tierpsychol.* **33:**461–491.

Yasargil, G. M., Greeff, N. G., Luescher, H. R., Akert, K., and Sandri, C., 1982, The structural correlate of saltatory conduction along the Mauthner axon in the tench (*Tinca tinca* L.): Identification of nodal equivalents at the axon collaterals, *J. Comp. Neurol.* **212:**417–424.

Zottoli, S. J., 1977, Correlation of the startle reflex and Mauthner cell auditory responses in unrestrained goldfish, *J. Exp. Biol.* **66:**243–254.

Zottoli, S. J., 1978, Comparative morphology of the Mauthner cell in fish and amphibians, in: *Neurobiology of the Mauthner Cell* (D. S. Faber and H. Korn, eds.), Raven Press, New York, pp. 13–45.

Zottoli, S. J., 1981, Electrophysiological and morphological characterization of the winter flounder Mauthner cell, *J. Comp. Physiol.* **143:**541–553.

Zottoli, S. J., and Faber, D. S., 1979, Properties and distribution of anterior VIIIth nerve excitatory inputs to the goldfish Mauthner cell, *Brain Res.* **174:**319–323.

Zottoli, S. J., and Faber, D. S., 1980, An identified class of statoacoustic interneurons with bilateral projections in the goldfish medulla, *Neuroscience* **5:**1287–1302.

9

Methodological Factors in the Behavioral Analysis of Startle

The Use of Reflex Modification Procedures and the Assessment of Threshold

HOWARD S. HOFFMAN

1. Introduction

It is common knowledge that for most organisms and for most stimulus modalities, almost any intense signal can elicit the rapid sequence of skeletal reactions that collectively constitute the overt startle reflex. Not so well known is the fact that with these same organisms and these same signals, almost any other stimulus can, if presented at an appropriate interval prior to the startle-eliciting stimulus, either prevent the elicited reflex or, at least in mammals, insure that it will occur with reduced amplitude. Since this effect occurs the very first time that the lead stimulus is presented, it is clear that it is not an example of classical conditioning or any other form of learning for that matter. In order to convey the idea that this effect entails a change in an elicited reaction, it has been described by the term, reflex modification.

The purpose of the present chapter is to examine certain of the methodological factors that require consideration if one is to employ reflex modification in the analysis of startle. Because the measurement of threshold is affected by reflex modification, a consideration of threshold is also presented. Reflex modification of startle has been studied almost exclu-

HOWARD S. HOFFMAN • Department of Psychology, Bryn Mawr College, Bryn Mawr, Pennsylvania 19010.

sively in mammals. It is hoped that these comments will be of use to those interested in neural mechanisms of startle in other types of organisms.

It is a curious fact that reflex modification has been independently discovered at least three times. This is not the place to attempt a detailed account of this historical anomaly other than to note that reflex modification was first described by Sechenov in the mid-1800s (Sechenov, 1965), that it apparently was independently discovered by Yerkes in the early 1900s (Yerkes, 1905), and that in the 1930s, Ernest Hilgard (1933) and Helen Peak (1936) both published important papers dealing with the phenomenon. For reasons that remain to be fully documented, in the 30 or so years following Hilgard's and Peak's work, the concept of reflex modification simply disappeared from the literature of both teaching and research and it did not again emerge until the 1960s when we began to publish studies of the phenomenon (Hoffman and Flesher, 1963). At present it is not clear why, during this 30-year period, such a basic and ubiquitous process as reflex modification should have been overlooked. Perhaps it was because at that time the behavioral sciences had become preoccupied with classical conditioning and this may have diverted the attention of the field. Another possibility is that reflex modification was somehow confused with classical conditioning. It is easy to see how this might have happened since the two processes entail identical operations, e.g., presenting a given stimulus just prior to a second stimulus that elicits a reflex. Whatever the reason, however, it is clear that the 30-year hiatus of reflex modification from the attention of behavioral scientists was an unfortunate event that requires further analysis. At the least, its interpretation seems likely to form an interesting chapter in the history of the behavioral sciences. In the meantime it is hoped that the present discussion of methodology will prompt interest into the basic factors that are responsible for reflex modification effects and thereby afford this interesting phenomenon more of the scientific attention that it clearly deserves.

Any experiment that entails reflex modification requires that one make a comparison between a response that is evoked both with and without a prior reflex-modifying signal. For this reason, it is important that one have good experimental control over the conditions under which the response is elicited. In particular, one must be able to accurately specify the stimulus that will elicit the response as well as the signal that is intended to modify that response. Moreover, it is important to be able to accurately measure the several parameters (e.g., the latency and the amplitude) of the response that is elicited.

In the early studies of startle, the stimulus consisted of an intense burst of noise produced by some form of shock wave, e.g., a gunshot, the snap of a mousetrap, a heavy weight falling on a surface. These bursts are short-duration complex signals, characterized by an extremely rapid onset and it is consequently difficult to accurately specify their physical attributes. Moreover, because they are produced either explosively or mechanically they are difficult to control. The contemporary solution to this problem is to employ sophisticated electronic equipment to generate signals and to carry out one's experiments in an acoustically treated (preferably anechoic) chamber.

The problem of response detection is a product of the rapidity of the events that compose overt startle. In man the entire sequence is typically completed within 200–300 msec, and its initial component (an eyeblink) often appears within 20–40 msec depending on the intensity and modality of the eliciting stimulus (Landis and Hunt, 1939). In the rat, the sequence occupies approximately 100 msec and the initial components of the skeletal response are detected approximately 10 msec after stimulation is applied (Hoffman and Searle, 1968; Chapter 10, this volume). Obviously such fleeting events as startle reflexes are difficult to observe with the unaided eye and, indeed, it is only through the use of specialized instruments with good time resolution that the details emerge. Many workers have used motion picture equipment capable of recording several hundred to several thousand frames per second (Landis and Hunt, 1939; Kimmel *et al.*, 1974). Because of the complexities in making quantitative measurements from film, contemporary workers on mammals and birds make use of instruments especially designed to measure the ballistic-type skeletal movements that characterize the response. Recently, direct computerized image analysis has become available for analysis of startle behavior (see Chapter 8, this volume).

What follows is a description of the essential features of the several instruments that are used to study reflex modification in the author's laboratory. Although one might not require access to all of these features in a given study, their description in the present context is intended to alert the reader to the kinds of instrumental issues that have required resolution in prior work and that seem likely to arise in future studies.

2. A Device for the Assessment of Startle

Figure 1 shows the device that is used to assess startle in small organisms such as rats and pigeons. It consists of a rigid cage attached to a superstructure of heavy gauge laminated epoxy and fiberglass. The cage is constructed of stainless steel rods to permit good penetration by acoustic and/or visual signals. It is suspended from the superstructure by flexible plates of laminated fiberglass and epoxy. Because the suspension

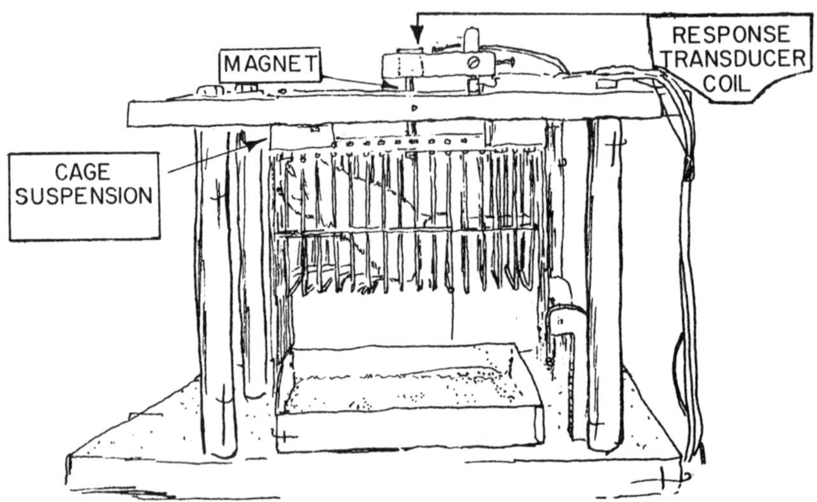

Figure 1. An apparatus for the assessment of startle in small organisms such as rats and pigeons (see text for details).

system is relatively stiff, oscillatory movements of the cage are heavily damped.

With this device, ballistic-type movements of the subject cause minute movements of its cage and these are transduced into an electrical signal. Firmly fixed to the top of the cage is an aluminum rod that has a magnet in its end. The magnet rides in a coil that is mounted rigidly on the superstructure in a manner permitting adjustment. Movements of the magnet in the coil produce a current that is proportional to the rate at which magnetic lines of flux cross the coil. For this reason, the device is highly sensitive to the sudden sharp movements involved in startle, but is relatively insensitive to the slow, although perhaps gross, movements involved in general activity. The output of the coil is passed through a bandpass filter to remove high-frequency artifacts resulting from the impact of intense acoustic signal on the apparatus. The output of the filter is subsequently amplified and passed to either a peak detector (the sample window of which is open for 100 msec following onset of the startle-eliciting stimulus) or to a high-speed analog recording device such as a storage oscilloscope.

This system is designed to respond linearly in proportion to the force of the subject's response. It is for this reason that a rigid cage suspension is used so that the travel of the magnet in the transducer coil is limited to a small range of uniform sensitivity. Stability and linearity are further assured by the use of materials unlikely to be affected by changes in humidity.

Figure 3 shows the traces from a dual channel storage oscilloscope during measurement of a rat's startle reaction to an intense burst of noise (top) and during measurement of a pigeon's startle reaction to an intense flash of light (bottom). The lower trace is the output from a transducer employed to detect the presentation of the startle-eliciting stimulus. The upper trace is the output from the coil of the response-detection unit. With both preparations, response latency is defined as the interval from the onset of the startle eliciting signal (point A) to the beginning of a deflection (point C) in the upper trace. In these circumstances, response amplitude is usually defined as the extent of the initial deflection of the response trace (e.g., the voltage difference between points C and D in the upper trace). Sometimes response amplitude is defined as the largest deflection within a fixed temporal interval that follows stimulus presentation, as for example when using a peak detector.

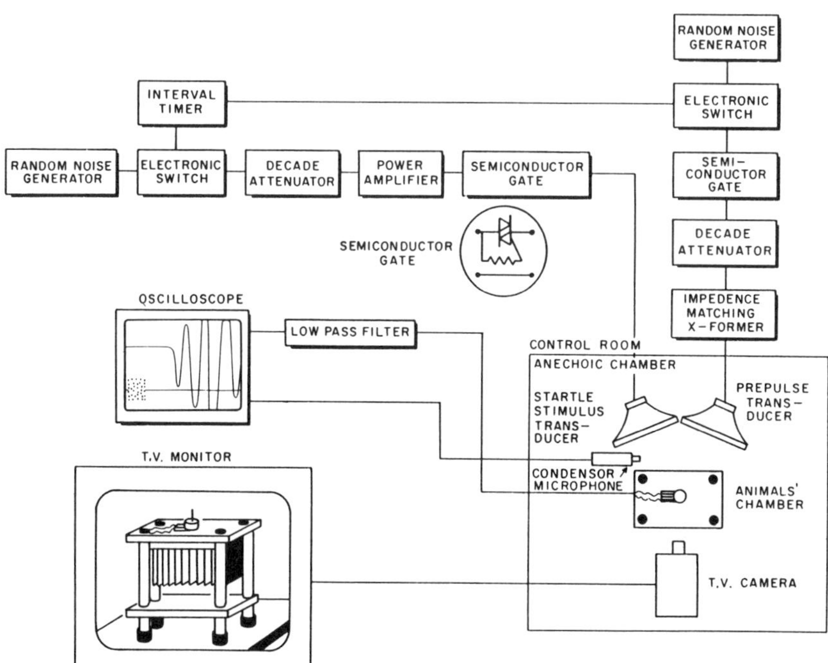

Figure 2. Schematic representation of the equipment for the production of stimuli and for the assessment of startle reactions.

2.1. Calibration of the Response Detection Unit

Calibration of this system is accomplished by dropping a 9.5-g lead slug from various heights onto a test anvil mounted on the top of the cage. Since the voltage generated by the transducer is proportional to velocity, the system's response is a direct function of the velocity of the test slug. This is calculated from the height of the test drop by the formula, velocity $= (2 \times g \times \text{distance})$, where g, the acceleration of gravity, is 908 cm/sec^2. In practice, the only source of "drift" in this system is the amplifier. Thus routine calibration checks are most conveniently performed by introducing a 70 Hz sine wave of known duration and amplitude to simulate the transducer output, and adjusting the gain of the amplifier as needed. Finally, it is relevant to note that because it is heavily damped, the sensitivity of this system is largely unaffected by minor variations in cage loading (200–400 g). This is important because it has been found that although rats exhibit reliable individual differences in the amplitude of their startle reactions, there is no significant correlation between body weight and the amplitude of startle to either acoustic signals (Brown *et al.*, 1951) or to electrical shocks (Hoffman *et al.*, 1964).

2.2. The Production of Acoustic Signals

In the author's laboratory acoustic startle stimuli are generated by feeding the output of a random noise generator (or an oscillator) to an electronic switch operated by an interval timer. This arrangement permits control over the duration of the signals as well as their rise-decay times. The amplitude control on the signal generator is used to set signal intensity. A final stage of amplification, provided by an audio power amplifier, is employed before feeding the signal through a semiconductor gate to a midrange driver with a 60-cm exponential horn. The semiconductor gate is a General Electric X12 Triac; this solid-state device appears as an open circuit until triggered and then permits current flow in both directions, a characteristic that allows the passage of AC signals. The triggering circuit is such that the appearance of a signal (the acoustical stimulus) will fire the Triac, but the unwanted internal noise of the equipment will not.

An auxiliary system is employed to produce the acoustic signals that precede the primary stimulus and that modify the reaction it elicits. Again, the output of the desired signal source is fed to an electronic switch that is under the control of an interval timer. In this case, however, stimulus intensity is controlled by a decade attenuator, the output of which is fed to an amplifier and from there to a second speaker mounted next to the subject's cage. Figure 2 illustrates the basic features of the experimental arrangement.

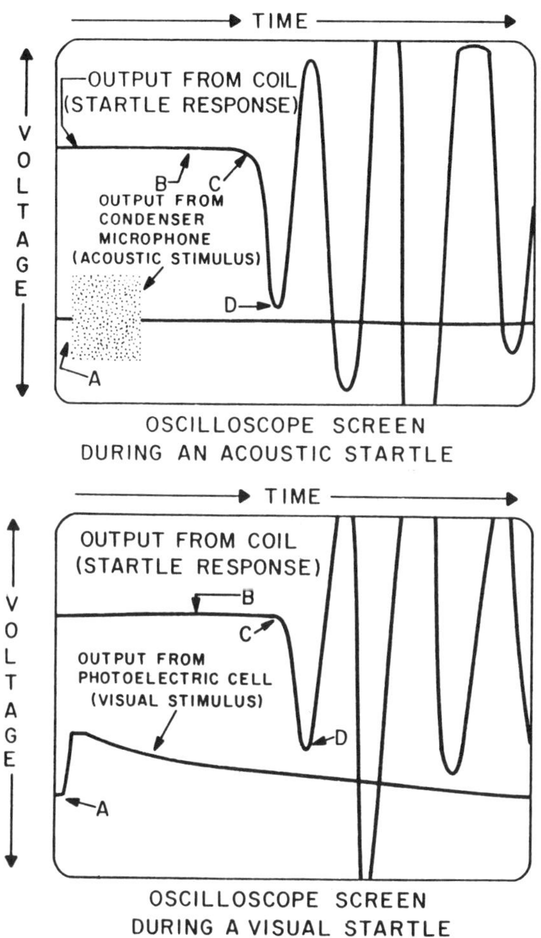

Figure 3. (Top) Oscilloscopic traces during the startle reaction of a rat to an intense sound and (Bottom) during the startle reaction of a pigeon to an intense flash of light (from Hoffman and Wible, 1970, *J. Acoust. Soc. Am.* **47**:490).

The effective part of the acoustic field emitted by a given transducer is that which penetrates the subject's cage; therefore, calibration of the acoustic signals requires that measurements be made within the cage. In assessing the intensity of a given signal, a calibrated condensor microphone is placed in the subject's cage and the desired signal is produced repeatedly. The output of the microphone is fed to a precision sound level meter and then displayed on one channel of a storage oscilloscope. The procedure is to take a large number of measurements, moving the microphone after each measurement until the entire space in the cage has been sampled. The intensity of the signal is defined as the mean of the resulting distribution of measurements.

2.3. The Production of Visual Signals

Although most mammals startle well to an acoustic signal, this author has yet to see a rat exhibit a startle reaction to a flash of light regardless of its intensity. The pigeon, on the other hand, startles well to a light flash, but rarely startles to an acoustic signal, regardless of its intensity. In view of these differences, it is of special interest that for both species, reflex modification effects are readily obtained when either an acoustic or visual signal is employed to modify the startle reaction to the appropriate reflex eliciting signal (Stitt *et al.*, 1976).

In the author's laboratory, startle-eliciting visual stimuli (used with pigeons) are produced by an electronic flash unit. To maintain these stimuli as pure visual events, it is crucial that they be presented silently. This poses a special problem because the flash unit produces a distinct "pop" when fired. In order to prevent subjects from hearing the pop, the flash unit is placed outside the acoustic chamber and is focused through its multilayered, sound-insulating window onto a translucent sheet attached to the side of the superstructure of the animal cage. A Fresnel lens mounted outside the acoustic chamber is used to ensure that the luminance of the translucent sheet will be uniform throughout its surface. The flash unit is operated by an electronic switch and other appropriate relay and electronic interfacing.

In those studies that require a lower intensity visual stimulus to precede a more intense startle-eliciting stimulus, a second electronic flash unit is employed. (It is not possible to use the same flash unit for both a visual reflex modifier and as a visual startle-eliciting stimulus due to its relatively long recharge time.) This second flash unit is also placed outside the acoustically treated chamber and is also focused with a Fresnel lens on the translucent sheet attached to the superstructure next to the animal's cage. With both units, stimulus intensity is adjusted through the use of neutral density filters placed between the flash unit and the Fresnel lens.

2.4. The Calibration of Visual Signals

The effective visual stimulus is the surface intensity of the subject's side of the translucent sheet next to its cage. For the flash stimuli employed in research into startle, direct luminance measurement is impossible due to the extremely short duration of the flash. Therefore, the luminance–time integral of the stimulus is determined by comparing photographic prints of the translucent sheet illuminated by a given flash stimulus with various exposure length photographic prints of the translucent sheet with a known surface intensity. The surface intensity can be determined by using a photometer to directly measure its brightness when it is illuminated by a 35-mm slide projector. Although this procedure is known to be fairly accurate, it should be noted that intensities derived in this manner must be considered as only best estimates due to spectral composition differences between the two light sources.

2.5. The Modification of Reflex Latency

When a signal that is too weak to produce measurable startle is presented, withdrawn, or otherwise changed just a few milliseconds before the presentation of an intense startle-eliciting signal, the latency of the reaction to the intense signal is often reduced. Usually the amount of reduction in latency approximates the lead time of the prior stimulus event, and response amplitude is neither reduced nor increased. Latency reduction has been reported for the rat's whole body startle reaction to an intense sound, using both visual and acoustical signals with lead times of 4 msec as reflex-modifying signals. It has also been found that the latency of the pigeon's whole body startle response to a flash of light can be reduced if either a weak flash of light or a weak tone is presented 4 msec prior to the intense flash. The same kind of latency reduction has been found when an otherwise continuously presented acoustic signal is terminated or its frequency components are changed 4 msec prior to the presentation of a startle-eliciting light flash in pigeons or to the presentation of a startle-eliciting acoustic signal in rats (Schwartz *et al.*, 1976; Stitt *et al.*, 1974; 1976).

2.6. The Modification of Reflex Amplitude

When a signal that is too weak to produce measurable startle is presented, withdrawn, or otherwise changed approximately 100 msec before an intense startle-eliciting stimulus, the amplitude of the reaction to the intense stimulus is reduced. The latency of that reaction, however, remains unchanged or, more often, is increased. In general, the amount of amplitude reduction is determined by the lead interval, by the intensity and nature of the stimulation employed, by the kind of change that occurs, and by the kind of organism tested. A common finding is that even with the most intense startle-eliciting stimulus, given an appropriate lead interval and a moderately intense reflex-modifying signal, the amplitude of the elicited response is no larger than 60% of the amplitude of the response when the startle-eliciting signal is presented alone. Given an appropriate lead interval, stimulus events that are at or near the thresholds for their detection produce measurable inhibition, and as the intensity of the reflex-modifying event increases, increasing amounts of inhibition are obtained. When a lead stimulus becomes sufficiently intense, one may begin to observe overt startle reactions to it, but the form of the function does not appear to change with this development.

The amplitude reduction effect is nicely illustrated by the results of an experiment (Hoffman and Searle, 1968) that was designed to determine what happens when an increasingly intense reflex-modifying stimulus is presented 100 msec prior to an intense startle-eliciting burst of noise. The subjects were nine, 200-day-old experimentally naive Wistar rats. The procedure involved the measurement of the amplitude of the startle response to each of two pulses of broad-band noise separated in time by 100 msec (measured from the onset of the first pulse to the onset of the second pulse). Both pulses had a rise-decay time of 2.5 msec and a duration of 20 msec at peak intensity. In a given stimulus configuration, the intensity of the second pulse (the main pulse) was always 140 dB SPL re: 0.0002 dyn/cm^2; the intensity of the first pulse (the prepulse), on the other hand, was set at one of several different levels: 50, 65, 80, 95, 110, 125, or 140 dB. Each subject was exposed to ten different series of the seven stimulus configurations. Trials occurred at approximately 20-sec intervals with the order of trials within each series determined by a random numbers table.

Figure 4 shows the mean amplitude (across subjects) of the initial deflections of the oscilloscope traces for the responses to both the prepulses and the main pulses. The several stimulus configurations are schematically illustrated below the graphs. With the data displayed in this fashion, one can evaluate the effects of a given prepulse on the reaction to the subsequent main pulse by using the amplitude of the responses to the 140-dB prepulse as a basis for comparison. The rationale derives from the fact that regardless of the prepulse intensity, the main pulse was always 140 dB. Consequently, the 140-dB prepulse was physically identical to the main pulse and the reactions to it are representative of responses to a main pulse alone; that is, a main pulse with no prior prepulse.

As seen in Figure 4, prepulses with intensities of 50, 65, and 80 dB partially inhibited the response to the main pulse, even though they themselves were too weak to yield measurable responses. As prepulse intensity

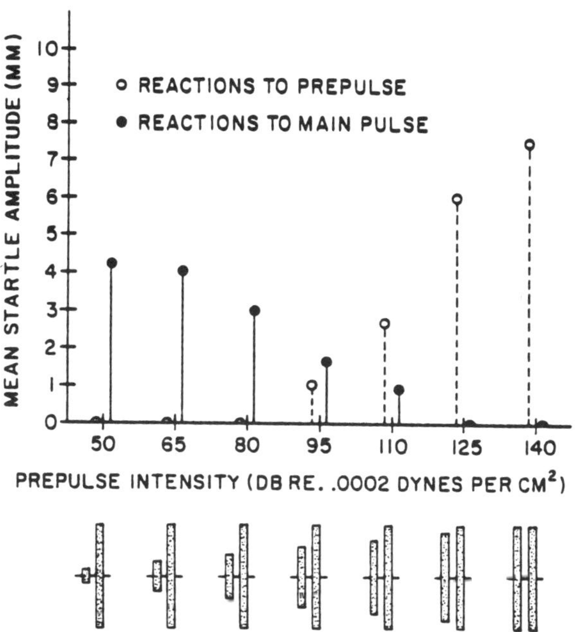

Figure 4. Mean amplitude of startle (across subjects) to each of two sequentially presented acoustic signals. The initial signal (the prepulse) led the second signal (the main pulse) by 100 msec. In all configurations main pulse intensity was 140 dB. Prepulse intensity is shown on the abscissa and the several stimulus configurations are schematically illustrated below the graph (from Hoffman and Searle, 1968, *J. Acoust. Soc. Am.* **43**:277).

increased, the inhibition effect increased, but at prepulse intensities of 95 dB and up, the prepulse itself tended to evoke a startle reaction.

2.7. General Procedural Considerations

In addition to illustrating certain details of the reflex modification effect, the previously described study illustrates the kind of approach that has proved most useful for the analysis of reflex modification. The usual procedure is to complete calibration of stimuli and of the startle-measuring device prior to each test session. The animal is then placed in the response-detection cage and allowed approximately 5 min to adapt to the situation. Once testing begins, stimuli are presented at random intervals with a mean of about 30 sec. Within these limitations, presentation of stimuli can either be under automatic control (by using solid-state programming modules interfaced with a microcomputer) or it can be controlled by the investigator. In most of our work, the experimental session contains approximately 50–80 stimulus presentations and therefore takes about 40 min or so to complete.

Because subjects vary in their tendency to startle, we have found it advantageous to try to compare control and treatment conditions within subjects wherever possible. To do so, we usually arrange to intersperse all control and experimental conditions in each trial block and to change the sequence of trials (using a random numbers table) from block to block. This insures that any effects of habituation or other temporal trends will be confounded with trial blocks but not with stimulus conditions. Statistical treatment of the results of such experiments is ordinarily quite straightforward. In most cases, standard repeated measures analysis of variance procedures are used. An index of the utility of this approach is that most of our experiments that have used it have yielded statistically significant results with samples that contained less than a dozen subjects.

3. Assessing the Threshold for Startle

3.1. The Method of Constant Stimuli

With startle, as with most other clearly identifiable reactions, it is a widely accepted practice to define its threshold as the intensity of a stimulus that, if repeatedly presented, would elicit the reaction on 50% of the trials. Defined in this way, the threshold is a statistical parameter that can at best only be estimated in a given study.

The usual procedure for assessing a threshold using the method of constant stimuli is to establish a criterion for deciding whether or not a response has occurred on a given trial and then to present a sequence of stimuli with intensities that seem likely to straddle the threshold for response evocation. Ordinarily the order of intensities within a sequence is random. Sequences are presented a dozen or so times, each time with a new random order of intensities.

With the method of constant stimuli, one need only keep track of the percentage of times that each of the stimuli in the sequence elicits a reaction that meets criterion. As will be seen, however, one can also derive some useful information by measuring the amplitude and latency of the reactions.

Figure 5 summarizes the data generated when the method of constant stimuli was used with each of four rats in an effort to assess their thresholds for startle to a brief acoustic signal. In this study, the startle-eliciting stimulus was a 20-msec burst of broad-band random noise (rise-decay

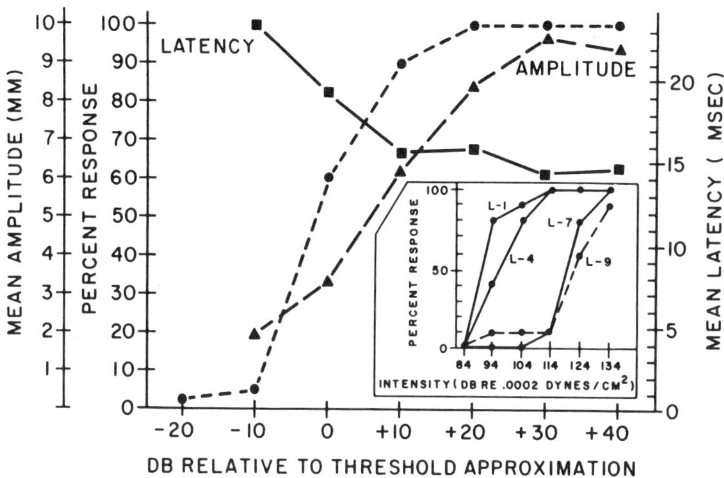

Figure 5. Frequency, amplitude, and latency of startle response during threshold assessment by the method of constant stimuli. The inset shows the percentage response (out of 20 trials) for each subject at each stimulus intensity. The larger curves were derived by assessing responses to the stimuli that spanned each subject's threshold (approximated to the nearest 10 dB). The unlabeled curve shows the percentage response (across subjects) to each of these stimuli. The curves labeled "amplitude" and "latency" are based on those trials during which a response occurred (from Hoffman and Searle, 1968, *J. Acoust. Soc. Am.* **43**:269–282).

time 0.1 msec). Stimulus presentations occurred at intervals of 1 min. On a given trial, stimulus intensity was either 84, 94, 104, 114, 124, or 134 dB re: 0.0002 dyn/cm^2. The order of stimuli was varied randomly within each sequence and trials proceeded until a given subject had been exposed to the entire sequence 20 times.

The inset for Figure 5 shows the percentage of trials (at each stimulus intensity) that yielded a response of at least 2 mm when recorded on a storage oscilloscope. These individual psychophysical functions were used to obtain approximations of the threshold for each subject (to the nearest 10 dB). The body of Figure 5 shows the composite psychophysical function that emerged when the individual functions were shifted upward or downward (so that these thresholds approximately coincided). The body of Figure 5 also shows the means (across subjects) of the latency and amplitude measures for those stimuli that yielded responses.

As can be seen, latency decreased and response amplitude increased as stimulus intensity crossed the psychophysical threshold. However, even when the stimulus was 40 dB above threshold, the average response latency was still approximately 15 msec.

3.2. The Up–Down Technique

One of the most useful procedures for assessing a subject's threshold for startle (or almost any other definable response for that matter) is entitled the up–down technique. This procedure involves a programming arrangement whereby the intensity of the test stimulus on successive trials is determined by the reactions of the subject. At the beginning of a threshold run, stimulus intensity is set below the level that will produce a measurable response. Thereafter, trials occur automatically at some fixed interval. Each time that the stimulus fails to elicit a measurable reaction, the intensity of the signal on the next trial is increased by some fixed amount (for example by 2 dB). However, if a given signal does evoke a response, stimulus intensity on the next trial is decreased by this amount (e.g., 2 dB). The run ends after some predetermined number of trials (for example 60) have elapsed following the subject's initial reaction. Under this procedure, the threshold is defined as the stimulus intensity that theoretically should evoke a response 50% of the time. Arithmetically, it is equal to the mean of the frequency distribution of stimulus intensities that yielded responses, shifted a half-step downward. Statistical procedures for dealing with data generated by the up–down technique have been discussed by Dixon and Mood (1948) and by Brownlee *et al.*, (1952).

Figure 6 shows the record of a rat's startle ractions to a 20-msec 4350 Hz tone (with a rise-decay time of 2.5 msec) during assessment of threshold using the up-down method. More specifically, Figure 6 provides a record of the sequence of stimulus intensities presented to the subject and it indicates which stimulus presentations yielded a measurable reaction. As can be seen, the up-down method is quite economical because the subject never encounters a stimulus that departs appreciably from its threshold. This can be important when one has reason to avoid exposing the subject to especially intense signals. Figure 6 also includes a table in which the data from the up-down run are summarized arithmetically. It shows the number of trials and the number of responses at each intensity (excluding trials that occurred before the first response). For this run, the subject's threshold for startle is calculated to be 107.6 dB.

Although the up-down technique provides little information about the psychometric function for an individual subject, it can be used to obtain an approximation of this function by pooling data across a group of subjects. The pooling procedure involves shifting each subject's distribution along the intensity scale so that the mean (rounded to the nearest integer) is approximately zero. The assumption underlying this pooling procedure is that differences between subjects reflect differences in the means of their functions, but not differences in the variances.

Figure 7 shows the composite cumulative function generated when

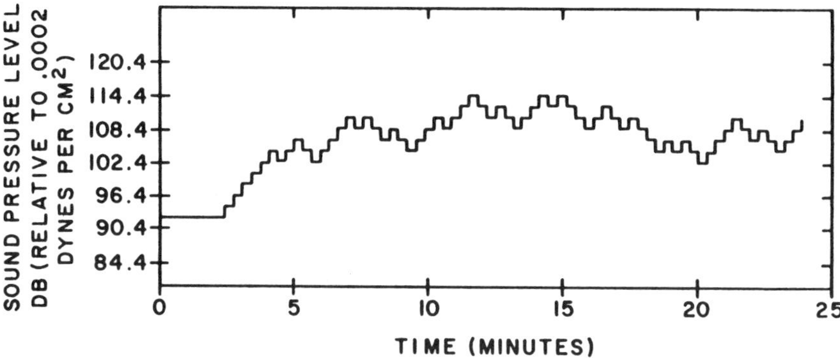

INTENSITY (I) IN DB	NO. OF TRIALS	NO. OF RESPONSES (R_I)	(R_I) x (I)
102.4	3	0	0
104.4	10	3	313.2
106.4	12	6	638.4
108.4	14	5	542.0
110.4	12	8	883.2
112.4	7	4	449.6
114.4	3	3	343.2
	61	29	3169.6

$$\text{TRESHOLD (DB)} = \frac{\sum(R_I)\times(I)}{\sum R_I} - 1 = \frac{3169.6}{29} - 1 = 107.6$$

Figure 6. Recording of successive stimulus intensities during assessment of a rat's threshold for startle to a 20-msec 4350-Hz tone using the up-down method. Each downward deflection of the trace indicates that the prior stimulus yielded a response of at least 2 mm on the oscilloscope trace. Upward deflections indicate that the prior stimulus failed to elicit a response that met this criterion. The table shows the number of responses at each stimulus intensity (excluding trials that occurred before the first response).

each of 16 rats was exposed to the up-down technique using a brief 4350 Hz tone as the startle-eliciting signal. By comparing Figure 7 with Figure 5, it becomes apparent that on a group level at least, the up-down technique and the method of constant stimuli can generate very nearly identical psychophysical functions.

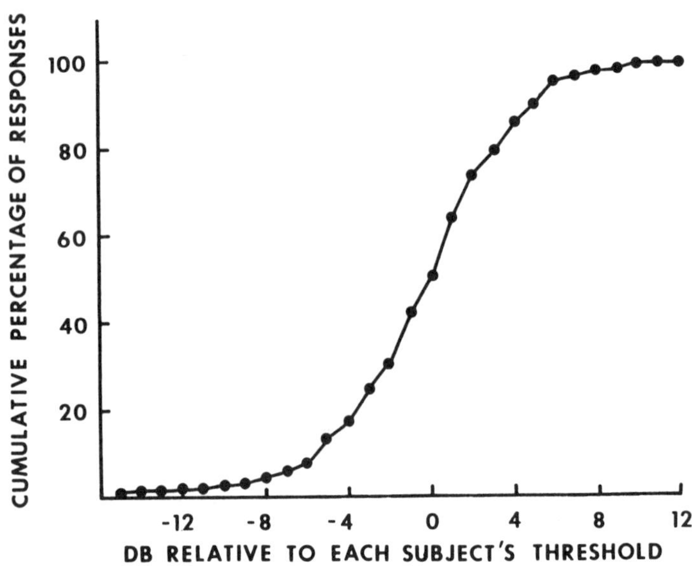

Figure 7. Composite cumulative percentage response across 16 rats. To obtain this function, subjects were exposed to an up-down procedure to assess their thresholds for startle to a 4350-Hz tone. Once threshold assessment was complete, the data from the 16 rats were pooled by shifting the response distribution for each subject along the intensity scale so that its mean approximated zero.

3.3. Reflex Modification Procedures to Assess Sensory Thresholds

Although both the up-down procedure and the method of constant stimuli can provide a good estimate of the threshold for a given target reflex, depending on the nature of that reflex, the obtained threshold is almost certain to be well above the organism's sensory threshold. This is clearly the case when the target reflex is startle. As was seen in Figures 5 and 6, for most of the rats tested, the threshold for startle was somewhere in the region of 100 dB. As noted earlier and as seen in Figure 4, however, a pulse of noise at 50 dB (an intensity that is well below the threshold for startle) was able to greatly reduce the amplitude of the startle reaction elicited by a 140 dB signal presented 150 msec later.

This observation, along with several others (Hoffman and Wible, 1970) suggested that reflex modification procedures might prove to be especially useful in assessing sensory thresholds in lower organisms. Their great advantage is that, unlike traditional approaches to animal psychophysics, they do not require that subjects be trained to respond in a given fashion.

The use of reflex modification procedures to assess sensory thresholds has received considerable attention in this and other laboratories (Hoffman and Wible, 1970; Reiter and Ison, 1977; Graham *et al.*, 1975). Perhaps the most straightforward (although not necessarily the most sensitive or economical) way to approximate a sensory threshold, using reflex modification procedures, is to present the same intense reflex-eliciting signal repeatedly and to arrange that on some of these trials the signal is preceded by a reflex-modifying stimulus at an interval appropriate to cause a reduction in reflex amplitude. With this procedure, one would try to select a sequence of reflex-modifying signals such that the weakest of them is somewhat below the anticipated threshold, whereas the strongest is somewhat above it. Under such circumstances, the threshold would be defined as the lowest-intensity reflex-modifying signal that produced a statistically significant reduction in the mean amplitude of response to the reflex-eliciting signal.

The approach is very nearly identical to the procedure that was used to generate the data shown in Figure 4. The major difference is that to assess a threshold, one must include trials in which only a startle-eliciting stimulus (the second of the two stimuli) is presented. Also, one would select the intensities of the first stimulus so that the weakest of them is below the anticipated threshold and the strongest of them is above it.

It is of interest that Young and Fechter (1982) have recently reported that with both rats and guinea pigs the detection thresholds for pure tones, as assessed with variations of these procedures, were in good agreement with the detection thresholds for the same pure tones when assessed with more traditional psychophysical techniques (Kelly and Masterton, 1977; Prosen *et al.*, 1978).

4. Conclusions

It was the purpose of this chapter to examine various methodological issues that must be resolved if one is to employ reflex modification procedures in the analysis of the startle reflex. In examining these issues, only a few of the known reflex modification effects could be described and little has been said about their implications. Fortunately detailed treatments of these effects are available elsewhere as are discussions of certain of their implications. (See Hoffman and Ison, 1980, for an outline of these effects and a list of references to their more extensive coverage.) For present purposes, it is perhaps sufficient to note that since all reflex modification effects are determined by the way in which the nervous system is arranged, their analysis can provide an avenue for the detailed investigation of this organization. This, coupled with the relative ease with which reflex modification can be studied with intact and unrestrained organisms, would seem to provide a compelling argument for its increasing use by contemporary investigators in the behavioral sciences.

ACKNOWLEDGMENTS. The research on which this chapter is based was supported by National Science Foundation Grant GB4348 and National Institutes of Health grant HD10511.

5. References

Brown, J. S., Kalish, H. I., and Farber, I. S., 1951, Conditioned fear as revealed by magnitude of startle response to an auditory stimulus, *J. Exp. Psychol.* **43**:317–378.

Brownlee, K. A., Hodges, J. L., and Rosenblatt, 1953, The up–down method with small samples, *J. Am. Stat. Assoc.* **48**:262–277.

Dixon, W. J., and Mood, A. M., 1948, A method for obtaining and analyzing sensitivity data, *J. Am. Stat. Assoc.* **43**:109–126.

Fletcher, L. D., and Young, J. S., 1983, Discrimination of auditory from nonauditory toxicity by reflex modulation audiometry: Effects of triethyltin, *Toxicol. Appl. Pharmacol.* **70**:216–227.

Graham, F. K., Putnam, L. E., and Leavitt, L. A., 1975, Lead stimulation effects on human cardiac orienting and blink reflexes, *J. Exp. Psychol Hum. Percept. Perform.* **1**:161–169.

Hilgard, E. R., 1933, Reinforcement and inhibition of eyelid reflexes, *J. Gen. Psychol.* **8**:85–111.

Hoffman, H. S., and Fleshler, M., 1963, Startle reaction: Modification by background acoustic stimulation, *Science* **141**:928–930.

Hoffman, H. S., and Ison, J. R., 1980, Reflex modification in the domain of startle: I. Some empirical findings and their implications for how the nervous system processes sensory input, *Psychol. Rev.* **87**:175–189.

Hoffman, H. S., and Searle, J. L., 1968, Acoustic and temporal factors in the evocation of startle, *J. Acoust. Soc. Am.* **43**:269–282.

Hoffman, H. S., and Wible, B., 1970, Role of weak signals in acoustic startle, *J. Acoust. Soc. Am.* **47**:489–497.

Hoffman, H. S., Fleshler, M., and Abplanalp, P. L., 1964, The startle reaction to electrical shock in the rat, *J. Comp. Physiol. Psychol.* **59**:53–58.

Kelly, J. B., and Masterton, B., 1977, Auditory sensitivity of the albino rat, *J. Comp. Physiol. Psychol.* **91**:930–936.

Kimmel, C. B., Patterson, J., and Kimmel, R. O., 1974, The development and behavioral characteristics of the startle response in the zebrafish, *Dev. Psychobiol.* **7**:47–60.

Landis, C., and Hunt, W. A., 1939, *The Startle Pattern,* Farrar and Rinehart, Inc., New York.

Peake, H., 1936, Inhibition as a function of stimulus intensity, *Psychol. Monogr.* **47**(2, Whole No. 212):135–147.

Prossen, C. A., Peterson, M., Moody, D., and Stebbins, W., 1978, Auditory thresholds and kanamycin-induced hearing loss in the guinea pig assessed by a positive reinforcement procedure, *J. Acoust. Soc. Am.* **63**:559–566.

Reiter, L. A., and Ison, J. R., 1977, Inhibition of the human eyeblink reflex: An examination of the Wendt-Yerkes method for threshold detection, *J. Exp. Psychol. Hum. Percept. Perform.* **3**:325–336.

Sechenov, I. M., 1965, *Reflexes of the Brain* (S. Belsky, trans.), M.I.T. Press, Cambridge, Massachusetts (Originally published St. Petersburg, Sushchinski, 1863.).

Schwartz, G. M., Hoffman, H. S., Stitt, C. L., and Marsh, R. R., 1976, Modification of the rat's acoustic startle responses by antecedent visual stimulation, *J. Exp. Psychol.* **2**:28–37.

Stitt, C. L., Hoffman, H. S., Marsh, R. R., and Boskoff, K. J., 1974, Modification of the rat's startle reaction by an antecedent change in the acoustic environment, *J. Comp. Physiol. Psychol.* **86**:826–836.

Stitt, C. L., Hoffman, H. S., Marsh, R. R., and Schwartz, G. M., 1976, Modification of the pigeon's visual startle reaction by the sensory environment, *J. Comp. Physiol. Psychol.* **90**:601–619.

Yerkes, R. M., 1905, The sense of hearing in frogs, *J. Comp. Neurol. Psychol.* **15**:279–304.

10

The Mammalian Startle Response

MICHAEL DAVIS

1. Introduction

One of the ultimate goals of neuroscience is to determine how the brain directs and changes behavior in man. Given the enormous complexity of this endeavor, an increasing number of neuroscientists have focused on invertebrates and lower vertebrates in an effort to simplify the problem. Eventually, however, it will be necessary to analyze the cellular basis of behavior in a complex mammalian brain. To approach the problem at this level, it would be useful to study a relatively simple behavior that can be elicited in mammals and that is sensitive to a variety of experimental treatments.

The mammalian acoustic startle reflex fulfills many of these criteria. Acoustic startle is under stimulus control so that inhibitory or excitatory effects can be measured. Acoustic startle can be measured at early ages in many different species, making it appropriate for developmental and comparative studies. Startle shows different types of plasticity such as habituation, sensitization, prepulse inhibition, and modification by prior associative learnings. Startle is also sensitive to a variety of drugs and is currently being used as a model system to analyze how drugs alter sensorimotor reactivity. Results gathered from studying startle plasticity have generalized to very different stimulus–response systems including those in man. In the rat the latency of acoustic startle is 8 msec when recorded electromyographically in the hindleg. This indicates that a very simple

MICHAEL DAVIS • Department of Psychiatry, Connecticut Mental Health Center, Yale University, New Haven, Connecticut 06508.

neuronal circuit must be involved. Acoustic startle represents, therefore, a sensitive vertebrate behavioral test system mediated by only a few central synapses.

2. The Behavioral Response

2.1. The Adequate Stimulus

The startle response consists of a characteristic sequence of muscular responses elicited by a sudden, intense stimulus. In mammals, intense acoustic stimulation is particularly effective in eliciting startle. Other things being equal, more intense stimuli produce larger responses. In fact, one of the most important features of the mammalian startle response is that it is highly graded in amplitude so that it is well suited for quantitative analysis in a single animal. This contrasts to startle in lower species that tends to be an all-or-none response. The graded amplitude of mammalian startle can be readily seen in direct muscle recording or in the output of cages sensitive to movements when whole body startle is measured. Change in startle amplitude is the major measure of startle plasticity in mammals. The critical parameter for an effective auditory stimulus is that it reaches some minimum intensity within a minimum of time. In the rat, an acoustic stimulus had to reach an intensity of about 90 dB (0.0002 dynes/cm^2) within 12 msec of its onset in order to elicit a measurable response (Fleshler, 1965). When longer rise times are used, startle will not occur even when very intense sound levels are reached (e.g., 140 dB). Beyond about 6–8 msec, increasing the duration of an auditory stimulus had little effect on either startle amplitude or startle threshold (Fleshler, 1965; Marsh *et al.*, 1973). For stimulus durations less than 8 msec, longer durations do produce greater startle amplitudes (Marsh *et al.*, 1973), indicating that some temporal summation does occur. Similar results have been reported for the eyeblink component of startle in humans (Berg, 1973, cited in Graham, 1975). Faster rise times were associated with lower thresholds to elicit eyeblinks using rise times of 0.3, 10, or 30 msec. Rapid temporal summation of the eyeblink also occurred using stimulus durations of 4, 8, 16, and 32 msec, with an asymptote around 16–32 msec for pure tones and somewhat longer for white noise bursts. In both rats and humans, therefore, startle represents the behavioral response to very short periods of sensory stimulation.

Anecdotally, rats startle more to high-frequency sounds than to low-frequency ones. Hence, the sound of a "kiss" or of two metal bars being tapped together are especially effective in eliciting startle, probably because each stimulus contains high-auditory frequencies. Consistent with this impression, Fleshler (1965) reported that the threshold to elicit startle in adult rats decreased monotonically as the frequency of the eliciting tones increased from 720 to 13,250 Hz. Similar results have been seen in adult mice over a range of 5–15 kHz. However, 20 kHz was actually less effective than 15 kHz in both adult and 15- to 17-day-old mice (Shnerson and Willott, 1980; Willott *et al.*, 1979). In mice, increased effectiveness of higher tone frequencies developed rapidly, since in 15- to 17-day-old mice a 15 kHz tone was more effective than a 7 kHz tone, whereas the 7 kHz tone was more effective in 13- to 14-day-old mice (Shnerson and Willott, 1980). In rats, auditory stimuli or air puffs are effective in eliciting startle, but visual stimuli are not. Light flashes also may be ineffective in humans (Meier-Ewert *et al.*, 1974). In contrast, in highly visual animals such as the pigeon, light flashes are very effective in eliciting startle but auditory stimuli are not (Stitt *et al.*, 1976).

2.2. The Form of the Response

As described by Strauss (1929, cited in Landis and Hunt, 1939), startle in humans consists of the following set of muscle movements: blinking of the eyes, forward head movement, a characteristic facial expression that includes a widening of the mouth and occasional baring of the teeth, raising and drawing forward of the shoulders, abduction of the upper arms, bending of the elbows, pronation of the lower arms, flexion of the fingers, forward movement of the trunk, contraction of the abdomen, and bending of the knees. The human response is illustrated in a highly stylized fashion in Figure 1. The response is primarily flexion, although extension of various limbs may frequently follow an initial flexion. Occasionally extension occurs in the absence of any prior flexion (Landis and Hunt, 1939). Electromyographic (EMG) recordings in leg and ankle muscles in man show that startle more often occurs in flexors than extensors, but that EMGs in extensors definitely do occur, and their latencies are slightly shorter than the EMG latencies of flexors (Rossignol, 1975). Whether or not extension or flexion occurs depends on the state of the nervous system at the time startle is elicited, since the probability of extension increases following voluntary extension or ongoing extension in the muscle being recorded (Rossignol, 1975).

The most striking feature of the startle reflex is its very short latency. This has been documented by recording EMG responses at various points along the body following a startle-eliciting stimulus. Table I summarizes the data from humans. The earliest detectable change in movement appears in the jaw muscles (14 msec from the onset of a loud click). The response then spreads down the neural axis to finally reach muscles in

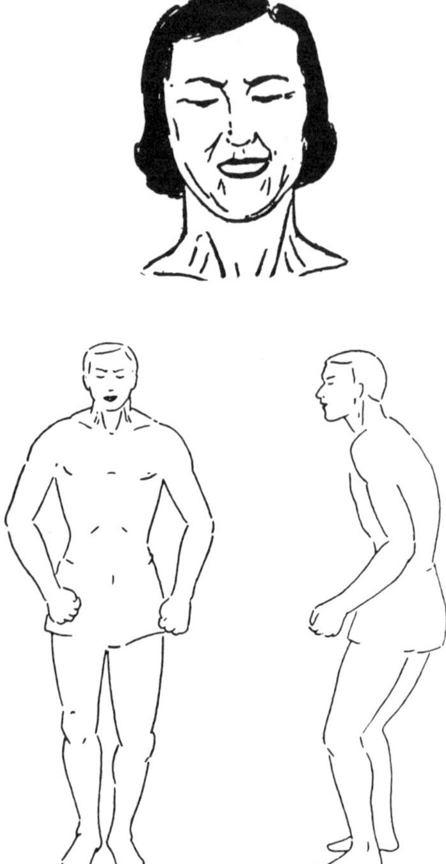

Figure 1. Schematic representation of facial and bodily pattern of the acoustic startle response in humans. Based on data from Landis and Hunt (1939) in which human startle was photographed at 64 frames/sec.

the legs, with longer latencies associated with longer distances from the ear. In children the response occurs much more rapidly in the legs, because of the smaller distance involved. Using high-speed photography, Landis and Hunt (1939) also report a progressive increase in latency as the response spread down the body. The longer average latencies in their work compared with the EMG data would be expected since they were measuring the actual movement of the muscle, rather than the arrival of the first afferent volley.

In rats and guinea pigs, startle also consists of a very rapid sequence of muscle movements (Figure 2). The earliest response seems to be a slight extension of the forepaws and hindpaws followed by a rapid gen-

Table I. Approximate Latencies of the Startle Reflex in Different Muscle Groups Recorded in Humans or Monkeys

Stimulus	Muscle recorded	EMG latency (msec)	Movement latency (msec)	Duration (msec)	Reference
Click or 750-Hz tone (85–100 dB)	M. masseter (jaw)	14		11	Meier-Ewert *et al.*, 1974
Noise burst (92 dB)	Obicularis oculi (eyelid)	20–40		160	Gogan, 1970
Click from castanet	Obicularis oculi (eyelid)	25[a]			Shimamura, 1973
Pistol shot	Trapezius (neck)	25–40			Jones and Kennedy, 1951
Click or 750-Hz tone	Trapezius	40		40	Meier-Ewert *et al.*, 1974
3000-Hz tone (97 dB)	Trapezius	17[b]			Mortimer, 1973
3000-Hz tone (97 dB)	Biceps	60		40	Meier-Ewert *et al.*, 1974
Click	Biceps	42[a]			Shimamura, 1973
1000-Hz tone (114 dB)	Tibialis anterior (leg flexor)	151		62	Rossignol, 1975
1000-Hz tone (114 dB)	Gastronemius (leg extensor)	123		—	Rossignol, 1975
Click	Gastronemius (leg flexor)	70[a]			Shimamura, 1973
Click	Head Forward		60–120		Landis and Hunt, 1939
	Neck		75–121		
	Shoulders		100–121		
	Arm		125–195		
	Hand		145–195		
	Knee		145–395		

[a]Children
[b]Monkeys

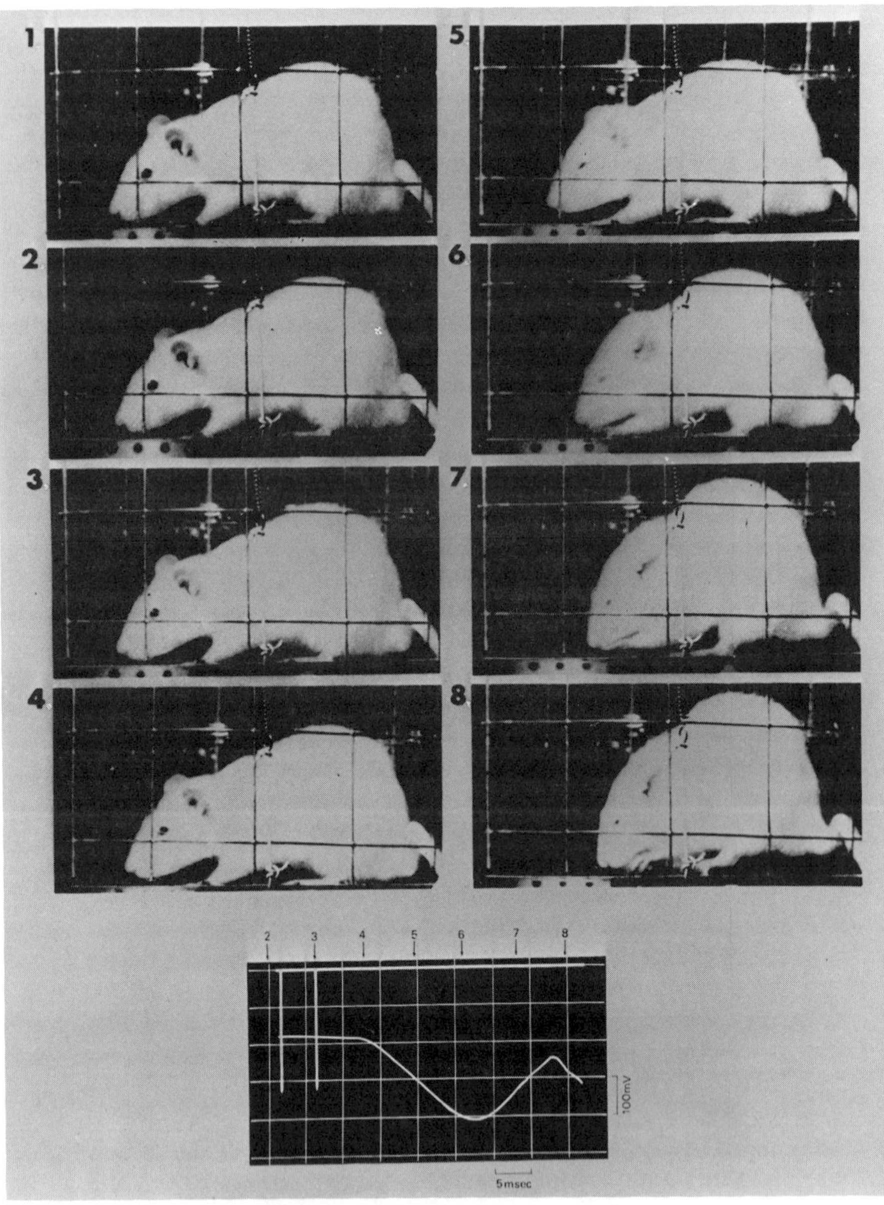

Figure 2. The acoustic startle pattern in rats. Photographs 1–8 are stills from a cine film (150 frames/sec) of an acoustic startle response. The photograph of the oscilloscope record shows on the top trace two signal spikes derived from a pulse unit. The first spike indicates the arrival of the sound burst at head height while the second indicates arrival of light from a flash unit for synchronization of the cine film with the accelerometer trace. The numbers along the top refer to the cine film stills and indicate the point in time at which each frame was exposed. The bottom trace is the output from the accelerometer, the downward deflection corresponding to a downward movement of the cage base (from Horlington, 1970).

eralized flexion that results in an overall shortening of the body and characteristic "hunched" position (Burnham, 1939; Dodge and Louttit, 1926; Horlington, 1968). The initial extension results in a downward movement of a cage, which can then be used to quantify startle magnitudes. Measured in this way, the average latency of acoustic startle in albino rats is about 12–13 msec, although movements as short as 10 msec can frequently be detected (Horlington, 1968). Using EMG recordings, latencies of 5–7 msec have been reported for the forelimbs (Ison *et al.*, 1973) and 8–10 msec for the hindlimbs (Davis *et al.*, 1982*a*; Ison *et al.*, 1973; Szabo, 1965). Similar to humans, extensors seem to have a slightly shorter latency than flexors (Ison *et al.*, 1973).

Although it is generally believed that startle is mainly a flexor reflex (based largely on Landis and Hunt, 1939), recent high-speed photography in our laboratories indicates that there is a great deal of extension in rat startle responses. In fact, in nonhabituated rats, startle is seen as a marked extension of both forelimbs and hindlimbs so that the rat actually jumps into the air. As habituation proceeds, extension seems to drop out leaving mostly flexion, similar to the response shown in Figure 2.

3. Organisms with the System

Startle can probably be elicited in all mammals. It is routinely studied in humans, rats, and mice. Startle reflexes have also been studied in cats, rhesus monkeys (Heninger and Davis, 1979), and guinea pigs (Dodge and Louttit, 1926). Landis and Hunt (1939) have noted startle responses in chimpanzees, orangutans, several different types of monkeys, the two-toed sloth, honey badger, ocelot, serval, Siberian lynx, binturong, South American wild dog, jackals, Andean wild dog, coyote, timber wolf, dingo, kinkaiou, meerkat, Cape ground squirrel, Tibetan bear, Alaska brown bear, and the Himalyanan tahr! Startle represents therefore an ideal behavior for comparative studies across a variety of mammalian species.

4. The Neural Network

4.1. Neural Mediation of Acoustic Startle

4.1.1. Literature Review

Early work on the acoustic startle circuit can be traced to Forbes and Sherrington (1914) who observed widespread muscle reflexes to acoustic stimuli in cats that had been decerebrated at the level of the superior colliculus. Similar observations have been made in decerebrate monkeys, dogs, rabbits, and even mesencephalic humans (cf., Szabo and Hazafi, 1965). These studies indicated that auditory structures rostral to the superior colliculus (i.e., the medial geniculate and the auditory cortex) are not required for acoustic startle, as one might expect given its very short latency. Decerebration at the midcollicular level would also eliminate the corticospinal tract, interstitial spinal tract, and the major part of the tectospinal tract, i.e., those fibers arising from the superior colliculus (Edwards, 1980). More recently, it has been reported that decerebrate and normal rats show similar changes in acoustic startle following administration of selected drugs (Davis *et al.*, 1977*a*; Davis *et al.*, 1980*a*) or various stimulus parameters (Davis and Gendelman, 1977; Fox, 1979; Hammond, 1973).

Strauss (1929, reported in Landis and Hunt, 1939) suggested that startle is mediated by activation of the red nucleus. However, this idea was not supported by the findings of Szabo and Hazafi (1965) that "after massive lesions at the mesencephalic level, involving the red nuclei, the entire mesencephalic reticular formation, caudally the rostral part of the pons, and rostrally the caudal portion of the diencephalon, no change could be observed in the excitability of the acoustic startle reaction." More recently, Groves *et al.* (1974*a*) also found that large lesions of the mesencephalic reticular formation, which sometimes involved destruction of the red nuclei, did not abolish startle.

Another influential paper reported that the startle response, as measured by the EMG in the rat gastrocnemius muscle, occurred with a latency between 15–25 msec (Prosser and Hunter, 1936). On the basis of this latency, Prosser and Hunter proposed the following circuit: cochlea, eighth nerve, cochlear nucleus, inferior colliculus, reticular nucleus in midbrain, reticulospinal tract, anterior horn cells, and motor nerves to limbs.

Evidence produced by Szabo and Hazafi (1965) and Groves *et al.* (1974*a*) challenged this pathway, since lesions of the inferior colliculus reduced but did not abolish startle in rats. On the other hand, Fox (1979) recently reported that destruction of the inferior colliculus did abolish startle in rats previously given a midcollicular brain transection, when startle was defined as an EMG response in the neck muscles having an average latency of about 11 msec. Moreover, Willot *et al.* (1979) have argued on the basis of electrophysiological work that the inferior colliculus is probably involved in mediating an acoustic startle reflex in mice with a latency of about 17 msec.

Part of the apparent confusion in the literature regarding the exact structures necessary for acoustic startle is that different studies have measured startle reflexes that have quite different latencies. Electromyographic recordings sometimes show two and occasionally three components of the acoustic startle reflex that arrive at the muscles at slightly different times (e.g., Szabo, 1965). Studies that implicate the inferior colliculus tend to report relatively long latencies (e.g., Fox, 1979; Prosser and Hunter, 1939; Willot *et al.*, 1979). However, as mentioned earlier, acoustic startle reflexes in the rat can be measured readily that have latencies of only 8 msec in the hindleg and about 5–6 msec in the forepaw (Ison *et al.*, 1973). These extremely short latencies, coupled with the inability of lesions of the inferior colliculus to always abolish startle, suggest that an acoustic startle circuit may exist that is completely subcollicular.

In contrast to the controversy concerning the role of the inferior colliculus, there is general agreement that the reticulospinal tract originating in the pontine reticular formation is required for acoustic startle. As early as 1926, Musken (cited in Larsson, 1956) observed that animals with injuries in the midbrain reticular formation had diminished startle responses; furthermore, startle disappeared "when the medial portion of the reticulum traversing the medulla and the pons was completely destroyed"; Szabo and Hazafi (1965) observed that startle was abolished by lesions of the ventromedial medullopontine reticular formation at the level of the trapezoid body. However, it is not clear whether this resulted from destruction of sensory fibers traversing the trapezoid body or from destruction of cells that form the reticulospinal tract. The involvement of the latter is implicated by the findings of Hammond (1973) that relatively small lesions in the region of the nucleus reticularis pontis caudalis, which would not have damaged the ventrally located trapezoid body, markedly attenuate startle. Groves *et al.* (1974*a*) also found that large lesions of the posterior regions of the reticular formation abolished startle although fibers crossing through the trapezoid body were frequently spared.

A number of studies emphasize the importance of the nucleus reticularis pontis caudalis as being the region of the reticular formation critical for the mediation of acoustic startle. Lesions of this structure abolish startle (Hammond, 1973; Groves *et al.*, 1974*a*; Leitner *et al.*, 1980), whereas lesions of more rostral areas of the reticular formation (Hammond, 1973; Groves *et al.*, 1974*a*) or more caudal areas (Hammond, 1973; Leitner *et al.*, 1980) did not. Finally, Leitner *et al.* (1980) reported that lesions of the nucleus reticularis pontis caudalis abolished or attenuated both acoustic and shock-elicited startle. The fact that both acoustic- and shock-elicited startle were impaired suggests that the lesions were effective in damaging the "motor" rather than the "sensory" limbs of these reflex arcs. Taken together, these studies support the involvement of the reticulospinal tract originating from the ventral pontine and medullary reticular formation in mediating the motor side of the startle circuit.

4.1.2. A Primary Acoustic Startle Circuit

Hence, a variety of studies have implicated different auditory or motor structures in mediating the acoustic startle response. What is lacking, however, is an effective means of determining the sequence of neural structures that are involved in translating an acoustic stimulus into a muscular response. Over the past four years, David Gendelman, Marc Tischler, Phillip Gendelman, and I have been trying to trace out the entire acoustic startle circuit using lesion, electrical stimulation, and anatomical tracing techniques (Davis *et al.*, 1982*a*). Figure 3 shows diagramatically those structures and pathways we believe are required to mediate the primary acoustic startle response defined as an EMG activity in the hindleg quadriceps femoris muscle complex beginning 8–9 msec after the onset of a loud click. This is the shortest latency that we can measure in the hindleg, hence, the designation, primary acoustic startle circuit. Sometimes, however, later responses (e.g., 15–25 msec) can be seen in EMG tracing after the initial 8-msec response, presumably via more complex neural pathways. These will not be addressed in this chapter. However, all of the effects described thus far for whole body startle (including potentiated startle) also occur (when startle is defined by the 8-msec EMG recordings) in this very early, 8-msec response, indicating its validity as a measure of startle.

4.1.2a. Posteroventral Cochlear Nucleus. The posteroventral cochlear nucleus (VCN) appears to be the first synapse in the primary acoustic startle circuit. Bilateral lesions of the VCN abolish acoustic startle. In contrast, lesions of the dorsal cochlear nucleus fail to abolish startle. In

Figure 3. Schematic diagram of a primary acoustic startle circuit. Abbreviations: A, aqueduct; CNIC, central nucleus of the inferior colliculus; CU, cuneate nucleus; DCN, dorsal cochlear nucleus; DLL, dorsal nucleus of the lateral lemniscus; DP, decussation of pyramids; DR, dorsal raphe nucleus; ENIC, external nucleus of the inferior colliculus; IO, inferior olive; LL, lateral lemniscus; LM, medial lemniscus, LV, lateral vestibular nucleus; MLF, medial longitudinal fasciculus; MTB, medial nucleus of the trapezoid body; MV, medial ventibular nucleus; nVII, nucleus of the seventh nerve; P, pyramids; RGI, nucleus reticularis gigantocellularis; RPC, nucleus reticularis pontis caudalis; RPO, nucleus reticularis pontis oralis; RST, reticulospinal tract; RSTm, medial reticulospinal tract; SO, superior olive; TSV, spinal tract of the fifth nerve; VAS, ventral acoustic stria; VCN, ventral nucleus; VII, seventh nerve; VLL, ventral nucleus of the lateral lemniscus (from Davis *et al.*, 1982*a*).

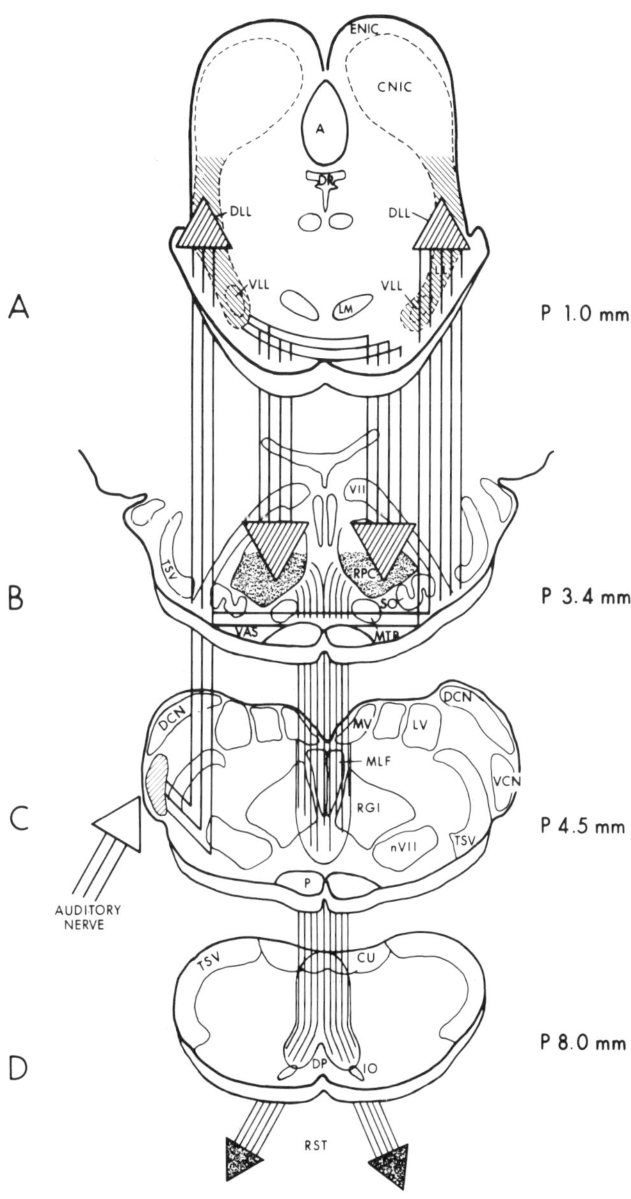
ENIC
CNIC
A
DR
DLL
DLL
VLL
VLL
LM
LLT
A
P 1.0 mm
VII
TSV
RPC
SO
VAS
MTB
B
P 3.4 mm
DCN
DCN
MV
LV
MLF
VCN
RGI
nVII
TSV
P
C
P 4.5 mm
AUDITORY
NERVE
TSV
CU
DP
IO
RST
D
P 8.0 mm

waking rats, bilateral, single-pulse stimulation (1-msec pulse width, 10–50 μA) elicits startlelike responses with a latency of 7.0–7.5 msec (Figure 4). In fact, this response looks so similar to acoustic startle that it is difficult to discriminate the two visually.

4.1.2b. The Dorsal and Ventral Nuclei of the Lateral Lemniscus. The second synapse in the primary acoustic startle circuit seems to occur in

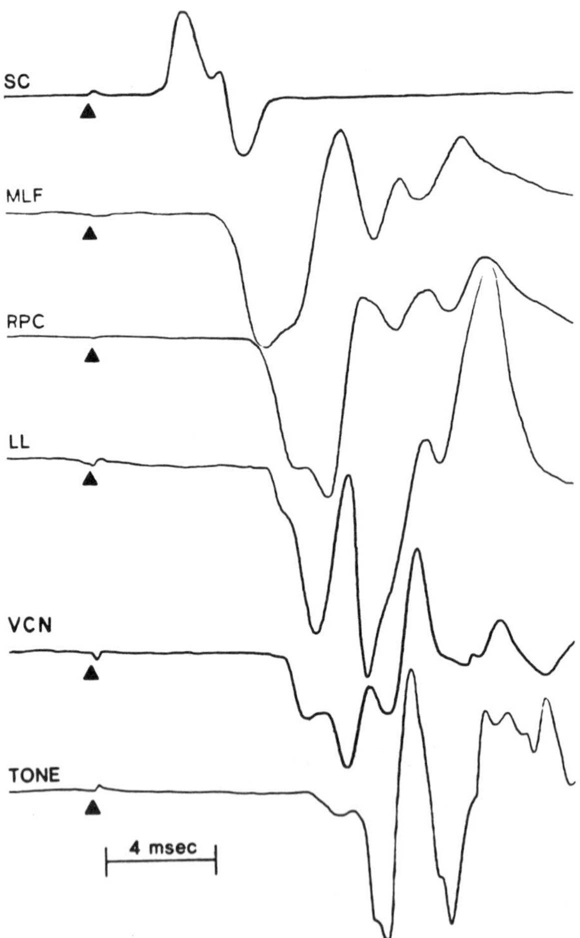

Figure 4. Electromyographic recording from the quadriceps femoris muscle complex of "startle" elicited by electrical stimulation in the spinal cord (SC), medial longitudinal fasciculus (MLF), nucleus reticularis pontis caudalis (RPC), nuclei of the lateral lemniscus (LL), or ventral cochlear nucleus (VCN), or by a tone (from Davis *et al.*, 1982*a*).

the dorsal and ventral nuclei of the lateral lemniscus, which are known to receive direct projections from the ventral cochlear nuclei. Bilateral lesions of these nuclei abolish acoustic startle. Both seem to be involved, since bilateral lesions of just the dorsal nuclei or just the ventral nuclei alone do not abolish startle. Moreover, both ipsilateral and contralateral nuclei are involved, since lesions of both sides are required to abolish startle in animals in which clicks are presented to just one ear. Finally, electrical stimulation of these nuclei elicits discrete startlelike responses with an average latency of about 6.0 msec (Figure 4). Stimulation of the dorsal nucleus elicits exclusively ipsilateral leg movements, whereas stimulation of the ventral nuclei elicits bilateral leg movements.

4.1.2c. The Nucleus Reticularis Pontis Caudalis. The next synapse probably occurs in a ventromedial region of the nucleus reticularis pontis caudalis (RPC). As outlined earlier, previous ablation work had pointed to the importance of this area for acoustic startle. These are the cells that form the reticulospinal tract, (cf., Peterson, 1979) long suspected of mediating acoustic startle in both mammals (e.g., Prosser and Hunter, 1936) and lower vertebrates (Chapter 8, this volume). We found that bilateral lesions of this area abolish acoustic startle when measured one day or two weeks later. Electrical stimulation of points within the RPC elicits startlelike responses with an average latency of about 5 msec (Figure 4). In contrast, lesions of areas more dorsal in the RPC or more rostral areas of the RPC or more caudally in the RGI do not abolish startle nor do they support electrical "startle" when reasonably low currents are used.

Although the RPC had been suspected to be important for acoustic startle, its connection with known auditory structures had not been described. To determine this, we iontophoresed horseradish peroxidase into the area of the RPC that seemed critical for startle. As suspected, labeling was found in both the dorsal and ventral nuclei of the lateral lemniscus, but not in other auditory areas such as the inferior colliculus. These results provided critical evidence for the connection between known auditory areas and this part of the reticular formation.

4.1.2d. Reticulospinal Tract. Cell bodies in the RPC send their axons to all levels of the spinal cord via the reticulospinal tract. This tract courses through the medial longitudinal fasciculus (MLF) on the midline and then bifurcates to form the ventral funiculi in the spinal cord. Midline lesions of the MLF abolish acoustic startle provided they have a great enough dorsal-ventral extent. At lower levels, bilateral lesions are required at the most ventral aspects of the medulla. In most cases these lesions are lethal, but in those animals that do survive, acoustic startle is abolished or markedly attenuated. Moreover, electrical stimulation on the midline through the MLF elicits leg movements with a latency of about 4.0–4.5 msec (Figure 4).

4.1.2e. Lower Motor Neurons in the Spinal Cord. Fibers from the reticulospinal tract synapse in the spinal cord, forming the last synapses before the neuromuscular junction. Both direct synapses as well as indirect ones through an interneuron in the cord are possible. To date our data are not able to determine if interneurons are involved.

4.2. Extrinsic Neural Systems That Modulate Startle

Although the neural circuitry that mediates acoustic startle is contained entirely within the brainstem, higher neural networks can modulate startle since lesions of various brain sites distant from the actual startle circuit do alter the response. Table II summarizes the effects of various brain lesions on acoustic and tactile (elicited by an air puff) startle amplitude per se, as well as different types of startle plasticity when they have been examined in the particular study.

Complete decerebration extending from the posterior splenium of the corpus collosum to the mammillary bodies had no obvious effect on startle amplitude per se (Davis and Gendelman, 1977). However, this is difficult to judge since at the short, 60-min posttransection interval used, the rats had not righted themselves in the test cage so that startle was transmitted to the cage in an abnormal way. It is certain, however, that complete decerebration did not alter auditory prepulse inhibition or the reduction in startle when short intertone intervals were used. Very similar results were found by Fox (1979) when startle was measured 24–48 hr after a complete transection. Davis and Gendelman (1977) also found that the decerebrate rats did not show evidence of habituation when relatively long intertone intervals were used, whereas the lack of effect on the intertone interval function would predict habituation when short intervals were used. Fox's (1979) data agreed with this conclusion, since pronounced habituation occurred when short intertone intervals were used, but not when a long, 150-sec interval was used. However, in Fox's study much longer intervals could still sustain habituation in the decerebrate rat, compared with our work, perhaps because Fox used a longer period of recovery following surgery. Finally, complete decerebration blocked the ascending limb of the nonmonotonic background noise function (see Section 7.3.1) without interfering with the descending limb (Davis and Gendelman, 1977). This indicates that structures rostral to the transection must be involved in mediating the excitatory effect of increased background noise levels.

Table II shows the results when various types of cortical tissues are removed or inactivated. Complete olfactory bulbectomy has been reported to increase both acoustic and tactile startle (Van Riezen *et al.*, 1977). Inactivation of the cerebral cortex via KCl increased acoustic startle, but did not alter either within- or between-session habituation, provided that conditions of cortical inactivation remained constant over test sessions. However, marked dishabituation did occur when the conditions of KCl infusion onto the cortex were changed from session to session (VanderStaak, 1976). Lesions of the frontal cortex tend to increase startle without interfering with either prepulse inhibition (Hammond, 1974) or the typical nonmonotonic background noise function (Ison and Silverstein, 1978). In contrast, lesions of the parietal cortex do not seem to alter startle, whereas lesions of the auditory cortex may depress it (Groves *et al.*, 1974*a*).

Lesions of the hippocampus have not consistently altered either startle amplitude per se or within- or between-session startle habituation or prepulse inhibition (Groves *et al.*, 1974*b*; Leaton, 1981; Kemble and Ison, 1971). Groves *et al.* (1974*a*) did report less between-session habituation after hippocampus lesions, but this was difficult to judge since an appropriate control was not included in this part of the study. Coover and Levine (1972) reported that hippocampal lesions increased acoustic startle, perhaps because in this study a much longer lesion-test interval was used than in the other studies.

On the other hand, lesions of the septal area markedly increase startle provided that testing occurs shortly after lesioning. In fact, the classic description of a septal rat is one of extreme hyperreactivity to sensory stimuli (Brady and Nauta, 1953). The hyperreactivity component of the "septal syndrome" seems to require inactivation of the lateral septal area, since intraseptal infusion of a local anesthetic into the lateral septal area induces hyperreactivity, whereas infusion into the medial septal area does not (Albert and Wong, 1978; Albert and Richmond, 1976). Septal lesions do not, however, impair within-session habituation when testing begins within 10–12 days after the lesion (Miller and Treft, 1979; Sagvolden and Wester, 1974). They also do not impair between-session habituation, although between-session habituation may not be as permanent in septal rats (Miller and Treft, 1979). This latter finding may explain why Sagvolden and Wester (1974) found impaired rehabituation in septal rats at 33 or 93 days postlesion relative to controls that may have had better retention and hence faster rehabituation.

Within the brainstem, lesions of the superior colliculus do not appear to alter startle (Groves *et al.*, 1974*a*), whereas lesions of the brachium of the inferior colliculus increase startle without altering within- or between-session habituation or the nonmonotonic background noise function (Jor-

Table II. Effects of Brain Lesions on Startle

Brain area lesioned	Time after lesion (days)	Effect of startle	Effect on habituation	Effect on other types of plasticity	References
Complete brain transection	1/24	Maybe none	Decrease when long intervals used	No effect on prepulse inhibition Blocks background noise enhancement	Davis and Gendelman, 1977
Complete brain transection	1–2	?	Maybe none	No effect on preinhibition	Fox, 1979
Olfactory bulbectomy	21	Increase	—	No effect on prepulse inhibition	Van Riezen *et al.*, 1977
Cerebral cortex	During KCl spreading depression	Small increase	None or decrease if KCl conditions changed	—	VanderStaak, 1976
Frontal cortex	4–7	None	None		Groves *et al.*, 1974*b*
Frontal cortex	11–17	Increase when high background noise used	—	No effect on noise function	Ison and Silverstein, 1978
Medial frontal cortex	3–14	Increase	—	No effect on prepulse inhibition	Hammond, 1974
Parietal cortex	4–7	None	None	—	Groves *et al.*, 1974*b*
Auditory cortex	4–7	Decrease	None	—	
Hippocampus	70	Increase	None	—	Coover and Levine, 1972
Hippocampus	21	None	None	—	Leaton, 1981
Hippocampus	6–10	None	—	—	Groves, *et al.*, 1974*a*
Parietal cortex		None			
Frontal cortex		None			
Hippocampus	4–7	None	Decrease		Groves *et al.*, 1974*a*
Septum	4–11	None		No effect on prepulse inhibition	Kemble and Ison, 1971
Amygdala					
Hippocampus					
Septal area	14	Increase	None	—	Miller and Treft, 1979
Septal area	1	Increase		—	

Septal area	10	None	None	None	Sagvolden and Wester, 1974
	33	None	Maybe decrease	Maybe increase	
	93	None	Maybe decrease		
Nucleus accumbens	7	None	—	—	Sorenson and Swerdlow, 1982
Medial forebrain bundle	42	None but increase in rats previously given shocks	None	—	Yunger and Harvey, 1977
Superior colliculus	2	None	None		Groves *et al.*, 1974*b*
	6	None	None	—	
Brachium of inferior colliculus	10 or more	Increase	None	No effect on background noise	Jordon and Leaton, 1982*b*
Inferior colliculus	2	Decrease	Maybe	—	Groves *et al.*, 1974*b*
	6	None or increase	Reduce		
Midbrain reticular formation	21–28	None	Decrease	—	Jordon and Leaton, 1982*a*
Midbrain reticular formation	15	None or up	Reduce	—	Capps and Stockwell, 1968
Mesencephalic reticular formation	4–7	Increase	None	—	Groves *et al.*, 1974*b*
Midbrain reticular formation	21	None or slightly up	Maybe Reduce	Marked attenuation of prepulse inhibition	Leitner *et al.*, 1981
Periaqueductal gray	1	Huge increase		—	Blair *et al.*, 1978
Dorsal plus median (B7) and raphe (B8)	14	Large increase	None	—	Davis and Sheard, 1974*a*
				No effect on noise function	
Median (B8) raphe	21	Large increase	None	None	Geyer *et al.*, 1976
Dorsal (B7) raphe	21	None	None	None	
Lateral (B9) raphe	21	None	None	None	
Locus coeruleus (electrolytic)	2–3	Decrease	None		Davis *et al.*, 1977*a*
Locus coeruleus (6-OHDA)	5	Decrease	Enhance	Maybe decrease sensitization	Adams and Geyer, 1981
Motor V Motor VII	6–10	Decrease plus large weight loss	—	Not clear on prepulse	Groves *et al.*, 1974*a*

don and Leaton, 1982*b*). When measuring whole body acoustic startle, lesions of the inferior colliculus markedly reduce startle when testing occurs shortly after the lesions, but may even increase startle when testing occurs several days later (Groves *et al.*, 1974*a*). This may explain why Fox (1979) and Wright and Barnes (1972) found severe reductions of startle after lesions of the inferior colliculus, since in each case testing occurred shortly after the lesions. If the nuclei of the lateral lemniscus are importantly involved in startle (Davis *et al.*, 1982*a*), then acute damage of the inferior colliculus may temporarily alter these nearby structures and hence temporarily alter startle.

The midbrain reticular formation is a particularly important area in the modulation of startle. Large bilateral lesions of the mesencephalic reticular formation do not seem to markedly increase initial startle amplitude, but they may decrease within session habituation (Capps and Stockwell, 1968), especially when short-interstimulus intervals are used (Leaton and Jordon, 1982*b*). Lesions in this area also interfere with prepulse inhibition (Leitner *et al.*, 1981). Large lesions on the midline, in the periaqueductal grey region or below in the dorsal and especially the medial raphe nuclei also increase acoustic and tactile startle (Blair *et al.*, 1978; Davis and Sheard, 1974*a*; Geyer *et al.*, 1976). We have also found that very low levels of electrical stimulation in the mesencephalic reticular formation, central grey, or raphe nuclei markedly reduce acoustic startle. Hence, this area is a particularly important one for startle inhibition. Perhaps, in part, this is because it contains the fibers that connect bilaterally the nuclei of the lateral lemniscus and inferior colliculus, which may mediate tonic inhibition from one side of the startle circuit to the other.

5. Pharmacology

5.1. Neurotransmitters That Mediate Acoustic Startle

Thus far there is no definitive evidence concerning the actual neurotransmitters that mediate startle from the auditory nerve to the spinal cord. It is almost certain that serotonin, norepinephrine, dopamine, glycine, and acetylcholine working at a muscarinic site are not involved, since depletion or antagonists of these transmitters do not prevent the occurrence of startle. Recent evidence suggests that aspartate may be the transmitter released from the auditory nerve onto the cochlear nucleus (Martin, 1980), although glutamate has not been entirely ruled out. It would be interesting, therefore, to apply aspartate antagonists locally onto the cochlear nucleus to determine if they would block startle.

Concerning the rest of the startle circuit, virtually nothing is known about the transmitters involved in mediating startle. Since complete spinal transection did not appreciably decrease spinal levels of glutamate, glycine, GABA, aspartate, alanine, serine, taurine, or methionine, it is unlikely that these substances would mediate transmission from the reticulospinal tract to the spinal cord (Singer *et al.*, 1981) provided that the pools being measured were ones relevant to neural transmission. Obviously, therefore, determining the actual neurotransmitters involved in mediating startle is an area that is in need of future research.

5.2. Neurotransmitters That Modulate Startle

5.2.1. Serotonin

A considerable amount of data leads to the conclusion that serotonin (5-HT) exerts a tonic inhibitory effect on startle and probably stimulus reactivity in general, since depletion of 5-HT is associated with increased startle.

5.2.1a. Lesions. Electrolytic lesions of the dorsal and median raphe nuclei, which contain the majority of cells in the brain that synthesize 5-HT, cause a pronounced increase in acoustic (Davis and Sheard, 1974a) or air-puff startle (Geyer *et al.*, 1976). For acoustic startle, lesions of both the dorsal and median raphe nuclei produce much larger effects than lesions of either nucleus alone (M. Davis, H. D. Bear, R. D'Aquila, and J. K. Walters, unpublished data). For air-puff startle, lesions of the median raphe produce the same size effect as combined lesions (Geyer *et al.*, 1976). More recently, Geyer *et al.*, 1982, have shown that repetitive presentation of startle-eliciting air puffs decrease intracellular content of 5-HT in the median but not the dorsal raphe nucleus, probably as a consequence of repetitive activation of cells in the median raphe nucleus by the startle stimulus.

5.2.1b. Drugs That Deplete 5-HT. Pharmacological depletion of 5-HT also usually results in increased startle, although the absolute magnitude of the effect generally is not as large as that achieved by mechanical lesions. *p*-Chlorophenylalanine (PCPA), a drug that blocks the synthesis of 5-HT, produces a small increase in acoustic startle (Carlton and Advokat, 1973; Conner *et al.*, 1970). However, some workers have not found significant changes after PCPA (Aghajanian and Sheard, 1968; Fechter, 1974*a*; Overstreet, 1977; Pohorecky *et al.*, 1976). When changes are found, they seem to emerge only after several test trials, probably because PCPA increases sensitization to repetitive stimulus exposure (Conner *et al.*, 1970), although even this is not clear (Fechter, 1974*a*; Overstreet, 1977). What is clear from the literature is that the magnitude of PCPA's effect on startle does not seem commensurate with the degree to which it depletes 5-HT. Why this might be the case will be discussed in Section 5.2.1g.

More striking results are obtained with *p*-chloroamphetamine (PCA). This drug has a number of actions of 5-HT neurons. Shortly after administration (15–30 min), PCA releases 5-HT. Thereafter, PCA causes a long-term depletion of 5-HT. When given to rats, PCA (5 mg/kg) depresses acoustic startle at the same time the drug releases 5-HT (Davis and Sheard, 1976). Later (2–15 hr), PCA has a marked excitatory effect on startle, during the period when 5-HT levels are rapidly falling. If these effects on startle depend on 5-HT, then prior depletion of 5-HT should make PCA ineffective. In fact, pretreatment with PCPA blocks both the early inhibitory and later depressant effects of PCA on startle (Davis and Sheard, 1976). In contrast, α-methyl-*p*-tyrosine (AMPT), a drug that blocks the synthesis of catecholamines, does not alter the effect of PCA on startle at a dose that completely blocks the excitatory effect of amphetamine on startle. One to four weeks after PCA, startle returns to control levels, despite the fact that 5-HT is still depleted by about 50%. Thus some system appears to compensate for the long-term depletion of 5-HT, perhaps by increasing the sensitivity of 5-HT receptors, although this has not been tested.

5.2.1c. Dietary Restrictions. More recently we have found a marked increase in acoustic startle when rats are fed a diet that lacks tryptophan, the precursor of 5-HT (Walters *et al.*, 1979). In our experiment, rats were intubated twice a day with either a tryptophan-free diet or the same diet with 0.5% L-tryptophan added. They were tested for startle every few days. Within 4–5 days, startle amplitudes of the tryptophan-free rats were about 60% higher than those of the control rats. At this point the diets were reversed. When this was done, startle performance of the groups also reversed. Parallel experiments indicated that 5-HT was depleted by about 60–70% within two days of being placed on the diet and then returned to normal levels within about two days after L-tryptophan was restored to the diet. Thus changes in startle lagged somewhat behind the changes in 5-HT. However, in a final experiment, systemic injection of L-tryptophan (125 mg/kg) in animals that had been on a tryptophan-free diet for several days was able to restore both startle amplitude and 5-HT to normal levels within 1 hr after injection.

5.2.1d. Hallucinogenic Drugs. 1. Indoleamine Hallucinogens. Another way of altering 5-HT transmission is by giving various doses of indoleamine hallucinogens, such as lysergic acid diethylamide (LSD) or N,N-dimethyltryptamine (DMT). In general, these drugs have biphasic dose-response effects on 5-HT transmission. Low doses decrease 5-HT transmission, whereas high doses enhance transmission (Haigler and Aghajanian, 1974; Aghajanian and Haigler, 1976; deMontigny and Aghajanian, 1977; Gallager and Aghajanian, 1975). If startle is modulated by 5-HT, hallucinogenic compounds should have biphasic dose-response effects; low doses should increase startle, whereas high doses should depress startle. This is what we have found. Low doses of DMT, psilocybin, or psilocin produce a small but consistent increase in startle, whereas higher doses depress startle (Davis and Bear, 1972; Davis and Sheard, 1974*c*; Davis and Walters, 1977). Bufotenin is more difficult to test since it does not pass the blood–brain barrier readily. Nonetheless, Geyer *et al.* (1975) found that bufotenin also produced biphasic dose-response effects on air-puff startle when given intraventricularly (D. B. Menkes, personal communication).

The pattern of results is similar for LSD, but more complicated when high doses are used. Low doses of LSD markedly increase acoustic startle (Davis and Sheard, 1974*b*; Miliaressis and St. Laurent, 1974). This may involve the raphe nucleus, since LSD no longer increases startle in rats with raphe lesions (Davis and Sheard, 1974*b*). Moreover, brom-LSD, which is relatively ineffective in depressing raphe cell firing rates, does not alter startle. High doses of LSD (greater than 300 μg/kg) do depress startle, although the depressant effect comes on rather slowly and follows an initial excitatory effect (M. Davis, unpublished data).

2. Phenylethylamines. It also is the case that phenylethylamine derivatives, which have hallucinogenic properties in humans, increase air-puff-elicited startle (Geyer *et al.*, 1978). The order of potency in increasing startle in the rat correlates positively with the order of potency in producing hallucinations in humans, DOM being the most potent. In addition, mescaline increases air-puff-elicited startle (Geyer, 1975; Geyer *et al.*, 1978). However, this effect does not seem dependent on the raphe nuclei, since Geyer *et al.* (1978) found that mescaline still increased tactile startle in rats with either dorsal or median raphe lesions. Furthermore, mescaline, DOM, and related compounds do not seem to have any consistent direct effect on the firing rate of raphe neurons (Haigler and Aghajanian, 1973).

None of these compounds has been systematically studied on acoustic startle. It was reported that neither mescaline nor 3,4-dimethoxyphenylethylamine altered acoustic startle (Bridger and Mandell, 1967). In contrast, mescaline does have excitatory effects on air-puff-elicited startle (Geyer *et al.*, 1978; 1979). However, very different test conditions were employed across these studies, so they are difficult to compare. Recently, I found that mescaline does increase acoustic startle using test procedures similar to (Geyer *et al.*, 1978).

Interestingly, LSD, DMT, and psilocybin do not appear to alter air-puff startle at doses that clearly have effects on acoustic startle (Geyer *et al.*, 1978). Hence, there seem to be differences in how acoustic and air-puff startle are affected by drug treatment. However, LSD will increase air-puff startle to the very first stimulus when a more intense air puff is used to elicit the response (Geyer *et al.*, 1978). Moreover, LSD may impair within-session habituation of tactile startle (Geyer *et al.*, 1978; Braff and Geyer, 1980). Perhaps similarly, levonatradol, a nonopiate analgesic related to cannabinoids also seems to impair habituation of tactile startle (Geyer, 1981). Thus the effects of the hallucinogens seem to be critically dependent on the characteristics of the stimuli used to elicit startle. In fact, with acoustic startle, the normal excitatory effect of LSD can be eliminated by testing at low levels of background noise (Davis and Sheard, 1974*b*) or by presenting tones at a rapid rate (Davis and Sheard, 1975). All of these data point to the conclusion that the hallucinogens alter startle by affecting the sensory rather than the motor side of the reflex arc, consistent with the profound effects these compounds have on perception in humans.

5.2.1e. Paradoxical Excitatory Effects of Serotonin. The data reviewed thus far point to the conclusion that 5-HT is inhibitory to startle. In fact, infusion of small amounts of 5-HT directly into the hippocampus (Geyer, 1975) or into the lateral ventricle (Geyer *et al.*, 1975) depresses startle. However, there are data that are difficult to reconcile with the conclusion that 5-HT depresses startle.

Fechter (1974*a*) reported that markedly increasing whole brain levels of 5-HT by administering a combination of 5-hydroxytryptophan (5-HTP, the immediate precursor of 5-HT), MK-486 (a peripheral decarboxylase inhibitor), and pargyline (a monoamine oxidase inhibitor that blocks the degradation of 5-HT) produced a large increase in acoustic startle. Similar results have also been found using a combination of L-tryptophan and pargyline (M. Davis, unpublished data). These results seem to depend on 5-HT since pretreatment with PCPA 72 hr before testing completely blocked the excitatory effect of L-tryptophan and pargyline.

Finally, the drug 5-methoxy,N,N-dimethyltryptamine (5-MeODMT), which acts as a 5-HT agonist in other behavioral tests, markedly increases acoustic startle (Davis *et al.*, 1980*a*). The excitatory effect of 5-MeODMT is completely blocked by the putative 5-HT antagonists cinanserin or cyproheptadine, but not by either haloperidol, a dopamine receptor blocker, or phenoxybenzamine, an α-adrenergic antagonist.

5.2.1f. Site of Action of Serotonin: A Dual Systems Theory. An overview of the literature reveals, therefore, a paradox. Depletion of 5-HT increases startle, but markedly elevating 5-HT also increases startle. To account for this I have proposed that 5-HT can have either inhibitory or excitatory effects on acoustic startle, depending on which 5-HT receptors are involved. More specifically, 5-HT in the forebrain may be inhibitory to startle, whereas 5-HT in the spinal cord may be excitatory. The threshold for the spinal excitatory effect may be higher, but once achieved it takes precedence over the forebrain inhibitory effect.

If 5-HT in the forebrain is inhibitory to startle, then direct infusion of 5-HT into the forebrain should depress startle. Conversely, if 5-HT in the spinal cord is excitatory to startle, then direct infusion of 5-HT into the spinal cord should increase startle. As mentioned earlier, administration of 5-HT into the lateral ventricles of the forebrain depresses tactile startle (Geyer, 1975). We have confirmed this with acoustic startle in rats pretreated with pargyline (Davis *et al.*, 1980*b*). Most striking however, is that 5-HT administered through chronic cannulae implanted in the subarachnoid space of the lumbar spinal cord, causes a marked increase in acoustic startle in rats pretreated with pargyline. More recently we have found that intrathecal administration of the 5-HT agonist 5-MeODMT produces a marked increase in startle (Astrachan and Davis, 1981; Commissaris and Davis, 1982; Davis *et al.*, 1980*a*), whereas intraventricular administration of the same drug depresses startle (Commissaris and Davis, 1982). These data provide further evidence that 5-HT can have either inhibitory or excitatory effects on acoustic startle depending on the place in the central nervous system where 5-HT acts. Moreover, the data are congruent with recent physiological evidence that 5-HT can have either inhibitory or facilitatory effects when applied directly to neurons in the forebrain or spinal cord, respectively (e.g., Haigler and Aghajanian, 1974; White and Neuman, 1980).

5.2.1g. Habituation and Sensitization. Depletion of 5-HT seems to increase acoustic startle by increasing sensitization without appreciably altering habituation. Connor *et al.* (1970) found that PCPA had no effect on startle to the first tone in the series, but instead increased responsivity to the second tone. Thereafter, PCPA-treated rats showed both within- and between-session response decrement, indicating that habituation could occur. *p*-Chloroamphetamine also did not alter habituation of acoustic startle (Davis and Sheard, 1976). Interestingly, PCPA-treated rats were much more dishabituated (a form of sensitization) when a novel change in background noise occurred or when the number of test tones were unexpectedly increased from 10 to 15 (Connor *et al.*, 1970). Rats with raphe lesions had startle amplitudes similar to controls on the first occurrence of a 115-dB tone, but had much higher amplitudes than controls to the next 115-dB tone when louder, 120-dB tones intervened (Davis and Sheard, 1974*a*). Thereafter, raphe-lesioned rats showed a response decrement curve that had a similar or even greater slope than that of the controls. Raphe lesions also fail to alter habituation of air-puff-elicited startle (Geyer *et al.*, 1976). Their effects on sensitization of startle to air puffs is less clear, since there is little information about the variables that are critical for producing sensitization of air-puff-elicited startle. Rats on tryptophan-free diets (Walters *et al.*, 1979) or rats given LSD (Davis and Sheard, 1974*b*) show exaggerated sensitization to background noise, since their response to the first tone after exposure to noise is greater than the response of controls. Thereafter, however, the rates of response decrement in the tryptophan-free or LSD rats are highly similar to controls. Indeed, the excitatory effect of LSD can be eliminated by testing at a low level of background noise. Similarly, the depressant effects of PCA on startle can also be eliminated by testing at a low level of background noise (Davis and Sheard, 1976). This is consistent with the idea that these compounds alter startle by interacting with sensitization. Removing the source of sensitization should block their behavioral effects.

More recent data indicate, however, that LSD may retard the rate of response decrement of air-puff-elicited startle when a large number of stimuli are given (Braff and Geyer, 1980). However, this effect was not seen following chronic administration, but instead LSD produced a marked increase in startle to the first stimulus in the series. Further work specifying the variables critical for habituation and sensitization of air-puff startle will be needed to explain the effects of LSD in this situation.

5.2.2. Dopamine

5.2.2a. Activation of Dopamine Transmission. Drugs that are thought to increase dopamine (DA) transmission increase startle. Apomorphine, a drug widely believed to be a direct DA agonist, increases acoustic startle (Davis and Aghajanian, 1976; Kehne and Sorenson, 1978) or air-puff-elicited startle (Geyer *et al.*, 1978). The excitatory effect occurs very rapidly after injection (2–4 min) and is essentially over within 30–50 min, depending on the dose. This probably explains why others (Fechter, 1974*b*;

Handley and Thomas, 1979) have failed to find an effect of apomorphine, since they did not begin testing until 40 min after injection.

The excitatory effect of apomorphine can be blocked by the putative DA antagonist haloperidol (Davis and Aghajanian, 1976) or pimozide (Kehne and Sorenson, 1978) at doses that do not have marked depressant effects on startle by themselves. Apomorphine also increases acoustic startle in the cat (M. Davis and G. R. Heninger, unpublished data) and this effect can be blocked by a very low dose of haloperidol but not by promethazine, a phenothiazine that lacks DA antagonist properties.

Consistent, but more complicated evidence comes from the finding that very high doses of *d*- or *l*-amphetamine also markedly increase acoustic startle in rats (Cladel *et al.,* 1966; Davis *et al.,* 1975; Kehne and Sorenson, 1978; Kirkby *et al.,* 1972); cats (M. Davis and G. R. Heninger, unpublished data) or mice (Kokkinidis and Anisman, 1978). The doses of *d*-amphetamine (greater than 3 mg/kg) or *l*-amphetamine (greater than 8 mg/kg) required to increase acoustic startle are in the same range as those required to produce stereotyped movements in these species, a behavior that is almost certainly mediated by increased dopamine transmission. However, the effect on startle cannot be explained by an artifact caused by stereotyped behavior but mistaken for startle, since sampling cage movement in the absence of a startle-eliciting tone revealed that the background level of movement induced by the drug was far below that induced by the tones (Davis *et al.,* 1975). Indeed, it is fascinating to observe these animals. Even though they are engaged in vigorous stereotyped movements and seem totally "preoccupied," they still show large startle responses when the tone is sounded. This indicates that what might look like competing motor movements do not necessarily interfere with startle under all circumstances.

The excitatory effects of amphetamine can be blocked by pretreatment with AMPT at a dose (100 mg/kg) that only slightly depresses startle by itself (Davis *et al.,* 1975). This is consistent with the general finding that the behavioral effects of amphetamine are dependent on newly synthesized catecholamines. Moreover, pimozide completely blocks the excitatory effect of either *d*-amphetamine or *l*-amphetamine (Kehne and Sorenson, 1978). Finally, rats treated with 6-hydroxydopamine intraventricularly and tested about three months later showed increased startle and decreased brain levels of DA relative to vehicle-treated rats (Sorenson and Davis, 1975). Since acute depletion of DA via AMPT depressed startle, it was concluded that the increased startle observed after chronic DA depletion reflected supersensitivity of DA receptors to remaining levels. However, direct tests of this hypothesis await further experiments.

5.2.2b. Inactivation of Dopamine Transmission. Although activation of dopamine transmission clearly increases startle, turning off dopamine transmission has only a small depressant effect on startle. Drugs that decrease dopamine synthesis, such as AMPT, slightly depress acoustic startle (Davis *et al.*, 1975; Sorenson and Davis, 1975). Similarly, the DA antagonists haloperidol or pimozide also depress acoustic startle in rats (Davis and Aghajanian, 1976; Kehne and Sorenson, 1978) or cats (M. Davis and G. R. Heninger, unpublished data; Pohorecky, 1976). However, whether these depressant effects are actually caused by a decrease in DA transmission or nonspecific sedative effects is not known. Fairly high doses of promethazine (1 mg/kg), which apparently do not block DA receptors, will depress acoustic startle in cats, probably because of its sedative effect. Furthermore, very low doses of apomorphine do not depress startle (Davis and Aghajanian, 1976) as might be expected, since low doses of apomorphine decrease DA release and preferentially depress firing rates of the presynaptic DA neurons (Skirboll *et al.*, 1979). In addition, a combination of haloperidol and AMPT, which should markedly depress DA transmission by preventing a feedback increase of DA (Carlsson *et al.*, 1973), also fails to alter startle significantly (M. Davis, unpublished data).

Based on this evidence, it is probably the case that DA does not exert a tonic excitatory effect on startle. If the system is driven via DA agonists, startle can be increased markedly. However, if the system is turned off, startle is only slightly affected, if at all.

Virtually nothing is known about which DA systems might be involved in mediating increased startle with increased transmission. We do know, however, that both apomorphine and *l*-amphetamine fail to augment acoustic startle in acutely decerebrate rats (Gendelman, 1979). Moreover, direct infusion of either DA itself or of apomorphine directly into the subarachnoid space around the spinal cord does not alter acoustic startle over a variety of doses (Astrachan and Davis, 1981).

5.2.3. Norepinephrine

5.2.3a. Activation of Norepinephrine Transmission. The role played by norepinephrine (NE) in modulating acoustic startle is less clear than that of DA. This stems primarily from the fact that there is no clean NE agonist that crosses the blood–brain barrier. Because of this, all of the experiments are somewhat indirect, but taken together they suggest that NE is excitatory to startle perhaps by activation of α-1-adrenoceptors in the spinal cord.

Low doses of *l*-amphetamine that may preferentially release NE

(Bunney *et al.*, 1975) increase acoustic startle significantly but only by about 25–30% (M. Davis, unpublished data). Drugs that increase the availability of central NE by blocking its reuptake, such as desipramine or chlordesipramine, increase acoustic startle somewhat at moderate doses (Davis *et al.*, 1977*b*). Yohimbine, a drug that increases the firing rate of cells in the locus coeruleus and increases NE turnover, also enhances acoustic startle (Davis and Astrachan, 1981).

5.2.3b. Spinal Norepinephrine and Startle. As with 5-HT, NE also facilitates the response of motor neurons to afferent stimulation (McCall and Aghajanian, 1979; White and Neuman, 1980). The spinal cord contains high levels of NE so that the excitatory effects of systemic drug administration might be mediated partially or fully by activation of NE in the spinal cord. Consistent with this, we have found that direct infusion of NE into the spinal cord or *d*-amphetamine, which releases NE, increases the amplitude of acoustic startle (Astrachan and Davis, 1981; Davis and Astrachan, 1981). This effect seems to be mediated by an α-1-adrenergic receptor, since intrathecal infusion of the α-1-adrenergic agonist, phenylephrine, also increases acoustic startle, and this effect can be blocked by systemic administration of the α-1-adrenergic agonist, WB-4101, but not by the 5-HT antagonist cyproheptadine or the β-adrenergic antagonist propranolol (Astrachan and Davis, 1981). Moreover, intrathecal infusion of the β-adrenergic agonist isoproterenol does not increase startle over a wide range of doses (Astrachan and Davis, 1981).

The excitatory effect of intrathecal phenylephrine can be markedly increased by depleting spinal NE with intrathecally administered 6-OHDA (Astrachan *et al.*, 1983). The increased behavioral response to phenylephrine seems to be explained by receptor supersensitivity since it is associated with an increase in the number of α-1-adrenergic receptor sites measured by [^{3}H]prazosin binding in lumbar spinal tissue with no change in receptor affinity or receptor occupation. In fact, Figure 5 shows that there was a remarkably high correlation between the increased effect of phenylephrine in facilitating startle and the number of binding sites at various times after denervation.

5.2.3c. Inactivation of Norepinephrine Transmission. Norepinephrine may tonically enhance startle, since various methods of decreasing NE transmission seem to decrease startle. Electrolytic lesions of the locus coeruleus or chemical destruction of these cells via 6-OHDA depress both acoustic and tactile startle, respectively (Davis *et al.*, 1977*a*; Adams and Geyer, 1981). High doses of the α-1-adrenergic antagonists phenoxybenzamine or prazosin depress startle (Davis *et al.*, 1980*a*; M. Davis, un-

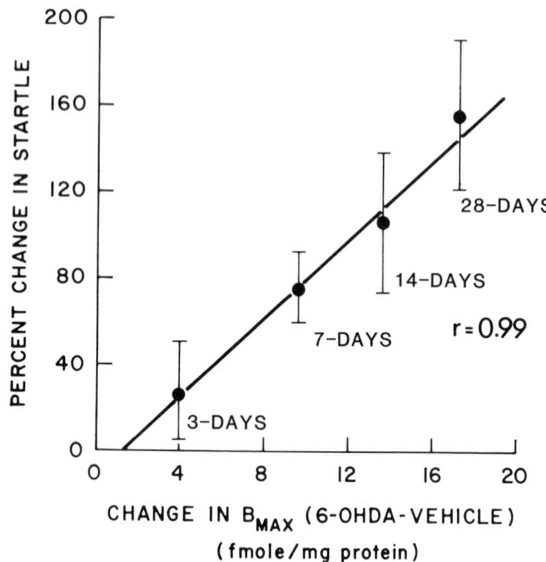

Figure 5. Correlation between increases in receptor number (Bmax) as described by Scatchard analysis of [^{3}H]prazosin binding in rat lumbar spinal cord and mean percent startle increase in response to 12.5 μg of intrathecal phenylephrine, [(startle averaged from 15 to 30 min postdrug infusion minus startle averaged over a 14-min predrug period) × 100] (from Astrachan *et al.*, 1983).

published data; Kehne and Sorenson, 1978). Finally, the drug clonidine, which decreases the firing rate of cells in the locus coeruleus and decreases NE release and turnover, markedly depresses acoustic and tactile startle at very low doses (Davis and Astrachan, 1981; Davis *et al.*, 1977*a*; Fechter, 1974*b*; Geyer *et al.*, 1978; Handley and Thomas, 1979). These effects can be blocked or attenuated by the α-2-adrenergic antagonists, piperoxane or yohimbine, but not by the α-1-antagonist WB-4101 (Davis and Astrachan, 1981; Davis *et al.*, 1977*a*). Clonidine still depresses startle in acutely decerebrate rats or in rats with lesions of the locus coeruleus (Davis *et al.*, 1977*a*). Clonidine also depresses startle when given intrathecally, and intrathecal administration of yohimbine attenuates the depressant effect of systemically administered clonidine (Davis and Astrachan, 1981). Although this suggests that clonidine may act within the spinal cord to depress startle, other areas may be involved since we have found that intrathecally administered clonidine apparently can get into the general circulation and act on supraspinal structures as well (Marwaha *et al.*, 1982). Moreover, clonidine also depresses startle after intraventricular administration (R. L. Commissaris and M. Davis, unpublished data).

In spinalized rats, high doses of clonidine act as α-1-agonists and increase the flexor reflex (Anden *et al.*, 1970). However, even very high doses of clonidine do not increase startle, as might be expected if it acted like other α-1-agonists in the spinal cord to increase startle. Recently, John Kehne in our laboratory has shown that clonidine increases the flexor reflex in spinalized rats, but actually depresses the same reflex in nonspinalized rats (Kehne, 1983). The reversal from clonidine inhibition to excitation occurs immediately on spinalization or cessation of impulse flow in the cord and does not seem attributable to supersensitivity of α-1-adrenergic receptors in the cord. All and all, Kehne's data indicate that systemic administration of clonidine must activate some supraspinal system that inhibits the flexor reflex to such a degree that it cannot be activated by clonidine's direct interaction with α-1-adrenoceptors in the spinal cord. It is possible that clonidine's depression of acoustic startle acts through the same supraspinal system. Ongoing studies in our laboratory are attempting to determine what this might be by making lesions in various supraspinal areas to see if these will make a normally inhibitory dose of clonidine excitatory to startle.

5.2.3d. Habituation and Sensitization. A particularly intriguing aspect of clonidine is that it appears to depress startle by improving habituation (Davis *et al.*, 1977*a*). This conclusion was based on the findings that clonidine accelerated the rate of response decrement during repetitive tone presentation, but did not attenuate sensitization to background noise, and therefore did not impair the ability of rats to startle. Accelerated response decrement could not be attributed to progressive absorption of clonidine since presentation of tones rather than simply previous drug injection was required to produce a response decrement. Moreover, prior exposure to several tones plus clonidine produced greater response attenuation than prior exposure to only a few tones plus clonidine. These results are indeed interesting, since other drugs that depress startle responsivity (e.g., high doses of DMT) do not appear to do so by accelerating habituation. Since there are only a handful of reports throughout the entire habituation literature that suggest that certain drugs accelerate habituation without impairing the ability to respond, the pharmacological mechanism by which clonidine affects startle is all the more important to discover.

More recently, we have found that decreases in NE transmission may block sensitization (Davis, 1981). In these studies startle was elicited electrically through the nucleus reticularis pontis caudalis, the part of the startle circuit that projects directly to the spinal cord. In these animals, repetitive elicitation of startle or exposure to background noise produced a marked increase in the response over a one-half-hour session. Systemic administration of the α-1-antagonist WB-4101 eliminated this sensitization, whereas administration of the 5-HT antagonist cyproheptadine or prior depletion of 5-HT by PCPA did not. Finally, Adams and Geyer (1981) found that 6-OHDA lesions of the locus coeruleus accelerated within-session response decrement of tactile startle and have argued that this change results from a decrease in sensitization.

5.2.4. Norepinephrine–Dopamine Interactions in Modulating Startle

A number of observations suggest that NE and DA interact in modulating startle. Kehne and Sorenson (1978) found that phenoxybenzamine markedly attenuated the excitatory effect of apomorphine on startle. Kokkinidis and Anisman (1978) found that FLA-63, a drug that depletes NE by blocking its conversion from DA, also blocked the *d*-amphetamine effect on acoustic startle or FLA-63 at a dose that had no significant effect by itself. Drugs that block α-1-adrenoceptors (systemic phenoxybenzamine, piperoxane, intrathecal phentolamine) also depress the excitatory effects of *d*-amphetamine on tactile startle (Handley and Thomas, 1979). We have found similar blocking actions on apomorphine using some different α-1-adrenergic antagonists. Thus WB-4101 or prazosin block the excitatory effects of apomorphine, whereas the peripherally acting NE antagonist phentolamine does not (M. Davis, R. L. Commissaris, and J. H. Kehne, unpublished data). Conversely, drugs that activate NE transmission, such as yohimbine or piperoxane, greatly augment the excitatory effects of apomorphine on acoustic startle. These effects cannot be explained by simple additivity, since the doses of yohimbine or piperoxane used only slightly increase startle by themselves. Measuring tactile startle, Handley and Thomas (1979) reported that intraventricularly administration of drugs that should enhance NE transmission (NE itself, α-methyl norepinephrine) as well as systemic administration of high doses of clonidine or yohimbine all potentiated the excitatory effects of systemically administered *d*-amphetamine.

Taken together, the data indicate that there are important interactions between NE and DA in modulating startle. In general the data indicate that DA and NE act in series, so that decreases in NE transmission prevent an expected DA-mediated activation of startle and increasing NE transmission augments excitatory effect of DA activation. This is particularly interesting in light of recent data showing that prior isolation or shock stress enhances the excitatory effect of *d*-amphetamine on acoustic startle

(Kokkinidis and MacNeill, 1982). Since stress activates NE as well as DA systems, this may represent another example of the interaction between catecholamines in increasing startle.

It should be pointed out, however, that there are some data that do not fit with this interpretation. The most striking example is the fact that the α-1-antagonist phenoxybenzamine did not block or even attenuate the excitatory effect of *d*-amphetamine on startle at a dose that essentially eliminated the effects of either *l*-amphetamine or apomorphine (Kehne and Sorenson, 1978). We have replicated this finding with phenoxybenzamine as well as with another α-1-antagonist, WB-4101. Furthermore, the excitatory effect of *d*-amphetamine is also not blocked by the β-adrenergic antagonist propranolol (C. A. Sorenson, personal communication) or the 5-HT antagonist cyrophetadine. It is important to reiterate, however, that the excitatory effect of *d*-amphetamine can be readily blocked by DA antagonists such as haloperidol (M. Davis, unpublished data) or pimozide (Kehne and Sorenson, 1978). Hence with *d*-amphetamine, activation of DA systems can increase startle somewhat independently from a link with NE systems. In contrast, with apomorphine, activation of DA systems seems to be linked to NE systems in increasing startle.

5.2.5. Acetylcholine

The role that acetylcholine (ACh) plays in modulating acoustic startle is not clearly understood. The most extensive analysis of this question has been made by Overstreet (1977), who concluded that "the cholinergic system modulates the basal startle response level, with activation of the system depressing and blockade of it facilitating startle." Supporting this conclusion was the fact that the muscarinic agonist, pilocarpine, depressed acoustic startle, whereas the muscarinic antagonist, atropine, increased acoustic startle, particularly over the first several test trials. However, within the same study, diisopropyl fluophosphate, an irreversible cholinesterase inhibitor that increases brain levels of ACh, slightly enhanced startle, perhaps by depressing within-session habituation. Moreover, physostigmine, a reversible cholinesterase inhibitor, did not alter baseline levels or habituation of startle. Finally, whereas doses of 1 or 5 mg/kg of atropine did appear to increase startle, the results were not statistically significant and a dose of 10 mg/kg had virtually no effect.

Handley and Thomas (1979) also concluded that activation of cholinergic transmission may depress tactile startle since the ACh agonists arecoline and carbacol (given intraventricularly) tended to depress startle, whereas the ACh antagonists atropine and hyoscine tended to elevate startle.

Warburton and Groves (1969) reported that scopolamine, a central muscarinic antagonist, which is generally more potent than atropine, enhanced acoustic startle at doses of 0.25–0.75 mg/kg. However, as the authors acknowledge, the design of this dose-response study did not permit an adequate conclusion. In a second study, scopolamine (0.5 mg/kg) produced about a 14% increase in startle. In contrast, Payne and Anderson (1967) reported that scopolamine (1.0 mg/kg) depressed acoustic startle in both albino and hooded rats. Finally, Williams *et al.* (1974) found no effect of either 0.5, 1, or 2 mg/kg scopolamine. In our own laboratory, we have not found consistent effects of scopolamine on startle.

Although the effects of ACh on baseline levels of startle are somewhat inconsistent, it seems safe to conclude that scopolamine, as well as other drugs that alter ACh, do not affect the rate of within-session habituation (Overstreet, 1977; Warburton and Groves, 1969; Williams *et al.*, 1974). On the other hand, scopolamine may depress between-session habituation (Warburton and Groves, 1969; Williams *et al.*, 1974). However, controls for state-dependent learning (state-dependent habituation) will have to be included to determine more precisely how scopolamine might attenuate between-session habituation.

Viewed together, it is still difficult to make any conclusion about the role of ACh in modulating startle. A systematic investigation of these variable-drug interactions will be required before definitive conclusions can be made. The possibility that the cholinergic system might interact with other transmitter systems in modulating startle could be a particularly fruitful way to approach this question.

5.2.6. CNS Depressants

5.2.6a. Alcohol. Pohorecky *et al.* (1976) found that ethanol produced a dose-dependent reduction in acoustic startle. At a dose of 1 g/kg, the effect was maximal 30 min after injection and lagged somewhat behind the time of peak blood ethanol levels. Rats maintained on liquid diets containing alcohol became tolerant to the depressant effects and then showed a pronounced increase in startle amplitude when alcohol was withdrawn. The peak increase in startle occurred about 9–12 hr after alcohol was removed, when alcohol was no longer detectable in the blood. Startle required about 4 days to return to normal. Gibbins *et al.* (1971) found a decrease in the flinch threshold elicited by electrical footshock during partial or complete withdrawal of ethanol, which could be quickly reversed by administration of alcohol.

More recently, Anadam *et al.* (1980) reported that rat offspring from mothers treated during pregnancy with alcohol show heightened startle reactivity in the absence of hyperactivity or impaired habituation.

5.2.6b. Sodium Pentobarbital. Hammond and Ison (1973) reported a very interesting sequence of events that occur after anesthetic doses (40 mg/kg) of sodium pentobarbital are injected. Very shortly after injection (about 2 min), at a time when rats become very active, startle was markedly increased over baseline. Thereafter, startle gradually declined to zero levels and was undetectable within about 7–8 min. Within about 1 hr, startle began to recover and reached a peak in about 75 min, which then returned to control levels within the next 2–3 hr. Most important, however, was the finding that prepulse inhibition was essentially absent during periods when startle was maximal (2 or 75 min after injection). However, as startle declined to low levels, just before narcosis (3–4 min after injection), prepulse inhibition was again evident. Although some explicit control for ceiling effects would have been comforting, such as using a weaker stimulus intensity to elicit startle, the results probably cannot be accounted for by ceiling effects. Thus in several individual instances, startle amplitudes were actually higher to tones preceded by the prepulse than on the control tones. This is very unusual and suggests that sodium pentobarbital interacts in an important way with whatever mechanism mediates prepulse inhibition.

5.2.6c. Opiates. The opiate agonist morphine (0.3–20 mg/kg) depresses acoustic startle (M. Davis and J. K. Walters, unpublished data). The effect is not very large (30%), is maximal once a dose of about 2.5 mg/kg is reached, and peaks about 30 min after an intraperitoneal injection. Very shortly after injection, morphine produces a transient increase in startle that is directly related to the dose. However, none of the doses tested was capable of making this excitatory effect last more than about 2–4 min. Both the early excitatory and later depressant effects of morphine were blocked by naloxone. By itself, naloxone (0.8–10 mg/kg) had a small excitatory effect on startle (20–25%) that is maximal about 20 min after injection. Based on this evidence it seems fair to conclude that stimulation of opiate receptors can produce a decrease in startle and that endogenous opiates may exert a tonic inhibitory effect on acoustic startle. It should be emphasized, however, that the effects are rather small and even large doses of these drugs do not have a very remarkable effect.

More recently, Warren and Ison (1982) have shown that startle elicited by a tone is minimally depressed by morphine, whereas comparable levels of startlelike responses elicited by a tail shock are markedly depressed by morphine. On the other hand, the inhibitory effect of a prior shock presentation on acoustic startle amplitude was not appreciably altered by morphine. As Warren and Ison point out, the data indicate first that the depressant effects of morphine on startle reactivity are not due to a general motor deficit for all reflex activity, and second that "morphine is not attenuating the simple sensory aspects of the noxious stimulus as much as reducing its nocioceptive properties, or aversiveness."

In contrast to these small effects on acoustic startle of morphine given systemically, injection of morphine directly into the periaqueductal gray region of the midbrain or intraventricularly produce explosive startle responses (Jacquet *et al.*, 1977; Jacquet and Lajtha, 1973; Sharpe *et al.*, 1974; Shizgal *et al.*, 1977). Lesions of the periaqueductal gray also produce explosive startle responses that look similar to the behavior elicited by infusion of morphine into this area (Blair *et al.*, 1980). The lesion effect subsides over time, so that 24 hr afterward explosive startles no longer occur. Importantly, morphine given intraventricularly at this time no longer produced heightened reactivity. This supports the idea that the periaqueductal gray region mediates the effect of morphine when given intraventricularly or directly into this region. It cannot be overemphasized how important this system must be. Without it animals literally startle themselves to death. If tested in a small cage, they will jump so hard that they will knock themselves unconscious on the top of the cage (Jacquet *et al.*, 1977). Since morphine injected into this region produces the same effect as a lesion of this area, morphine may somehow inhibit this tonic inhibitory system.

Clearly, however, morphine given systemically does not produce explosive motor behavior as expected if it only acted on this critical region in the central gray. In fact, it slightly depresses startle. This means that systemic morphine must also act on some other system, which can cancel the effect expected by morphine's action on the central gray. In fact, Jacquet *et al.* (1977) mentioned that pretreating animals with systemic morphine blocked the explosive behavior normally elicited by infusing morphine into the central gray. They have also shown that the unnatural morphine isomer (+)-morphine also produces explosive motor behavior when injected into the central gray, but is much weaker than the natural (−)-morphine on virtually all other tests. Moreover, naloxone fails to block the explosive motor effects of either isomer, whereas it does of course block many other effects of (−)-morphine.

To account for this, Jacquet *et al.* (1977) have suggested that there are two different receptors on which morphine acts. One, the opiate receptor, mediates analgesia and other effects. Another, the receptor associated with the central gray, mediates heightened stimulus reactivity. Here, naloxone is not an effective antagonist. By acting on both receptors, morphine given systemically does not produce explosive motor behavior since this is somehow canceled by its action on other neuronal systems where naloxone may be the antagonist. However, when an animal that had been given morphine chronically was then given naloxone, morphine would no longer act on the naloxone-sensitive sites, but still could act on the naloxone-insensitive sites. If the naloxone sites normally inhibit the heightened reactivity expected by morphine interaction with the latter, central gray, then morphine could now produce explosive motor behavior. In fact, a classic syndrome of animals made dependent on morphine and then challenged with naloxone is heightened reactivity to sensory stimuli, including explosive jumping (Way *et al.*, 1969).

However, naloxone does not produce explosive motor behavior in rats given morphine acutely. This is what one would expect if naloxone blocked morphine's interaction with systems that protect the organism from heightened reactivity. Hence there is something about chronic morphine treatment that brings out this effect. Future studies using other antagonists after acute morphine will be necessary to fully uncover the several ways in which the opiates may modulate startle.

5.2.7. Amino Acids

5.2.7a. Glycine. Glycine appears to tonically inhibit acoustic startle. Thus the glycine antagonist strychnine causes a dose-dependent increase in startle at doses that did not elicit convulsions (Kehne *et al.*, 1981). These effects seem to be mediated in the spinal cord and brainstem, since intrathecal or intracisternal infusion of strychnine also increased startle, whereas intraventricular infusions had no effect or actually depressed the reflex. Thus far, strychnine is the most efficacious drug in increasing startle we have found. If its effects are indeed due to blockade of glycine, this indicates that startle is always tonically inhibited in a normal animal. Given that the central gray and mesencephalic reticular formation seem to be very important areas for tonically inhibiting startle, it would be important to test if glycine were the transmitter that mediated this inhibition.

5.2.7b. GABA. The role played by GABA is less clear than glycine. Recently we have suggested that GABA may inhibit startle in the spinal cord, but functionally increase startle in supraspinal areas (Gallager *et al.*, 1983), since the GABA antagonist picrotoxin increases startle when given intrathecally, but depresses startle when given intraventricularly. After systemic administration, picrotoxin depresses startle at subconvulsant doses. Even doses that cause convulsions in 50% of the rats still depress startle. Moreover, we have found that other convulsants such as pentylenetetrazol depress startle in the rat (M. Davis and D. W. Gallager, unpublished data). Interestingly, however, both picrotoxin and pentylenetetrazol increase acoustic startle in rat pups below the age of about 21 days (Gallager *et al.*, 1983). Hence it is possible that the system that mediates a functionally excitatory effect of GABA on startle does not mature until this time, so that only the inhibitory system is involved. It is even conceivable, therefore, that different species could differ in the degree to which these functionally excitatory or inhibitory GABA systems predominated. If the inhibitory system tended to predominate in cats, this would explain why cats seem to show an increase in startle after convulsant doses of pentylenetetrazol (Faingold *et al.*, 1981) in contrast to rats.

5.2.8. Effects of Antianxiety Drugs on the Potentiated Startle Effect

The magnitude of the startle reflex can be augmented by presenting the eliciting stimulus in the presence of a cue that has previously been paired with shock (Brown *et al.*, 1951). This phenomenon has been termed the potentiated startle effect and has been replicated a number of times (cf., Davis and Astrachan, 1978). The effect has been used as evidence that startle is increased by fear and could provide a useful model system to study how drugs alter fear.

If potentiated startle does reflect fear, then it should be attenuated or blocked by drugs that are thought to reduce fear in other situations or anxiety clinically. Consistent with this expectation, Chi (1965) found that sodium amytal produced a dose-related reduction in potentiated startle. D. R. Williams (reported in Miller and Barry, 1960) found a similar reduction using alcohol. In both cases the effect was somewhat selective since the same doses did not alter baseline levels of acoustic startle. In contrast, amphetamine, mescaline, or dimethoxyphenylethylamine have been reported to increase the potentiated startle effect when administered before a combined training and testing session (Bridger and Mandel, 1967).

More recently, the benzodiazepines diazepam and flurazepam have also been found to cause a dose-dependent reduction of the potentiated startle effect when given prior to testing (Davis, 1979*a*). The minimum effective dose of diazepam was about 0.3 mg/kg, which makes the potentiated startle effect more sensitive to diazepam than the widely used operant conflict test (cf., Cook and Sepinwall, 1975). Moreover, a 2 × 2 design in which rats were trained and tested after injection of either diazepam or saline indicated that its attenuation of potentiated startle could not be explained by state-dependent generalization decrement and that it did not interfere with fear acquisition. Morphine also causes a decrease in potentiated startle with a minimum effective dose of about 2.5 mg/kg (Davis, 1979*b*). Importantly, each of these compounds at the doses used is selective in the same sense that they decrease potentiated startle without appreciably decreasing baseline levels of startle. Finally, we have found that drugs that reduce NE transmission and reduce anxiety in humans (low doses of clonidine, propranolol) decrease potentiated startle, whereas drugs that increase NE transmission and produce anxiety in humans (piperoxane, yohimbine) actually increase the magnitude of potentiated startle (Davis *et al.*, 1979). For the latter drugs, doses were used that had no consistent effect on baseline levels of startle. Taken together, the data indicate that potentiated startle can be a more sensitive measure for certain drugs than unconditioned startle. Moreover, the paradigm may be especially useful for analyzing how drugs alter fear and perhaps anxiety, since it is sensitive to drugs that either increase or decrease anxiety in humans.

6. Development

In rats and mice, acoustic startle can first be seen at postnatal day 12 (Parisi and Ison, 1979; Shnerson and Willott, 1980). This point is delayed by 1 day if rats are treated neonatally with the neurotoxin 6-OHDA (Parisi, 1979). The fact that the external auditory meatus also opens at this time (Gottlieb, 1971) may account for the abrupt onset of startle, since auditory systems do appear to develop somewhat earlier (cf., Rubel, 1978). Over the next several days, startle increases in amplitude both visually and as measured by cage movement. However, determining the relative magnitude of startle in rat pups and adults is exceedingly difficult, since body weight changes so rapidly over early development, it is difficult to decide if an increase in startle with age occurs because startle is actually greater or simply that heavier weights are able to displace the cage more effectively.

On the other hand, latency measurements suggest that the responses are probably very similar in pups and adults. Parisi (1979) found that 14-day-old rats had a startle latency about 3.5 msec longer than 18-day-old rats when startle was measured electromyographically in the neck. We have found very similar latencies for acoustic startle recorded in cages equipped with accelerometers in 16-day-old pups (about 14 msec) and adults (about 12 msec). Hence, the very rapid nature of startle seen at early ages strongly suggests that the same neural circuit is involved in pups and adults.

It is also the case that habituation of startle occurs very early; in fact, habituation seems to occur as soon as one can measure the response

(Williams *et al.*, 1975). Although there may be some differences in the rate of habituation at different ages (Hamilton and Timmons, 1979), this is a tricky issue since weights and initial startle levels are also varying across age, which can make interpretation of differences in habituation rates very difficult. In contrast, prepulse inhibition (see Chapter 9, this volume) does seem to develop over time so that adult levels are not reached until about 20 days of age using an auditory prepulse or 30 days of age using a visual prepulse (Parisi and Ison, 1979).

Kellogg *et al.*, (1980) found that rats exposed to several consecutive doses of either 2.5, 5.0, or 10.0 mg/kg diazepam in utero during the last week of gestation showed abnormal patterns of locomotor activity from days 14 to 20, when no detectable amounts of diazepam could be measured in their brains. During this period, these rats failed to show the usual increase in acoustic startle when background noise levels were increased from 60–70 dB, in contrast to control rats whose mothers had been treated with the diazepam vehicle. The authors conclude that prenatal exposure to diazepam interfered with the development of normal arousal.

In humans, elements of the startle response (eyeblink and contraction of the abdomen) are present at very early ages, although they are not identical to the full adult pattern and may be masked by the more prominent Moro reflex (Landis and Hunt, 1939). The Moro reflex also occurs to sudden auditory stimuli. However, it is primarily an extension response where the arms extend straight out from the trunk, the fingers extend, and the trunk is arched backward with head extended. This reflex is prominent in the first month of life but then gradually decays over time so that it is not seen beyond the fourth month. As the Moro reflex decays, the adult startle pattern emerges more prominently, although the exact relationship between the two is not clear. Landis and Hunt (1939) suggest that from 6 weeks on, both the Moro and startle reflex occur together, with the startle pattern preceding the Moro reflex in terms of latency. With development the startle pattern remains, whereas the Moro reflex gradually decays.

7. Modification of Startle by Prior Experience

7.1. Prepulse Facilitation and Inhibition

Small changes in the sensory environment (prepulses) that precede intense startle-eliciting stimuli produce highly reproducible changes in startle (cf., Hoffman and Ison, 1980; Chapter 9, this volume). At very short intervals (e.g., 5–10 msec) prepulse exposure reduces the latency of startle. At longer intervals (15–2000 msec) prepulse exposure reduces the amplitude of startle. These effects are cross modal and occur using prepulses that are near threshold. Prepulse inhibition develops from 12–30 days of age (Parisi and Ison, 1979), and can be attenuated by lesions in the mesencephalic reticular formation at the level of the cuneiform nucleus (Leitner *et al.*, 1981). The elegant and extensive work done by Ison and his colleagues and Hoffman and his colleagues is thoroughly reviewed by Howard Hoffman in Chapter 9 of this volume.

7.2. Habituation

Practically everyone who has measured startle in any systematic fashion has observed a decline in this behavior with repeated presentations of the eliciting stimulus. In humans this may occur very rapidly, after a few stimuli, particularly when whole body movements are being measured (Landis and Hunt, 1939). Head movements and especially the eyeblink may never habituate totally, although their amplitudes do decline (e.g., Davis and Heninger, 1972). Occasionally people are found who show extremely slow rates of startle habituation to the point where their lives are even disrupted (e.g., Thorne, 1944). In rats, startle habituation does not seem to occur as readily as it does in humans, monkeys, or cats (e.g., Heninger and Davis, 1979). Nonetheless, both within- and between-session startle habituation very definitely does occur in the rat, and may even last over long retention intervals (Leaton, 1976). Hoffman and Stitt (1969) showed that startle habituation still occurs when rats are trained to stay in the same position during testing, indicating that habituation does not result because the rat locates itself in a less intense part of the acoustic field. In fact, startle habituation in the rat has been used extensively to study the process of habituation and findings derived from the study of startle habituation have proven to generalize to a variety of other stimulus–response systems in other organisms.

7.2.1. Stimulus Intensity and Startle Habituation

It has frequently been asserted that the rate or degree of habituation is inversely related to the intensity of the eliciting stimulus. This conclusion was based on observations from an experimental procedure in which a given subject has been repetitively presented with the same stimulus intensity. However, like most other behaviors, startle is more probable when elicited by an intense versus a weak stimulus. Hence it might be concluded from previous studies that the more intense a test stimulus that is presented, the less likely will a prior training series have rendered that stimulus ineffective, independent of the intensity employed during training. To evaluate differential changes in responsiveness that may be attributed to the history of intensities experienced, it is necessary to arrange separate training and testing phases in which subjects are exposed during a habituation series to different stimulus intensities, but are then tested with a common intensity or common series of intensities.

To evaluate this, startle was measured in rats presented with 20 exposures to each of five tone intensities before and after 300 or 700 presentations of either a 108- or 120-dB tone (Davis and Wagner, 1968). Hence, all rats received identical pretests and posttests, whereas the number and intensity of the stimuli differed among the four groups during the "habituation training session."

Figure 6 shows the mean percentage startle responses of the four groups at each of the five test intensities, before and after the different habituation treatments. As expected, there was a highly reliable overall increase in startle frequency with increasing intensity of the test stimulus in all groups. During the treatment phase (not shown), all groups showed a decreasing frequency of startle, with repeated exposures to the selected tone intensity. Of primary interest, however, was performance following the separate habituation treatments, when each group was again presented with the same sequence of test intensities. Figure 6 indicates that at each

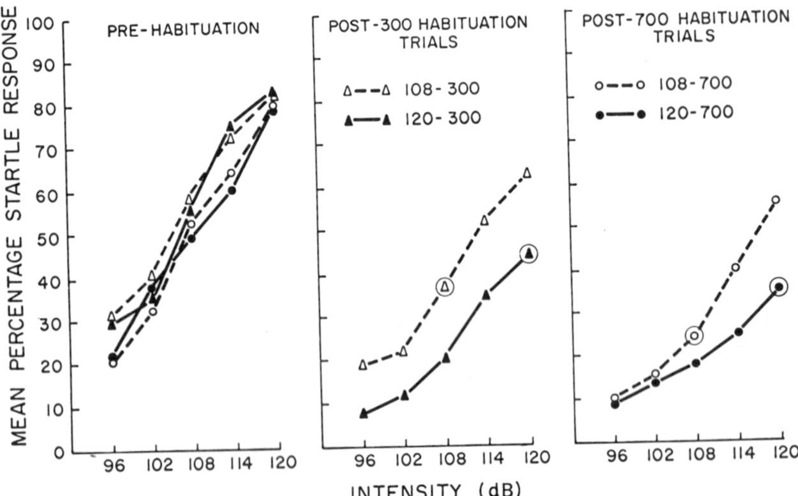

Figure 6. Mean percentage startle responses at each of five tone intensities, before and after habituation, for each of four groups, receiving either of two tone intensities (108 or 120 dB) and either of two numbers of stimulations (300 or 700) during habituation (from Davis and Wagner, 1968).

test intensity, and following either 300 or 700 habituation trials, the groups habituated with 120-dB tones responded less than the groups habituated with 108-dB tones. These findings indicated, therefore, that exposure to the more intense, 120-dB stimulus, during habituation produced a greater decrease in startle responsiveness than did exposure to the 108-dB stimulus.

In a subsequent series of experiments, the differences between the intensity of the eliciting tone and the intensity of the background noise was found to be an important parameter for startle (Davis, 1974*a*). This suggested that the greater effectiveness of the 120-dB tone in producing habituation in the Davis and Wagner (1968) study might be attributed to its greater signal–noise ratio, rather than to its intensive properties per se. To evaluate this it would be necessary to use a variety of tone intensities in combination with several background noise intensities in some training session to see if the degree of habituation is most closely related to tone intensity, or background intensity, or signal–noise ratio.

To test this, six different groups of rats were exposed to either 110- or 120-dB tones at background levels of either 60, 70, or 80 dB. Thus different groups deliberately had equivalent signal–noise ratios (e.g., groups 80–120 and 70–110 or groups 70–120 and 60–110), but differed both in terms of tone intensity and noise intensity. Before and after these training sessions, the animals were given identical test sessions in which a series of all combinations of background noise and tone intensities were presented. This insured that the test conditions under which habituation was evaluated would be comparable for all animals.

The details of the procedures and results have been described elsewhere (Davis, 1974*b*). The most striking finding, however, was that when the various groups were viewed in terms of the signal–noise ratio in the training session, then signal–noise ratio was a better predictor of the degree of habituation than either tone intensity alone or background noise intensity alone (Figure 7). These data indicate, once again, that habituation is generally greater following training with strong versus weak tone intensities, but suggest that this may be a special case of the more general law that habituation is a direct function of the signal–noise ratio used in training.

It should be emphasized, however, that these conclusions are based on an experimental design in which all animals have been tested with a full series of test intensities prior to and following a specific habituation training session. It is possible that the multiple-intensity test procedures are important for demonstrating these effects. In another study, we found that startle amplitude during a block of 120-dB tones was lower after a series of tones that gradually increased in intensity than after a series of 120-dB tones (Davis and Wagner, 1969). The gradual sequence seemed important, since another group that received a weak stimulus intensity

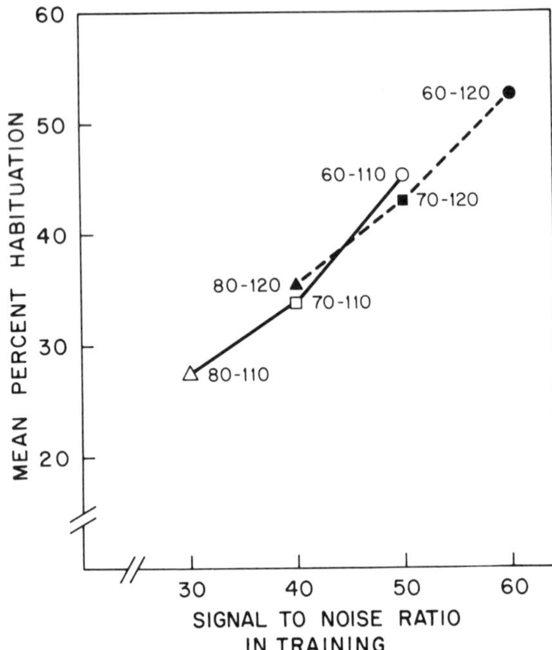

Figure 7. Mean percent habituation [(pretest amplitude − posttest amplitude)/pretest amplitude × 100)] for six groups as a function of the signal–noise ratio in training. The first number in each group refers to the level of background noise in training (e.g., 60, 70, or 80 dB), the second number refers to the intensity of the eliciting stimulus in training (110 or 120 dB). All animals were tested with all combinations of background noise and tone intensity (from Davis, 1974*a*).

during training equivalent to the average intensity experienced by the gradual group was the least habituated during the test. Nonetheless, this study did demonstrate that exposure to intense tones may not always be the most effective way to reduce subsequent startle propensity.

In an effort to explain the superior performance of the gradual group, Davis and Wagner (1969) proposed that "simple stimulus exposure, in addition to producing habituation may also produce any of a variety of state changes depending on the degree of prior habituation to that stimulus. Thus, exposure to an unhabituated stimulus may produce arousal which could persist and potentiate the response measurement on subsequent exposure occasions." A similar explanation was offered by Groves and Thompson (1970) in which they suggested that intense stimuli would produce more sensitization that would then elevate subsequent response levels. If sensitization is a relatively transient effect, whereas habituation is more permanent (e.g., Davis, 1972), then the interval between training and testing may be critical for measuring the effects of stimulus intensity on habituation. Short retention intervals would allow prior arousal or sensitization to carry over into testing, which could elevate response levels following loud versus weak stimuli. Hence, to address this issue, a careful analysis of habituation and stimulus intensity should include several retention intervals between training and testing.

7.2.2. "Refractorylike" Effects

Many studies have shown that the amplitude of startle can be reduced by the presentation of a single, prior startle-eliciting stimulus. For example, Dodge and Louttit (1926) reported that the magnitude of the guinea pig's startle response, elicited by the bang of a mousetrap, was markedly reduced if preceded up to 3 sec by a single exposure to that same stimulus, and that there was a greater reduction the shorter the interval between the two. In rats, this effect decays gradually over about 60 sec (Davis, 1970*a*; Wilson and Groves, 1973). It still occurs in a completely decerebrate rat (Davis and Gendelman, 1977) or after adrenalectomy (Davis and Zolovick, 1972). The effect also occurs when "startlelike" responses are elicited by electrical stimulation in the nucleus reticularis pontis caudalis, suggesting that the spinal cord is sufficient to mediate the effect (Davis, 1981). However, a greater depression of startle responsivity is seen with pairs of acoustic stimuli or when startle is elicited electrically in the cochlear nucleus. This probably means that refractory effects occur at multiple points along the startle circuit that then summate when more and more parts of the circuit are activated.

7.2.3. Interstimulus Interval and Habituation

It has consistently been asserted that the rate of habituation is inversely related to the interval between successive stimuli (interstimulus interval, ISI), using a design where one group of animals is presented with a series of stimuli at long ISIs, while another group receives stimuli at short ISIs. As described in Section 7.2.1 regarding stimulus intensity, in these designs each stimulus serves both as a "test" stimulus, to which response likelihood is evaluated at that particular exposure, and as a "training" stimulus, by which further response decrements may be produced. The ISI, therefore, may be thought of as both the interval between successive training stimuli (training interval) and the interval between each test stimulus and an immediately prior stimulus exposure (test interval). Such procedures do not allow, therefore, any easy separation of the effects of training at different ISIs from those of testing at different ISIs. In order to distinguish between the two, a design would have to be used in which different groups of animals would be trained with different ISIs but then later tested with a common ISI.

To evaluate this question, rats were exposed to a series of tones at ISIs of 2, 4, 8, or 16 sec before and after presentation of 1000 tones at either a 16-sec or 2-sec ISI (Davis, 1970*a*). Figure 8 shows that during the prehabituation and posthabituation tests, rats showed a smaller likelihood of responding the shorter the time since an immediately previous stimulus. During the "habituation training" session, the degree of responding was also much lower when stimuli were presented at the shorter ISI. Most striking, however, was that when the ISI test conditions were subsequently equated for both groups, overall startle levels were lower for the rats that had been trained with long rather than short ISIs. Thus, when conditions were arranged so as to unconfound training and test ISIs, habituation was found to be greater following long versus short ISIs. Moreover, a subsequent study found that the superior habituation follow-

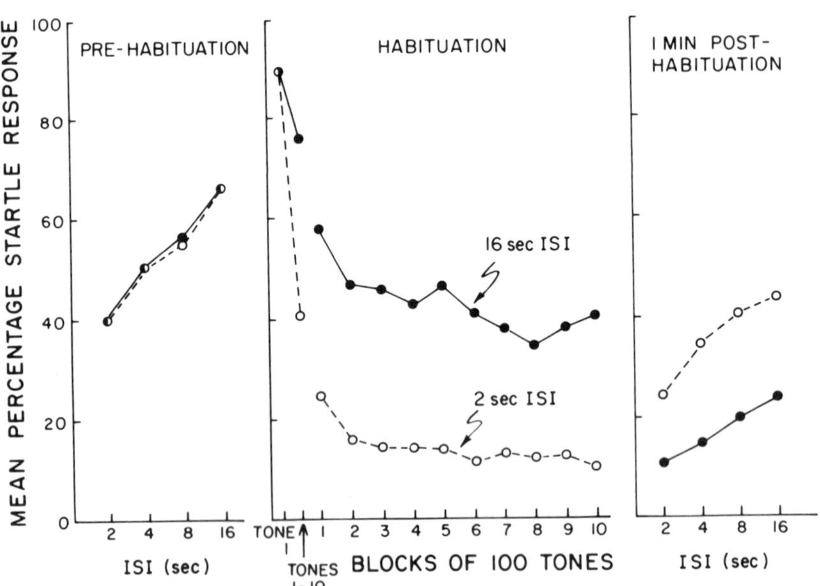

Figure 8. Mean percentage of startle responses of the 2- and 16-sec groups at each of the four test intervals prior to and 1 min following habituation training and over each block of 100 tones during habituation training. Data for the first tone and the average of tones 1–10 in training are also shown (from Davis, 1970*a*).

ing the longer ISI could not be attributed to the longer total time in the experimental situation that this group necessarily had to spend prior to test (Davis, 1970*b*).

These ISI data indicate, therefore, that there are at least two types of startle habituation. One type is akin to the refractorylike effect where response amplitude is dependent on the presentation of an immediately prior stimulus and is inversely related to ISI. Another type is dependent on a history of several prior stimuli and is directly related to the ISI between those stimuli. This general finding has been replicated in other stimulus–response systems and has provided the foundation for a new theory of habituation (Wagner, 1976).

7.2.4. Changes in the Properties of a Habituated Stimulus

Given that habituation of startle occurs, the next question concerns the mechanism underlying habituation. This question has been addressed in part by asking about the changes in the properties of a habituated stimulus. One suggestion offered for characterizing the habituation process is that stimulus repetition reduces the perceived intensity of the eliciting stimulus (Davis, 1970*b*, 1974*a*). Thus following habituation to a 120-dB tone, this stimulus now acted as a less intense stimulus in terms of how it was affected by varying the level of background noise (Davis, 1974*a*). Similarly, inferior habituation at rapid rates of stimulus repetition may result because a stimulus presented within the refractory period of the prior stimulus acts as a weaker stimulus intensity leading to less startle and less habituation (Davis, 1970*b*). However, Russo *et al.* (1975) showed that repetition of a mild shock led to a reduction in its ability to elicit startle but not in its ability to act as a prepulse to inhibit startle to an auditory stimulus. Since other data indicated that a reduction in the intensity of this shock reduced its effectiveness as a prepulse, these results were taken as evidence that stimulus repetition did not reduce the effective intensity of the stimulus. One assumption in this test, however, is that changes in the stimulus attributes that alter the ability of that stimulus to elicit startle are similar to those that allow that stimulus to produce prepulse inhibition. However, as Russo *et al.* (1975) suggest, these different consequences of stimulation may "represent progressively complex transformations of stimulus input." Consistent with this idea, we are finding that drugs such as diazepam can markedly depress the startle-eliciting effects of a white noise burst without altering in any degree the ability of that noise burst to produce prepulse facilitation or inhibition of electrically elicited startle reflexes (M. Davis, K. M. Berg, and D. W. Gallager, unpublished data). Perhaps similarly, exposure to a prepulse can markedly decrease the ability of a loud stimulus to elicit startle, but not affect the ability of that stimulus to inhibit startle elicited by a subsequent loud stimulus (Ison and Krauter, 1974).

Another characterization of habituation is that it might result in a prolongation of the usual refractory period seen after a single stimulus presentation (Cohen, 1929; Davis, 1970*a*; Ratner, 1970). However, direct tests of this idea have shown that stimulus repetition can markedly reduce the startle-eliciting capability of that stimulus without altering the time course of the refractory period induced by that same stimulus (Wilson and Groves, 1973).

More recently it has been suggested that habituation may result from conditioning to background cues in the experimental situation (Wagner, 1976). However, in an extensive series of experiments, Marlin and Miller (1981) could find no evidence in support of this regarding startle habituation. We also have not found much support for this idea (Davis and File, 1984).

Although the mechanism underlying startle habituation is far from understood, we are beginning to determine the location within the acoustic startle circuit where it might occur. In these studies, startlelike responses have been elicited electrically from either the ventral cochlear nucleus or the nucleus reticularis pontis caudalis (RPC) (Davis *et al.*, 1982*b*). When using relatively slow rates of stimulus repetition (one stimulus every 30 sec), startle elicited through the VCN showed consistent habituation across the session, similar to acoustic startle. In contrast, startle elicited through the RPC showed a marked increase in amplitude over the session. Other experiments indicated that the increase in startle in the RPC rats was caused by exposure to background noise, a potent sensitizer of acoustic startle. Taken together the data suggest that sensitization of startle is ultimately mediated in the spinal cord whereas habituation of startle occurs in supraspinal parts of the startle pathway. Currently we are trying to localize where this point might be.

7.3. Sensitization

In addition to habituation, it is also the case that startle may show abrupt increases in strength during stimulus repetition. In fact, a typical curve of startle amplitude across trials for any individual rat is remarkably variable from one trial to another even under apparently invariant test conditions. In general, startle is greatest when a rat is sitting quietly in a cage and less when the animal is engaged in any kind of activity (turning around, rearing). However, startle can still be shown to vary appreciably from trial to trial in a rat that remains motionless through the test. Hence, startle is extremely sensitive to a variety of state changes, some of which are not obviously under stimulus control. On the other hand, a number of variables can be arranged that consistently increase acoustic startle and hence can be used to measure the effects of arousal or state changes on the reflex.

7.3.1. Nonmonotonic Effects of Background Noise during Startle Elicitation

A particularly potent environmental factor that influences startle is the level of background noise that exists at the time the reflex is elicited. As originally reported by Hoffman and Fleshler (1963) and extended by Hoffman and Searle (1965), startle amplitude increased as the level of background white noise increased from 50 to 90 dB. In another study, in which a wider variety of intermediate background noise intensities were sampled, it was found that the relationship between startle amplitude and background noise was actually a nonmonotonic one. Startle increased as the background noise level was raised from 65 to 75 dB, but then decreased as the background noise was raised from 75 to 90 dB (Ison and Hammond, 1971).

To account for this nonmonotonic function, Ison and Hammond (1971) offered two alternative hypotheses. In the first, startle amplitude was suggested to be a joint function of arousal (which increased with an increase in the level of background noise and thus enhanced startle) and the signal–noise ratio (which decreased with an increase in the level of background noise and thus reduced startle). To generate the nonmonotonic function, it was assumed that over the ascending limb of the function the enhancing effect of the change in arousal outweighed the reducing effect of the change in signal–noise ratio, whereas over the descending limb the reverse was true. In the second hypothesis it was suggested that the entire nonmonotonic function was the product of a single underlying process, namely, arousal. In this case startle amplitude at different levels of background noise was said to reflect the apparently U-shaped character of the "arousal function" similar to those reported in other behavioral situations. The peak of the startle function would then represent optimal arousal surrounded by lower levels of startle representing underarousal and overarousal.

In an effort to distinguish between these alternatives, rats were presented with several different tone intensities at several different background noise levels (Davis, 1974*a*). If a single process were operating, one would expect to see parallel changes in startle over different noise backgrounds, irrespective of the intensity of the eliciting stimulus. If signal–noise ratio was also important, one would expect to see an interaction between tone intensity and background noise. The data clearly supported the later, since background noise and tone intensity interacted. When very loud tones were used, increases in background noise produced a monotonic increase in startle. When weak tones were used, increases in background noise produced a monotonic decrease in startle. Other data also support the idea that increases in background noise will enhance startle unless maskinglike effects interfere with the expected increase. Thus background noise has a monotonic facilitatory effect on startle elicited by bright light flashes in pigeons, where masking presumably would be minimized by this cross-modal comparison (Stitt *et al.*, 1976). Indeed, when background illumination was varied, a nonmonotonic function on visual startle was obtained.

Although these results supported the hypothesis that the descending limb of the nonmonotonic noise function was akin to something similar to masking, more recent data suggest this may not be entirely correct. In a very clever experiment, Ison and Reiter (1980) tested whether diminished startle in the presence of high background noise behaved similar to startle elicited by a weak tone intensity. In this test, a visual prepulse inhibited startle markedly when a weak startle-eliciting stimulus was used, but had little inhibitory effect when a comparable baseline level of startle was produced by presenting a higher stimulus intensity against a very high level of background noise. As Ison and Reiter put it, "This difference in susceptibility of these two phenotypically similar reflexes to preliminary stimuli must have significance for our understanding of the terminal decrement in the acoustic startle reflex produced by intense background noise. In particular, it must be that the earlier masking hypothesis (Davis, 1974*a*; Ison and Hammond, 1971) is inappropriate. If the two reflexes were of the same strength because the high background noise works through a process functionally equivalent to that engendered by a weak eliciting stimulus, then the reflexes should be similarly affected by initial stimuli. That they are not reveals the inadequacy of this hypothesis and suggests, instead, that the apparently similar responses are produced by different underlying mechanisms."

In view of these findings, plus a number of studies that would suggest that masking alone cannot account for all the data, Ison has gone back to the "overarousal" hypothesis to account for the descending limb of the nonmonotonic function. In essence he suggests that noise tends to activate or depolarize motor neurons in the startle circuit. "At a low noise level, motor units are excited and more readily fire, given the eliciting stimulus. At high noise levels, motor units spontaneously discharge and then less readily fire, given the eliciting stimulus" (Ison and Reiter, 1980).

An interesting pharmacological test of this hypothesis would be to try and activate motor neurons in the startle circuit pharmacologically to see how this might interact with background noise. For example, background noise and drugs that depolarize motor neurons should synergize in enhancing or depressing startle, depending on the dose of the drug and level of background noise that were used.

7.3.2. Monotonic Effects of Noise Prior to Startle Elicitation

Another potent variable for acoustic startle is the length of time animals have been exposed to background noise prior to the introduction of startle-eliciting stimuli. Figure 9 shows data from an experiment in which rats were presented with two tones either 5, 15, 30, or 75 min after

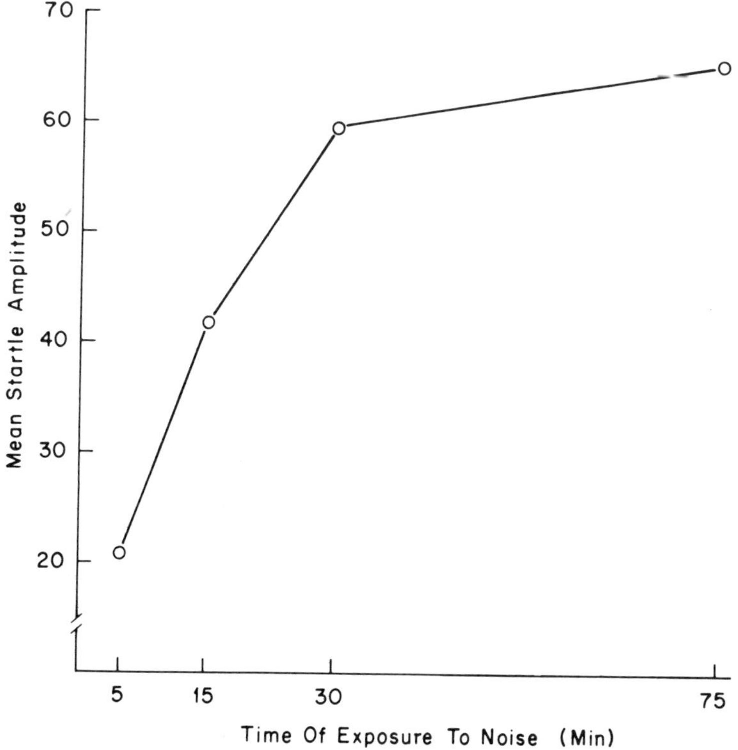

Figure 9. Mean amplitude of the startle response at various times after placing rats in the startle chamber and exposing them to a constant level of 80 dB background noise (from Davis, 1974*b*).

being placed in a startle test cage where a constant level of 80-dB background noise was maintained throughout the session. Startle amplitude showed a very marked increase over this test period. Figure 10 shows that the magnitude of the effect was directly related to the intensity of the intervening noise level. Here rats were placed in the chamber and tested either 5 or 30 min later. During testing, a constant level of 70 dB was maintained for all groups. However, during the rest of the session, background noise levels were either 60, 70, or 80 dB. The fact that a greater increase in startle was seen following exposure to higher noise levels rules out the possibility that the facilitatory effect of noise results only from a depression of startle in the early part of the session following handling (Wedeking and Carlton, 1979), since all groups were treated identically except for the level of background noise prior to the 30-min

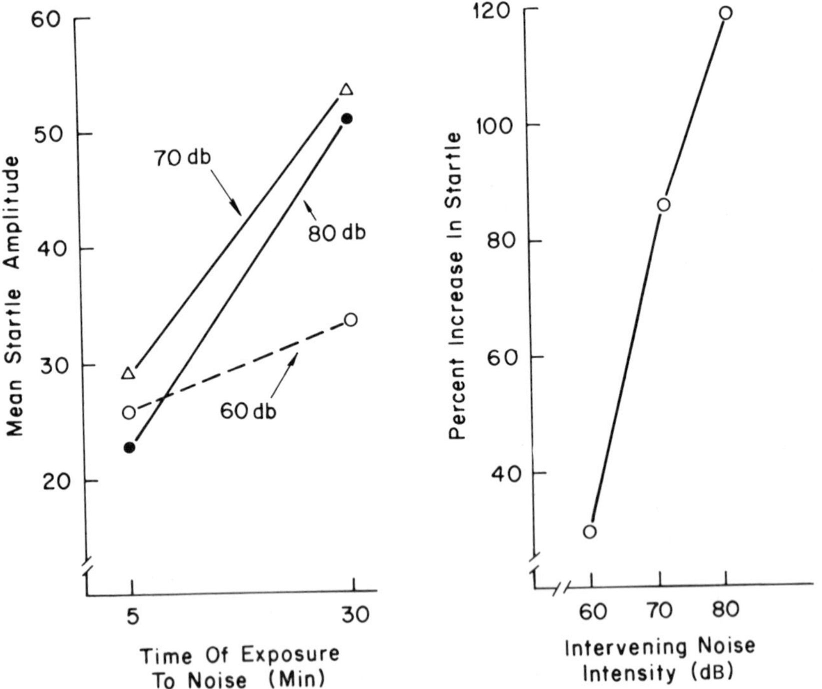

Figure 10. (A) Mean amplitude of the startle response 5 min and 30 min after being exposed to either a 60, 70, or 80-dB level of background noise. (B) The mean percentage increase in startle amplitude from the 5- to the 30-min test points as a function of the intensity of the intervening white noise level. All animals were tested using a 70-dB background noise at the time the startle tones were presented (from Davis, 1974*b*).

test point. Other experiments showed that the "sensitizing effect" of exposure to background noise dissipated in about 15 min, once the noise level was reduced from 80 to 60 dB, again under test conditions that were equated for all animals.

7.3.3. "Arousal" and Startle in Rats

Thus far the tacit assumption has been that procedures that increase arousal, increase startle. However, experiments designed to test this assumption directly have produced contradictory results. Early experiments tended to support the idea that increased arousal increased startle. Prosser and Hunter (1936), Moyer and Bunnell (1960*b*) and Korn and Moyer (1965) reported that acoustic startle was increased when elicited shortly after a series of inescapable tail- or footshocks. Meryman (1952, cited in Brown and Farber, 1968) reported an increase in startle with an increase in food deprivation. Mellgren (1969) reported an increase in startle following either 24 or 48 hr of water deprivation, as well as an increase in startle in the presence of a cue that had previously been paired with the absence of water in water-deprived rats. Earlier, Wagner (1963) also showed that startle was elevated in the presence of a cue associated with frustrative nonreward, and Hoffman and Stitt (1969) found an elevation of startle during a test session in which the normal delivery of intermittent food reinforcements were withheld. Conversely, a few reports indicate that startle may be decreased if elicited in the presence of a cue associated with reward. Thus Mellgren (1969) reported that startle in water-deprived rats was somewhat reduced in the presence of a stimulus associated with a presentation of water. Armus *et al.* (1964) report a small decrease of startle in food-deprived rats in the presence of a cue that signaled food, provided that cue had previously been paired with food.

However, there are a good deal of data that fail to find a simple relationship between startle and arousal produced by deprivation. In the same study where Mellgren (1969) reported that water deprivation increased startle, two later experiments failed to replicate this effect. Ison and Krauter (1975) report large decreases in startle following water deprivation. Trapold (1962) found no effect on startle of a cue paired with food. Szabo (1967) reported a monotonic decrease in startle as a function of the time of food deprivation using intervals of 0, 24, 48, or 72 hr. Finally, Fechter and Ison (1972) reported a series of experiments all showing that food deprivation depressed acoustic startle that could not be explained by a simple change in body weight.

Other manipulations designed to increase arousal or emotionality using other types of stress have also not been found to increase startle, but again appear to decrease it. Brown *et al.* (1956) reported that prior footshocks tended to depress startle, with a greater depression, the shorter the time since an immediately previous shock. However, since footshocks elicit startlelike responses (Hoffman *et al.*, 1964), this probably reflects refractorylike effects induced by the shocks that inhibit subsequent startle reactions to the auditory stimulus. Stern (1971, p. 32) found that prior stress or sleep deprivation markedly depressed acoustic startle, contrary to expectation. However, he also found that these animals weighed less and were much more active during the startle test sessions, and suggested that this activity "masked the occurrence of startle responses or induced postures which were incompatible with the performance of a startle response."

7.4. Conditioned Fear and Startle

In contrast to exposure to shocks themselves, prior fear conditioning does markedly increase startle, provided that appropriate training and test parameters are used. Brown *et al.* (1951) reported that startle was augmented when the eliciting stimulus occurred in the presence of a cue previously paired with footshock. This phenomenon has been termed the potentiated startle effect and has been replicated a number of times (Anderson *et al.*, 1969*a,b*; Bridger and Mandel, 1967; Chi, 1965; Galvani, 1970; Kurtz and Siegel, 1966; Wagner *et al.*, 1967). The effect has been used as evidence that startle is increased by fear.

Other data indicated, however, that the relationship between startle amplitude and fear may not be as simple as originally thought. Kurtz and Siegel (1966) reported that startle was increased in the presence of a cue previously paired with a footshock, but not when the cue previously had been paired with a backshock, despite the fact that shock levels at the two loci were adjusted to support equivalent avoidance responding in another situation. Chalmers *et al.* (1974) found that previous exposure to intense footshocks did not alter acoustic startle in spite of pronounced effects on other measures sensitive to fear, such as general activity, corticosteroid levels, and heart rate.

In reviewing those studies that had reported potentiated startle, we found that relatively low-intensity or short-duration shocks had been used. In contrast, Chalmers *et al.* (1974) failed to find an effect after prolonged and intense shocks. These observations suggested that startle potentiation might bear a nonmonotonic relationship to prior shock experience and, by inference, to fear.

To evaluate this, rats were given light-shock pairing in which the intensity of the footshock was either 0.2, 0.4, 0.8, or 1.6 mA, using a different group of 12 rats at each of the four shock intensities. One day later the rats were tested for potentiated startle by presenting tones in the presence or absence of the light. Four other groups were treated identically except in the training the lights and shocks were presented in a "random" relationship to each other.

Figure 11 shows that startle potentiation bore a nonmonotonic relationship to the shock intensity used in training in rats that had light and shock consistently paired. Random pairings led to no significant potentiation. In another experiment in which backshocks instead of footshocks were used, startle potentiation also was related nonmonotonically to the shock intensity in training. In fact, the magnitude of the potentiated startle effect at the peak of the two nonmonotonic curves was quite comparable for the two shock loci indicating that robust and consistent startle potentiation does occur with backshocks, provided an appropriate shock in-

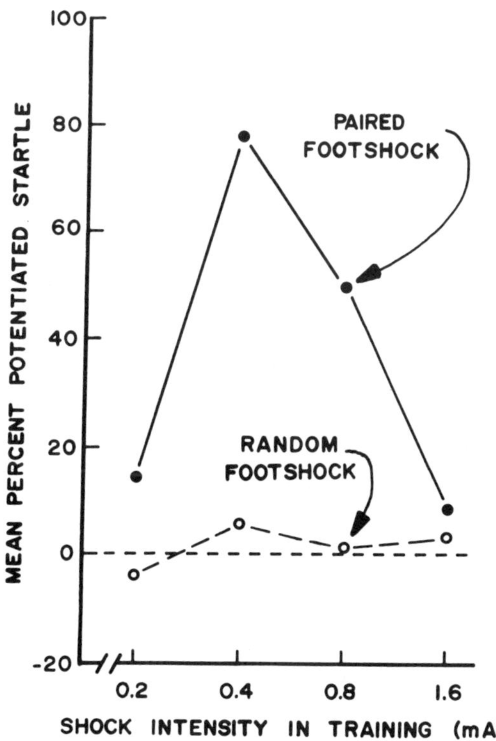

Figure 11. Mean percent potentiated startle as a function of the footshock intensity used in training for the paired (——) and random (–––) groups (from Davis and Astrachan, 1978).

tensity is used. It is possible, therefore, that the failure of the Kurtz and Siegel (1966) study to find potentiated startle using backshocks resulted because the effective shock intensity used was either too high or too low.

The lack of potentiated startle at the high-shock intensity could result because conditioning was simply inferior at this intensity. However, in other situations the literature is generally consistent in showing a direct function between unconditioned stimulus intensity and degree of classical conditioning (cf., Mackintosh, 1974). If this were so, it would follow that startle potentiation shows a nonmonotonic relationship to the level of fear in testing. Moreover, it would predict that high-shock animals should manifest potentiated startle if their level of fear were reduced somewhat. One way to reduce fear would be to give an extended extinction session in which the conditioned stimulus was presented without the unconditioned stimulus. At some point during extinction, high-shock rats should show potentiated startle as fear gradually dissipated from a level initially too high to produce potentiated startle to a level optimal for increasing startle.

To evaluate this, two groups of rats were given light-shock pairings, using a 0.4 mA shock in one group (moderate shock group) and a 1.6 mA shock in the other group (high shock) and then given an extended test session one day later. Figure 12 shows that the moderate-shock group had a high level of startle potentiation on the first block of extinction test trials (which represents the full test session used the first experiment), whereas the high-shock group had little potentiation. With further extinction test trials, the moderate-shock group showed a decline in potentiated startle, which reached baseline in about three blocks, as would be expected if fear dissipated over extinction. In contrast, the high-shock group showed a progressive increase in potentiated startle over the first several test blocks, which then declined to baseline by the end of testing.

These results showed, therefore, that the lack of potentiated startle after training with a high-shock level did not result simply because conditioning was inferior at this intensity. When a prolonged extinction-test session was given, these rats showed a progressive increase in potentiated startle, which peaked in the middle of extinction and dissipated thereafter. Under the appropriate conditions, therefore, prior conditioning could be measured. Moreover, assuming that fear in these animals progressed from an initial high level to a final low level during extinction, their startle performance was exactly what one would expect as they "moved to the left" on the inverted U-shaped function relating startle to fear. More generally, the data indicate that startle can be enhanced in the presence of a fearful stimulus, provided that appropriate training and testing parameters are used.

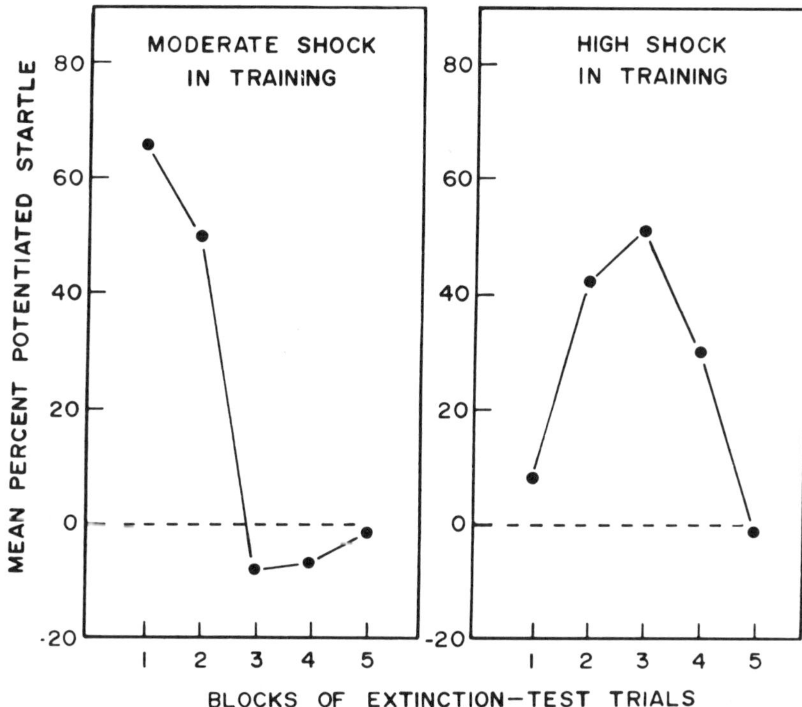

Figure 12. Mean percent potentiated startle over successive blocks of extinction-test trials for groups previously trained with either a moderate (left) or high (right) shock (from Davis and Astrachan, 1978).

8. Summary and Conclusions

The mammalian startle response consists of a rapid sequence of muscle movements that begin in the head and move down the neural axis. Startle responses are highly graded in amplitude being directly related to stimulus intensity and the interval between successive stimuli. Startle does not occur to stimuli that have slow rise times. Startle can be elicited in all mammals that have been studied. The most unique characteristic of startle is its very short latency. In humans, electromyographic responses in jaw muscles can be recorded within 14 msec of stimulus onset. In rats, acoustic startle reflexes measured in the forepaw occur with a latency of about 6 msec. This extraordinary short latency indicates that only a few central synapses are involved in the neural circuit that mediates startle. We have provided evidence that this neural pathway consists of the ventral cochlear nucleus, ventral nucleus of the lateral lemniscus, nucleus reticularis pontis caudalis, and motor neurons in the facial motor nucleus and spinal cord. In the rat, startle-like responses can be elicited electrically from each of these loci.

Startle reflexes show habituation and sensitization and many results concerning these processes originally derived from startle experiments have generalized to other stimulus-response systems. Startle can also be modified by small changes in the experimental environment (pre-pulse facilitation or inhibition). In addition, startle amplitude can be increased if elicited in the presence of a cue previously paired with shock. This latter paradigm is currently being used to analyze the neural bases of classical conditioning as well as to investigate the mechanism of action of antianxiety drugs.

Startle is modulated by a number of different neurotransmitters. In the spinal cord and facial motor nucleus, norepinephrine and serotonin increase acoustic startle. Glycine tonically inhibits startle in the spinal cord. GABA may have a similar effect. Supraspinally, serotonin depresses startle, while activation of dopamine, and perhaps GABA, receptors functionally enhances startle. Startle is also modulated by several brain regions distant from the primary startle pathway itself. Thus, lesions of the olfactory bulbs, frontal cortex, septal area, brachium of the inferior colliculus, mesencephalic reticular formation, periadqueductal gray, and the median raphe nuclei all increase startle amplitude. Lesions of the locus coeruleus tend to depress startle.

Startle represents, therefore, a unique experimental tool to investigate brain mechanisms involved in behavioral plasticity. By eliciting startle electrically at different points along its neural pathway, or by recording single unit activity at each of the various nuclei, it should be possible to determine where different types of behavioral plasticity or different drugs ultimately act to alter neural transmission. With these points as an anchor, it should then be possible to work backwards through the nervous system (e.g., using retrograde tracing techniques) to determine the participation of distant brain structures in mediating various behavioral or pharmacological effects. Eventually, this information will allow us to determine the cellular processes involved in behavioral plasticity in a complex vertebrate.

9. References

Adams, L. M., and Geyer, M. A., 1981, Effects of 6-hydroxydopamine lesions of locus coeruleus on startle in rats, *Psychopharmacology* **73**:394–398.

Aghajanian, G. K., and Haigler, H. J., 1976, Hallucinogenic indoleamines: Preferential action upon presynaptic serotonin receptors, *Psychopharmacol. Commun.* **1**:619–629.

Aghajanian, G. K., and Sheard, M. H., 1968, Behavioral effects of midbrain raphe stimulation: Dependence on serotonin, *Commun. Behav. Biol.* **1**:37–41.

Albert, D. J., and Richmond, S. E., 1976, Hyperreactivity and aggressiveness following infusion of local anesthetic into the lateral septum or surrounding structures, *Behav. Biol.* **18**:211–226.

Albert, D. S., and Wong, R. C. K., 1978, Hyperreactivity, muricide, and intraspecific aggression in the rat produced by infusion of local anesthetic into the lateral septum or surrounding areas, *J. Comp. Physiol. Psychol.* **92**:1062–1073.

Anandam, N., Felegi, W., and Stern, J. M., 1980, In utero alcohol heightens juvenile reactivity, *Pharmacol. Biochem. Behav.* **13**:531–536.

Anden, N. E., Corrodi, H., Fuxe, K., Hokfelt, B., Hokfelt, T., Rydin, C., and Svensson, T., 1970, Evidence for a central noradrenaline receptor stimulation by clonidine, *Life Sci.* **9**:513–523.

Anderson, D. C., Johnson, D., and Kempton, H., 1969*a*, Second-order fear conditioning as revealed through augmentation of a startle response: Part I, *Psychonomic Sci.* **16**:5–7.

Anderson, D. C., Johnson, D., and Kempton, H., 1969*b*, Second-order fear conditioning as revealed through augmentation of a startle response: Part II, *Psychonomic Sci.* **16**:7–9.

Armus, H. L., Carlson, H. R., Guinan, J. F., and Crowell, R. A., 1964, Effect of secondary reinforcement stimulus on the auditory startle response, *Psychol. Rep.* **14**:535–540.

Astrachan, D. I., and Davis, M., 1981, Spinal modulation of acoustic startle: The role of norepinephrine, serotonin and dopamine, *Brain Res.* **206**:223–228.

Astrachan, D. I., Davis, M., and Gallager, D. W., 1983, Behavior and binding: Correlations between alpha-1-adrenergic stimulation of acoustic startle and alpha-1-adrenoceptor occupancy and number of rat lumbar spinal cord, *Brain Res.* **260**:81–90.

Berg, K. M., 1973, Elicitation of acoustic startle in the human, Ph.D. thesis, University of Wisconsin.

Blair, R., Liran, J., Cytryniak, H., Shizgal, P., and Amit, Z., 1978, Explosive motor behavior, rigidity, and periaqueductal grey lesions, *Neuropharmacology* **17**:205–210.

Blair, R., Cytryniah, H., Shizgal, P., and Amit, Z., 1980, Heroin, but not levorphanol produces explosive motor behavior in naloxone-treated rats, *Psychopharmacology* **69**:313–314.

Brady, J. V., and Nauta, W. J. H., 1953, Subcortical mechanisms in emotional behavior: Affective changes following septal forebrain lesions in the albino rat, *J. Comp. Physiol. Psychol.* **46**:339–346.

Braff, D. L., and Geyer, M. A., 1980, Acute and chronic LSD effects on rat startle: Data supporting an LSD-rat model of schizophrenia, *Biol. Psychiatr.* **15**:909–916.

Bridger, W. H., and Mandel, I. J., 1967, The effects of di-methoxy-phenylethylamine and mescaline on classical conditioning in rats as measured by the potentiated startle response, *Life Sci.* **6**:775–781.

Brown, J. S., and Farber, I. E., 1968, Secondary motivation systems, *Ann. Rev. Psychol.* **19**:99–134.

Brown, J. S., Kalish, H. I., and Farber, I. E., 1951, Conditioned fear as revealed by magnitude of startle response to an auditory stimulus, *J. Exp. Psychol.* **41**:317–327.

Brown, J. S., McRyman, J. W., and Marzocco, F. N., 1956, Sound-induced startle response as a function of time since shock, *J. Comp. Physiol. Psychol.* **49**:190–194.

Bunney, B. S., Walters, J. R., Kuhar, M. J., Roth, R. H., and Aghajanian, G. K., 1975, D & L amphetamine stereoisomers: Comparative potencies in affecting the firing of central dopaminergic and noradrenergic neurons, *Psychopharmacol. Commun.* **1**:177–190.

Burnham, R. W., 1939, Repeated auditory stimulation of the startle response in the guinea pig, *J. Psychol.* **7**:79–89.

Capps, M. J., and Stockwell, C. W., 1968, Lesions in the midbrain reticular formation and the startle responses in rats, *Physiol. Behav.* **3**:661–665.

Carlsson, A., Ross, B. E., Walinder, J., and Skott, A., 1973, Further studies on the mechanisms of antipsychotic action: Potentiation by alpha-methyl tyrosine of thioridazine effects in chronic schizophrenia, *J. Neural Trans.* **34**:125–132.

Carlton, P. L., and Advokat, C., 1973, Attenuated habituation due to parachlorophenylalanine, *Pharmacol. Biochem. Behav.* **1**:657–663.

Chalmers, D. V., Hohf, J. C., and Levine, S., 1974, The effects of prior aversive stimulation on the behavioral and physiological responses to intense acoustic stimuli in the rat, *Physiol. Behav.* **12**:711–718.

Chi, C. C., 1965, The effect of amobarbital sodium on conditioned fear as measured by the potentiated startle response in rats, *Psychopharmacologia* **7**:115–122.

Cladel, C. E., Cho, M. H., and McDonald, R. D., 1966, Effects of amphetamine and catecholamines on startle response and general motor activity in albino rats, *Nature* **210**:864–865.

Cohen, L. H., 1929, The relationship between refractory period and negative adaptation in reflex responses, *J. Comp. Psychol.* **9**:1–16.

Commissaris, R. L., and Davis, M., 1982, Opposite effects of N,N-dimethyltryptamine (DMT) and 5-methoxy-N,N-dimethyltryptamine (5-MeODMT) on acoustic startle: Spinal vs. brain sites of action, *Neurosci. Biobehav. Rev.* **6**:515–520.

Conner, R. L., Stolk, J. M., Barchas, J. D., and Levine, S., 1970, Parachlorophenylalanine and habituation to repetitive auditory startle stimuli in rats, *Physiol. Behav.* **5**:1215–1219.

Cook, L., and Sepinwall, J., 1975, Behavioral analysis of the effects and mechanisms of action of benzodiazepines, in: *Mechanisms of Action of Benzodiazepines* (E. Costa and P. Greengard, eds.), Raven Press, New York, pp. 1–28.

Coover, G. D., and Levine, S., 1972, Auditory startle response of hippocampectomized rats, *Physiol. Behav.* **9**:75–78.

Davis, M., 1970*a*, Effects of interstimulus interval length and variability on startle response habituation in the rat, *J. Comp. Physiol. Psychol.* **72**:177–192.

Davis, M., 1970*b*, Interstimulus interval and startle response habituation with a "control" for total time during training, *Psychonomic Sci.* **20**:39–41.

Davis, M., 1972, Differential rates of decay of sensitization and habituation of the startle response in the rat, *J. Comp. Physiol. Psychol.* **78**:260–267.

Davis, M., 1974*a*, Signal-to-noise ratio as a predictor of startle amplitude and habituation in the rat, *J. Comp. Physiol. Psychol.* **86**:812–825.

Davis, M., 1974*b*, Sensitization of the rat startle response by noise, *J. Comp. Physiol. Psychol.* **87**:571–581.

Davis, M., 1979*a*, Diazepam and flurazepam: Effects on conditioned fear as measured with the potentiated startle paradigm, *Psychopharmacology* **62**:1–7.

Davis, M., 1979*b*, Morphine and naloxone: Effects on conditioned fear as measured with the potentiated startle paradigm, *Eur. J. Pharmacol.* **54**:341–347.

Davis, M., 1981, Habituation and sensitization of a startle-like response elicited by electrical stimulation at different points in the acoustic startle circuit, *Adv. Physiol. Sci.* **16**:67–78.

Davis, M., and Aghajanian, G. K., 1976, Effects of apomorphine and haloperidol on the acoustic startle response, *Psychopharmacology* **47**:217–223.

Davis, M., and Astrachan, D. I., 1978, Conditioned fear and startle magnitude: Effects of different footshock or backshock intensities used in training, *J. Exp. Psychol. Anim. Behav. Process.* **4**:95–103.

Davis, M., and Astrachan, D. I., 1981, Spinal modulation of acoustic startle: Opposite effects of clonidine and d-amphetamine, *Psychopharmacology* **75**:219–225.

Davis, M., and Bear, H. D., 1972, Effects of N,N-dimethyltryptamine on retention of startle habituation in the rat, *Psychopharmacologia* **27**:29–44.
Davis, M., and File, S. E., 1984, Intrinsic and extrinsic habituation and sensitization: Implications for the design and interpretation of experiments, in: *Habituation in the Behaving Organism* (H. V. S. Peeke and L. Petrinovich, eds.), Academic Press, New York, pp. 287–323.
Davis, M., and Gendelman, P. M., 1977, Plasticity of the acoustic startle reflex in acutely decerebrate rats, *J. Comp. Physiol. Psychol.* **91**:549–563.
Davis, M., and Heninger, G. R., 1972, Comparison of response plasticity between the eyeblink and vertex potential in humans, *Electroenceph. Clin. Neurophysiol.* **33**:283–293.
Davis, M., and Sheard, M. H., 1974*a*, Habituation and sensitization of the rat startle response: Effects of raphe lesions, *Physiol. Behav.* **12**:425–431.
Davis, M., and Sheard, M. H., 1974*b*, Effects of lysergic acid diethylamide (LSD) on habituation and sensitization of the startle response in the rat, *Pharmacol. Biochem. Behav.* **2**:675–683.
Davis, M., and Sheard, M. H., 1974*c*, Biphasic dose-response effects of N,N-dimethyltryptamine on the rat startle reflex, *Pharmacol. Biochem. Behav.* **2**:827–829.
Davis, M., and Sheard, M. H., 1975, Effects of lysergic acid diethylamide (LSD) on temporal recovery (pre-pulse inhibition) of the acoustic startle response in the rat, *Pharmacol. Biochem. Behav.* **3**:861–868
Davis, M., and Sheard, M. H., 1976, *p*-Chloroamphetamine (PCA): Acute and chronic effects on habituation and sensitization of the acoustic startle response in the rat, *Eur. J. Pharmacol.* **35**:261–273.
Davis, M., and Wagner, A. R., 1968, Startle responsiveness following habituation to different intensities of tone, *Psychonomic Sci.* **12**:337–338.
Davis, M., and Wagner, A. R., 1969, Habituation of the startle response under an incremental sequence of stimulus intensities, *J. Comp. Physiol. Psychol.* **67**:486–492.
Davis, M., and Walters, J. R., 1977, Psilocybin: Biphasic dose response effects on the acoustic startle reflex in the rat, *Pharmacol. Biochem. Behav.* **6**:427–431.
Davis, M., and Zolovick, A. J., 1972, Habituation of the startle response in adrenalectomized rats, *Physiol. Behav.* **8**:579–584.
Davis, M., Svensson, T. H., and Aghajanian, G. K., 1975, Effects of *d*- and *l*-amphetamine on habituation and sensitization of the acoustic startle response in rats, *Psychopharmacologia* **43**:1–11.
Davis, M., Cedarbaum, J. M., Aghajanian, G. K., and Gendelman, D. S., 1977*a*, Effects of clonidine on habituation and sensitization of acoustic startle in normal, decerebrate and locus coeruleus lesioned rats, *Psychopharmacology* **51**:243–253.
Davis, M., Gallager, D. W., and Aghajanian, G. K., 1977*b*, Tricyclic antidepressant drugs: Attenuation of excitatory effects of *d*-lysergic acid diethylamide (LSD) on the acoustic startle reflex, *Life Sci.* **20**:1249–1258.
Davis, M., Redmond, D. E., Jr., and Baraban, J. M., 1979, Noradrenergic agonists and antagonists: Effects on conditioned fear as measured by the potentiated startle paradigm, *Psychopharmacology* **65**:111–118.
Davis, M., Astrachan, D. I., Gendelman, P. M., and Gendelman, D. S., 1980*a*, 5-methoxy-N,N-dimethyltryptamine: Spinal cord and brainstem mediation of excitatory effects on acoustic startle, *Psychopharmacology* **70**:123–130.
Davis, M., Astrachan, D. I., and Kass, E., 1980*b*, Excitatory and inhibitory effects of serotonin on sensorimotor reactivity measured with acoustic startle, *Science* **209**:521–523.
Davis, M., Gendelman, D. S., Tischler, M., and Gendelman, P. M., 1982*a*, A primary acoustic startle circuit: Lesion and stimulation studies, *J. Neurosci.* **2**:791–805.

Davis, M., Parisi, T., Gendelman, D. S., Tischler, M. D., and Kehne, J. H., 1982*b*, Habituation and sensitization of electrically elicited 'startle' reflexes, *Science* **218**:688–690.

deMontigny, C., and Aghajanian, G. K., 1977, Preferential action of 5-methoxytryptamine and 5-methoxydimethyltryptamine on presynaptic serotonin receptors: A comparative iontophoretic study with LSD and serotonin, *Neuropharmacology* **16**:811–818.

Dodge, R., and Louttit, C. M., 1926, Modification of the pattern of the guinea pig's reflex response to noise, *J. Comp. Psychol.* **3**:267–285.

Edwards, S. B., 1980, The deep cell layers of the superior colliculus: Their reticular characteristics and structural organization, in: *The Reticular Formation Revisited* (J. A. Hobson and M. A. B. Brazier, eds.), Raven Press, New York, pp. 193–209.

Faingold, C. L., Hoffman, W. E., and Skittsworth, J. D., Jr., 1981, Pentylenetetrazol-induced enhancement of reticular-formation auditory evoked potential and acoustic startle responses in the cat, *Neuropharmacology* **20**:787–790.

Fechter, L. D., 1974*a*, Central serotonin involvement in the elaboration of the startle reaction in rats, *Pharmacol. Biochem. Behav.* **2**:161–172.

Fechter, L. D., 1974*b*, The effect of L-dopa, clonidine, and apomorphine on the acoustic startle reaction in rats, *Psychopharmacologia* **39**:331–344.

Fechter, L. D., and Ison, J. R., 1972, The inhibition of the acoustic startle reaction in rats by food and water deprivation, *Learn. Motivation* **3**:109–124.

Fleshler, M., 1965, Adequate acoustic stimuli for startle reaction in cat, *J. Comp. Physiol. Psychol.* **60**:200–207.

Forbes, A., and Sherrington, C. S., 1914, Acoustic startle reflexes in the decerebrate cat, *Am. J. Psychiatr.* **35**:367–376.

Fox, J. E., 1979, Habituation and prestimulus inhibition of the auditory startle reflex in decerebrate rats, *Physiol. Behav.* **23**:291–298.

Gallager, D. W., and Aghajanian, G. K., 1975, Effects of chlorimipramine and lysergic acid diethylamide on efflux of precursor-formed 3H-serotonin: Correlations with serotonergic impulse flow, *J. Pharmacol. Exp. Ther.* **193**:785–795.

Gallager, D. W., Kehne, J. H., Wakeman, E. A., and Davis, M., 1983, Developmental changes in pharmacological responsivity of the acoustic startle reflex: Effects of picrotoxin, *Psychopharmacology* **79**:87–93.

Galvani, P. F., 1970, Air-puff elicited startle: Habituation over trials and measurement of a hypothetical emotional response, *Behav. Res. Methods Instrumentation* **2**:232–233.

Gendelman, P. M., 1979, The effect of psychotropic drugs on acoustic startle modulation in the decerebrate rat, M.D. thesis, Yale University School of Medicine.

Geyer, M. A., 1975, Functional heterogeneity within neurotransmitter systems. *Psychopharmacol. Commun.* **1**:675–686.

Geyer, M. A., 1981, Effects of levonantradol on habituation of startle in rats, *J. Clin. Pharm.* **21**:2355–2395.

Geyer, M., Warbritton, A. J. D., Menkes, D. B., Zook, J. A., and Mandell, A. J., 1975, Opposite effects of intraventricular serotonin and bufotenin on rat startle responses, *Pharmacol. Biochem. Behav.* **3**:687–691.

Geyer, M. A., Puerto, A., Menkes, D. B., Segal, D. S., and Mandell, A. J., 1976, Behavioral studies following lesions of the mesolimbic and mesostriatal serotonergic pathways, *Brain Res.* **106**:257–270.

Geyer, M. A., Peterson, L. R., Rose, G. J., Horwitt, D. D., Light, R. K., Adams, L. M., Zook, J. A., Hawkins, R. L., and Mandell, A. J., 1978, The effects of lysergic acid diethylamide and mescaline-derived hallucinogens on sensory-integrative function: Tactile startle, *J. Pharmacol. Exp. Ther.* **207**:837–847.

Geyer, M. A., Rose, G. J., and Peterson, L. R., 1979, Mescaline increases startle responding equally in normal and raphe-lesioned rats, *Pharmacol. Biochem. Behav.* **10**:293–298.

Geyer, M. A., Flicker, C. E., and Lee, E. H. Y., 1982, Effect of tactile startle on serotonin content of midbrain raphe neurons in rats, *Behav. Brain Res.* **4**:369–376.

Gibbins, R. J., Kalant, H., LeBlanc, A. E., and Clark, J. W., 1971, The effects of chronic administration of ethanol on startle thresholds in rats, *Psychopharmacologia* **19**:95–104.

Gogan, P., 1970, The startle and orienting reactions in man. A study of their characteristics and habituation, *Brain Res.* **18**:117–135.

Gottlieb, G., 1971, Ontogenesis of sensory function in birds and mammals, in: *The Biopsychology of Development* (E. Tobach, L. R. Aronson, and E. Shaw, eds.), Academic Press, New York, pp. 67–129.

Graham, F. K., 1975, The more or less startling effects of weak prestimulation, *Psychophysiology* **12**:238–248.

Groves, P. M., and Thompson, R. F., 1970, A dual process theory, *Psychol. Rev.* **77**:419–450.

Groves, P. M., Boyle, R. D., Welker, R. L., and Miller, S. W., 1974*a*, On the mechanism of prepulse inhibition, *Physiol. Behav.* **12**:367–375.

Groves, P. M., Wilson, C. J., and Boyle, R. D., 1974*b*, Brain stem pathways, cortical modulation and habituation of the acoustic startle response, *Behav. Biol.* **10**:391–418.

Haigler, H. J., and Aghajanian, G. K., 1973, Mescaline and LSD: Direct and indirect effects on serotonin-containing neurons in brain, *Eur. J. Pharmacol.* **21**:53–60.

Haigler, H. J., and Aghajanian, G. K., 1974, Lysergic acid diethylamide and serotonin: A comparison of effects on serotonergic neurons and neurons receiving a serotonergic input, *J. Pharmacol. Exp. Ther.* **188**:688–699.

Hamilton, L. W., and Timmons, C. R., 1979, Maturation of startle reflex habituation in rats, *Bull. Psychon. Sci.* **14**:427–430.

Hammond, G. R., 1973, Lesions of pontine and medullary reticular formation and prestimulation inhibition of the acoustic startle reaction in rats, *Physiol. Behav.* **10**:239–243.

Hammond, G. R., 1974, Frontal cortical lesions and prestimulus inhibition of the rats acoustic startle reaction, *Physiol. Psychol.* **2**:151–156.

Hammond, G. R., and Ison, J. R., 1973, Stimulus-produced reflex inhibition in the rat during induction of and recovery from barbiturate anesthesia, *J. Comp. Physiol. Psychol.* **84**:436–444.

Handley, S. L., and Thomas, K. V., 1979, Potentiation of startle response by *d*- and *l*-amphetamine: The possible involvement of pre- and postsynaptic alpha-adrenoceptors and other transmitter systems in the modulation of a tactile startle response, *Psychopharmacology* **64**:105–112.

Heninger, G. R., and Davis, M., 1979, Delayed effects of lithium on the somatosensory evoked response and the acoustic startle reflex in cats and monkeys, in: *Lithium Controversies and Unresolved Issues* (S. Gershon, M. Schou, and N. S. Kline, eds.), Excerpta Medica, Amsterdam, pp. 841–853.

Hoffman, H. S., and Fleshler, M., 1963, Startle reaction: Modification by background acoustic stimulation, *Science* **141**:928–930.

Hoffman, H. S., and Ison, J. R., 1980, Reflex modification in the domain of startle. I. Some empirical findings and their implications for how the nervous system processes sensory input, *Psychol. Rev.* **87**:175–189.

Hoffman, H. S., and Searle, J. L., 1965, Acoustic variables in the modification of the startle reaction in the rat, *J. Comp. Physiol. Psychol.* **60**:53–58.

Hoffman, H. S., and Stitt, C., 1969, Behavioral factors in habituation of acoustic startle reactions, *J. Comp. Physiol. Psychol.* **68**:276–279.

Hoffman, H. S., Fleshler, M., and Abplanalp, P. L., 1964, Startle reaction to electric shock in the rat, *J. Comp. Physiol. Psychol.* **58**:132–139.

Horlington, M., 1968, A method for measuring acoustic startle response latency and mag-

nitude in rats: Detection of a single stimulus effect using latency measurements, *Physiol. Behav.* **3:**839–844.
Ison, J. R., and Hammond, G., 1971, Modification of the startle reflex in the rat by changes in the auditory and visual environment, *J. Comp. Physiol. Psychol.* **75:**435–452.
Ison, J. R., and Krauter, E. E., 1974, Reflex-inhibiting stimuli and the refractory period of the acoustic startle reflex in the rat, *J. Comp. Physiol. Psychol.* **86:**420–425.
Ison, J. R., and Reiter, L. A., 1980, Reflex inhibition and reflex strength, *Physiol. Psychol.* **8:**345–350.
Ison, J. R., and Silverstein, L., 1978, Acoustic startle reactions, activity and background noise intensity, before and after lesions of medial cortex in the rat, *Physiol. Psychol.* **6:**245–248.
Ison, J. R., McAdam, D. W., and Hammond, G. R., 1973, Latency and amplitude changes in the acoustic startle reflex of the rat produced by variation in auditory prestimulation, *Physiol. Behav.* **10:**1035–1039.
Jacquet, Y. F., and Lajtha, A., 1973, Morphine action at central nervous system sites in rat: Analgesia or hyperalgesia depending on site and dose, *Science* **182:**490–492.
Jacquet, Y. F., Klee, W. A., Rice, K. C., Iijima, I., and Minamihawa, J., 1977, Stereospecific and nonstereospecific effects of (+) and (−) morphine: Evidence for a new class of receptors, *Science* **198:**842–845.
Jones, F. P., and Kennedy, J. L., 1951, An electromyographic technique for recording the startle pattern, *J. Psychol.* **32:**63–68.
Jordan, W. P., and Leaton, R. N., 1982*a*, Effects of mesencephalic reticular formation lesions on habituation of startle and lick suppression responses in rat, *J. Comp. Physiol. Psychol.* **96:**170–183.
Jordan, W. P., and Leaton, R. N., 1982*b*, Startle habituation in rats after lesions in the brachium of the inferior colliculi, *Physiol. Behav.* **28:**253–258.
Kehne, J. H., 1983, The effects of clonidine on the flexor reflex in intact and spinalized rats, unpublished Ph.D. thesis, University of Massachusetts.
Kehne, J. H., and Sorenson, C. A., 1978, Effects of pimozide and phenoxybenzamine pretreatments on amphetamine and apomorphine potentiation of the acoustic startle response in rats, *Psychopharmacology* **58:**137–144.
Kehne, J. H., Gallager, D. W., and Davis, M., 1981, Strychnine: Brainstem and spinal mediation of excitatory effects on acoustic startle, *Eur. J. Pharmacol.* **76:**177–186.
Kellogg, C. K., Tervo, D., Ison, J. R., Parisi, T., and Miller, R. K., 1980, Prenatal diazepam exposure alters behavioral development in rats, *Science* **207:**205–207.
Kemble, E. D., and Ison, J. R., 1971, Limbic lesions and the inhibition of startle reactions in the rat by conditions of preliminary stimulation, *Physiol. Behav.* **7:**925–928.
Kirkby, R. J., Bell, D. S., and Preston, A. C., 1972, The effects of methylamphetamine on stereotyped behavior, activity, startle and orienting response, *Psychopharmacologia* **25:**41–48.
Kokkinidis, L., and Anisman, H., 1978, Involvement of norepinephrine in startle arousal after acute and chronic d-amphetamine administration, *Psychopharmacology* **59:**285–292.
Kokkinidis, L., and MacNeil, E. P., 1982, Stress-induced facilitation of acoustic startle after d-amphetamine administration, *Pharmacol. Biochem. Behav.* **17:**413–418.
Korn, J. H., and Moyer, K. E., 1965, The effects of pre-shock and handling on the startle response in the rat, *Psychonomic Sci.* **3:**409–410.
Kurtz, K. H., and Siegel, A., 1966, Conditioned fear and magnitude of startle response: A replication and extension, *J. Comp. Physiol. Psychol.* **62:**8–14.
Landis, C., and Hunt, W., 1939, *The Startle Pattern,* Farrar and Rinehart, New York.
Larsson, L., 1956, The relation between the startle reaction and the non-specific EEG

response to sudden stimuli with a discussion of the mechanism of arousal, *Electroenceph. Clin. Neurophysiol.* **8**:631–644.

Leaton, R. N., 1976, Long-term retention of the habituation of lick suppression and startle response produced by a single auditory stimulus, *J. Exp. Psychol. Anim. Behav. Process.* **2**:248–259.

Leaton, R. N., 1981, Habituation of startle response, lick suppression, and exploratory behavior in rats with hippocampal lesions, *J. Comp. Physiol. Psychol.* **95**:813–826.

Leitner, D. S., Powers, A. S., and Hoffman, H. S., 1980, The neural substrate of the startle response, *Physiol. Behav.* **25**:291–297.

Leitner, D. S., Powers, A. S., Stitt, C. L., and Hoffman, H. S., 1981, Midbrain reticular formation involvement in the inhibition of acoustic startle, *Physiol. Behav.* **26**:259–268.

McCall, R. B., and Aghajanian, G. K., 1979, Denervation supersensitivity to serotonin in the facial nucleus, *Neuroscience* **4**:1501–1510.

Mackintosh, N. J., 1974, *The Psychology of Animal Learning*, Academic Press, London.

Marlin, N. A., and Miller, R. R., 1981, Associations to contextual stimuli as a determinant of long-term habituation, *J. Exp. Psychol. Anim. Behav. Process.* **7**:313–333.

Marsh, R., Hoffman, H. S., and Stitt, C. L., 1973, Temporal integration in the acoustic startle reflex of the rat, *J. Comp. Physiol. Psychol.* **82**:507–511.

Martin, M. R., 1980, The effects of iontophoretically applied antagonists on auditory nerve and amino acid evoked excitation of anteroventral cochlear nucleus neurons, *Neuropharmacology* **19**:519–528.

Marwaha, J., Commissaris, R. L., Kehne, J. H., and Davis, M., 1982, The effects of intrathecal administration of clonidine on neuronal activity in nucleus locus coeruleus, *Soc. Neurosci. Abst.* **8**:726.

Meier-Ewert, K., Gleitsmann, K., and Reiter, F., 1974, Acoustic jaw reflex in man: Its relationship to other brain stem and microreflexes, *Electroenceph. Clin. Neurophysiol.* **36**:629–638.

Mellgren, R. L., 1969, Magnitude of the startle response and drive level, *Psychol. Rep.* **25**:187–193.

Meryman, S. W., 1952, Magnitude of startle response as a function of hunger and fear, unpublished master's thesis, University of Iowa.

Miliaressis, T. E., and St. Laurent, J., 1974, Effets de lamide de lacid lysergique-25-sur la reaction de sursant chez le rat, *Can. J. Physiol. Pharmacol.* **52**:126–129.

Miller, N. E., and Barry, H., III., 1960, Motivational effects of drugs: Methods which illustrate some general problems in psychopharmacology, *Psychopharmacology* **1**:169–199.

Miller, S. W., and Treft, R. L., 1979, Habituation of the acoustic startle response following lesions of the medial septal nucleus, *Physiol. Behav.* **23**:645–648.

Mortimer, J. A., 1973, Temporal sequence of cerebellar purkinje and reticular activity in relation to the acoustic startle response, *Brain Res.* **50**:457–462.

Moyer, K. E., and Bunnell, B. N., 1960*a*, Effect of adrenal demedullation, operative stress, and noise stress on emotional elements, *J. Genet. Psychol.* **96**:375–382.

Moyer, K. E., and Bunnell, B. N., 1960*b*, Relationship between emotional elimination, startle response, and blood count after stress, *J. Genet. Psychol.* **97**:237–243.

Overstreet, D. H., 1977, Pharmacological approaches to habituation of the acoustic startle response in rats, *Physiol. Psychol.* **5**:230–238.

Parisi, T., 1979, The ontogeny of modification of the acoustic startle reflex in normal rats and in rats treated systemically with 6-hydroxydopamine, unpublished Ph.D. thesis, University of Rochester.

Parisi, T., and Ison, J. R., 1979, Development of the acoustic startle response in the rat:

Ontogenic changes in the magnitude of inhibition by prepulse stimulation, *Dev. Psychobiol.* **12:**219–230.
Payne, R., and Anderson, D. C., 1967, Scopolamine-produced changes in activity and in the startle response: Implications for behavioral activation, *Psychopharmacology* **12:**83–90.
Peterson, B. W., 1979, Reticulo-spinal projections to spinal motor nuclei, *Annu. Rev. Physiol.* **41:**127–140.
Pohorecky, L. A., Cagan, M., Brick, J., and Jaffe, C.S., 1976, Startle response in rats: Effects of ethanol, *Pharmacol. Biochem. Behav.* **4:**311–316.
Prosser, C. L., and Hunter, W. S., 1936, The extinction of startle responses and spinal reflexes in the white rat, *Am. J. Physiol.* **17:**609–618.
Ratner, S., 1970, Habituation: Research and theory, in: *Current Issues in Animal Learning* (J. Reynierse, ed.), Lincoln, University of Nebraska Press, pp. 55–84.
Rossignol, S., 1975, Startle responses recorded in the leg of man, *Electroenceph. Clin. Neurophysiol.* **39:**389–398.
Rubel, E. W., 1978, Ontogency of structure and function in the vertebrate auditory system, *Handbook of Sensory Physiol.* **IX:**135–234.
Russo, J. M., Reiter, L. A., and Ison, J. R., 1975, Repetitive exposure does not attenuate the sensory impact of the habituated stimulus, *J. Comp. Physiol. Psychol.* **88:**665–669.
Sagvolden, T., and Wester, K., 1974, Habituation of the startle reflex in rats with septal lesions, *Behav. Biol.* **12:**413.
Sharpe, L., Garnett, J. E., and Cicero, T. J., 1974, Analgesia and hyperreactivity produced by intracranial microinjections of morphine into the periaqueductal gray matter of the rat, *Behav. Biol.* **11:**303–313.
Shimamura, M., 1973, Neural mechanisms of the startle reflex in cerebral palsy, with special reference with spino-bulbo-spinal reflexes, in: *New Development in Electromyography*, Vol. 3 (J. E. Desmedt, ed.), Karger, Basel, pp. 761–766.
Shizgal, P., Brown, Z. W., Amit, Z., and Sklar, L., 1977, Differential motor effects of intraventricular infusions of morphine and etonitazene, *Pharmacol. Biochem. Behav.* **6:**17–20.
Shnerson, A., and Willott, J. F., 1980, Ontogeny of the acoustic startle response in C57B1/6J mouse pups, *J. Comp. Physiol. Psychol.* **94:**36–40.
Singer, H. S., Coyle, J. T., Frangia, S., and Price, D. L., 1981, Effects of spinal transection on presynaptic markers for glutamanergic neurons in the rat, *Neurochem. Res.* **6:**485–496.
Skirboll, L. R., Grace, A. A., and Bunney, B. S., 1979, Dopamine auto- and postsynaptic receptors: Electrophysiological evidence for differential sensitivity of dopamine agonists, *Science* **206:**80–83.
Sorenson, C. A., and Davis, M., 1975, The effects of 6-hydroxydopamine and alpha-methylparatyrosine on the acoustic startle in rats, *Pharmacol. Biochem. Behav.* **3:**325–329.
Sorenson, C. A., and Swerdlow, N., 1982, The effect of tail pinch on the acoustic startle response in rats, *Brain Res.* **247:**105–113.
Stern, W., 1971, Effects of desynchronized sleep deprivation upon startle response habituation in rats, *Psychonomic Sci.* **23:**31–32.
Stitt, C. L., Hoffman, H. S., Marsh, R. R., and Schwartz, G. M., 1976, Modification of the pigeon's visual startle reaction by the sensory environment, *J. Comp. Physiol. Psychol.* **90:**601–619.
Strauss, H., 1929, "Oas Zusammenschrenken," *J. Psychol. U. Neurol.* **39:**111–321.
Szabo, I., 1965, Analysis of the muscular action potentials accompanying the acoustic startle reaction, *Acta Physiol. Acad. Sci. Hung.* **27:**167–178.
Szabo, I., 1967, Positive correlation between the magnitude of the acoustic startle reaction

and the performance of the active avoidance reaction, *Acta Physiol. Acad. Sci. Hung.* **31:**191–198.

Szabo, I., and Hazafi, K., 1965, Eliciability of the acoustic startle reaction after brain stem lesions, *Acta Physiol. Acad. Sci. Hung.* **27:**155–165.

Thorne, F. C., 1944, Startle neurosis, *Am. J. Psychiatr.* **101:**105–109.

Trapold, M. A., 1962, The effect of incentive motivation on an unrelated reflex response, *J. Comp. Physiol. Psychol.* **55:**1034–1039.

VanderStaak, C., 1976, Habituation and sensitization of the acoustic startle response during cortical spreading depression in rats, *Physiol. Behav.* **16:**681–687.

Van Riezen, H., Schnieden, H., and Wren, A. F., 1977, Olfactory bulb ablation in the rat: Behavioral changes and their reversal by antidepressant drugs, *Br. J. Pharmacol.* **60:**521–528.

Wagner, A. R., 1963, Conditioned frustration as a learned drive, *J. Exp. Psychol.* **66:**142–148.

Wagner, A. R., 1976, Priming in STM: An information processing mechanism for self-generated or retrieval-generated depression in performance, in: *Habituation: Perspectives from Child Development, Animal Behavior and Neurophysiology* (T. J. Tighe and R. N. Leaton, eds.), Lawrence Erlbaum Association, New Jersey, pp. 95–128.

Wagner, A. R., Siegel, L. S., and Fein, G. S., 1967, Extinction of conditioned fear as a function of percentage of reinforcement, *J. Comp. Physiol. Psychol.* **63:**160–164.

Walters, J. K., Davis, M., and Sheard, M. H., 1979, Tryptophan free diet: Effects on the acoustic startle reflex in rats, *Psychopharmacology* **62:**103–109.

Warburton, D. M., and Groves, P. M., 1969, The effects of scopolamine on habituation of acoustic startle in rats, *Commun. Behav. Biol.* **3:**289–293.

Warren, P. H., and Ison, J. R., 1982, The selective action of morphine on reflex expression to nociceptive stimulation in the rat: A contribution to the assessment of analgesia, *Pharmacol. Biochem. Behav.* **16:**869–874.

Way, E. H., Loh, H. H., and Shen, F. H., 1969, Simultaneous quantitative assessment of morphine tolerance and physical dependence, *J. Pharmacol. Exp. Ther.* **167:**1–8.

Wedeking, P., and Carlton, P. L., 1979, Habituation and sensitization in the modulation of reflex amplitude, *Physiol. Behav.* **22:**57–62.

White, S. R., and Neuman, R. S., 1980, Facilitation of spinal motoneurone excitability by 5-hydroxytryptamine and noradrenaline, *Brain Res.* **188:**119–128.

Williams, J. M., Hamilton, L. W., and Carlton, P. L., 1974, Pharmacological and anatomical dissociation of two types of habituation, *J. Comp. Physiol. Psychol.* **87:**724–732.

Williams, J. M., Hamilton, L. W., and Carlton, P. L., 1975, Ontogenetic dissociation of two classes of habituation, *J. Comp. Physiol. Psychol.* **89:**733–737.

Willott, J. F., Shnerson, A., and Wilson, G. P., 1979, Sensitivity of the acoustic startle response and neurons in subnuclei of the mouse inferior colliculus to stimulus parameters, *Exp. Neurol.* **65:**625–644.

Wilson, C. J., and Groves, P. M., 1973, Refractory period and habituation of acoustic startle response in rats, *J. Comp. Physiol. Psychol.* **83:**492–498.

Wright, C. G., and Barnes, C. D., 1972, Audio-spinal reflex responses in decerebrate and chloralose anesthetized cats, *Brain Res.* **36:**307–331.

Yunger, L. M., and Harvey, J. A., 1973, Effect of lesions in the medial forebrain bundle on three measures of pain, sensitivity, and noise-elicited startle, *J. Comp. Physiol. Psychol.* **83:**173–183.

11

Escapism

Some Startling Revelations

MICHAEL V. L. BENNETT

1. Introduction

The large fiber systems have been historically important in providing material for relatively convenient study of membrane properties. The giant axon of the squid (which escaped inclusion in this volume) continues to be an important experimental subject, although patch clamping may make large size less important in the future (Hammill *et al.*, 1981). The Mauthner cell has provided important information on synaptic transmission and is still under active investigation. The thrust of this volume, however, is in organizational aspects. What do we learn from comparing different startle responses and escape systems? Are there generalizations to be formed and predictions to be made?

We are fond of classifying, that is, grouping things together according to common traits. This activity is useful in suggesting interrelations and general principles of operation. However, there are pitfalls, such as that of confusing homology and analogy, an error pointed out to us when we are very young, probably by third grade in the post-Sputnik era. We can accept without difficulty the possibility of phylogenetic classifications, because a family tree must exist (with a slight reservation about the most primitive microorganisms). When we define functional sets we deal with analogy, and as Bullock (Chapter 1, this volume) points out, the sets may be overlapping. This feature is perfectly reasonable for functional categories and common in family trees, indeed embedded in the definition of sexually reproducing species. However, overlapping is only occasionally

MICHAEL V. L. BENNETT • Division of Cellular Neurobiology, Department of Neuroscience, Albert Einstein College of Medicine, Bronx, New York 10461.

found in phylogenetic trees, for example in allotetraploidy. When one comes to questions of startle responses, giant fibers, command neurons, and the like, the meaning of the classifications must be carefully examined.

A set can be defined by naming its members or it can be defined by an attribute they hold in common. Except when working with random samples, grouping by common attribute is more useful in suggesting generalizations about other attributes within the set. When defining a set, we generally give it a name related to the common attribute of interest. It is important that the name has an appropriate connotation. Command and grandmother neurons have a nice ring and imply the common attribute in aesthetically pleasing ways. Inducer is a nice embryological term, but the set remains embarrassingly small. To name a set by an attribute does not mean that the set contains any members.

Most of the systems discussed in this book are independently evolved and to varying degrees are examples of convergent evolution. Therefore, any generalizations that can be arrived at will concern modes of operation and modes of evolution. We can ask what is easy evolutionarily, that is, what kinds of neural changes occur frequently and does this tell us anything about the genetic and developmental mechanisms. Also are there features in common that give insights into the selection pressures or the best strategies for the individual's response?

2. Large Size, Electrical Transmission, and Speed of Response

Large-diameter axons have evolved convergently in many instances. One reason, meaning one survival advantage they confer, is more rapid conduction, where the utility in escape is obvious. The saving provided by giant fiber systems need not be great (a few milliseconds between fastest Mauthner fiber and non-Mauthner fiber mediated startle responses; see Chapter 8, this volume) and the advantage provided by this saving when muscle contraction time is an order of magnitude greater has not been shown experimentally. Nevertheless, we generally feel quite comfortable with this explanation.

The matching of fiber diameter and conduction velocity to the requirements of the task performed is about as common as matching the size of a muscle to the mechanical work to be done. Fast-twitch muscle fibers are innervated by more rapidly conducting axons; priority sensory information is carried in faster pathways. Diameter, then, appears to be under easy evolutionary control, implying the existence of size or rate of growth determining genes that are readily mutable or exist in many alleles. A size gene may also be pleiotropic in that it may be expressed at many sites. The electric organ of the sternarchids consists of giant myelinated nerve fibers that are processes of the homologs of spinal electromotor neurons in other gymnotids (Bennett, 1971). In one genus, *Adontosternarchus,* an accessory organ consists of similarly enlarged sensory axons, the same morphology being expressed at an entirely new site. The molecular biology behind these apparently saltatory changes should become clear as more is learned of the control of eukaryotic differentiation.

Synchronization of effector organs at different distances from the controlling neurons offers other examples of precisely determined size. Prime examples are provided by the squid mantle, where more distant muscle is innervated by faster fibers so that contraction is uniform (Pumphrey and Young, 1938). Other equally refined examples are found in the control of electric organs (Bennett, 1971; Meszler *et al.,* 1974). Comparison of auditory information from the two ears must also involve very precisely controlled conduction times.

Increasing diameter is only one method of increasing conduction velocity. The vertebrate myelinated axon is a marvelous invention representing a whole new approach, an evolutionary breakthrough. The function, however, is not to make conduction saltatory, but to reduce membrane capacity, and having many membranes in series to reduce capacity can be, and is, done in more than one way. The mechanism may be clarified by example. Granted that the wavelength of an action potential is long compared with internodal distance (duration times velocity is several centimeters for large fibers compared with a 2-mm internodal distance), conduction really cannot be considered saltatory in the sense of jumping between adjacent nodes. Conduction velocity would be the same if the nodal area were distributed differently, e.g., doubling the number of nodes and halving their area, or even distributing the active membrane in a continuous narrow strip running along the axon. Obviously there would be problems of seal at the margin of the myelin and active membrane and of forming the myelin layers, but the point is that conduction velocity is little affected by the degree of separation of active membrane areas below a certain value. Thus, there is no reason why the Mauthner axon, whose myelin formation is as yet undescribed, should not be able to conduct quite well with active areas (or functional nodes) that are also its synaptic terminals and much more closely distributed than they should be compared with nodes of ordinary vertebrate myelin. However, there may be so many terminals that capacity is increased beyond that for maximum velocity.

The foregoing electrical point makes shrimp and earthworm myelin seem perfectly reasonable, for active membrane at a branch is just as good as an annular node in terms of generating action currents (see Chapter 1, this volume). The large periaxonal space of the shrimp axons also presents little problem for local circuit propagation provided current can easily enter and leave the axon just inside its passages through the sheath. Myelination is readily "available" to vertebrates (except *Agnatha*) and widely utilized. A few invertebrates have found another way to do essentially the same thing. One wonders whether other invertebrate axons with extensive sheaths but no compact myelin, for example the medial giant of the crayfish, are at an intermediate stage of myelin development. Histochemical techniques revealing the location of active membrane areas may ultimately aid in demonstrating such intermediates (Ellisman and Levinson, 1982).

In many giant fiber systems there are also electrically transmitting synapses (cf. Bennett, 1977). Here, too, one explains their occurrence in terms of reduced latency; at 20°C perhaps half a millisecond per synapse. In septate axons that saving can add up, in annelids more than in arthropods (and even more in vertebrate heart). The mode of transmission at synapses in giant fiber systems points out some of the evolutionary questions. In the Mauthner system, for example, there are four synapses from saccular hair cell to axial muscle (and five to pectoral fin muscle). Only one synapse, the club ending on the Mauthner cell lateral dendrite, is electrical (two in the second pathway, the club ending and the giant fiber-motoneuron synapse). Whether there are cellular and molecular biological reasons for failure to develop electrical transmission at more synapses, or whether the reasons are simply historical is undecided. Increase in size seems to be under relatively simple control, so fast escape systems without giant fibers may be there because the benefits of increase in speed are too little to justify the costs of increase in size. Particularly in large structures with large inertias, e.g., the flukes of a blue whale, the latency savings at even many electrical synapses in series would lead to negligible increase in speed of response.

One price paid for giant size is increase in space occupied. Without myelin we couldn't pack our spinal columns with anything like as many fast fibers. If neural signals are complex, there may not be room for giant fibers to conduct them. The cockroach giant fibers that carry information on wind direction are the most numerous discussed in this volume. If there were only a few more they might just be considered a tract of large fibers.

Electrical transmission and large size are not associated only with short latency. Precision of synchronization in the mormyrid electric organ involves rapid electrical transmission, but neither short latency of response nor giant fibers (Bennett, 1971). The giant supramedullary neurons of the puffer fish are the largest vertebrate somata known, 0.3 mm in diameter, size may substitute for large number and serve to increase synchronization of firing, which is also provided by the reciprocal property of the electrotonic coupling between them (Bennett, 1977; Bennett *et al.*, 1959). However, their axons conduct slowly, and although each cell generates the same number of impulses, they occur at quite different times in different regions of the neuronal cluster. Thus, speed of response appears of little importance in this effector system of unknown function. Much faster communication could be produced with myelinated axons and chemical transmission. Apparently homologous systems in many other fishes involve more numerous neurons of more normal size, but still electrotonically coupled and synchronously active (Bennett, 1960). The inherent synchronization associated with a small number of elements is also found in the R2 and $P1_1$ neurons in *Aplysia* (Rayport *et al.*, 1983), but their role in mucus secretion does not require speed, and indeed R2's giant axon conducts relatively slowly because of invaginations that increase its capacity per unit length.

Computers can be better when faster, but only recently has switching become so fast that conduction time in the connecting leads is important. Interneuronal computations within nuclei or cortical regions may not be limited by conduction times, but communication between nuclei may require a more rapidly conducting axon. Furthermore, a small neuron can receive fewer synapses and is intrinsically more noisy because of its higher resistance. Larger size of a soma may represent a need to maintain a larger axon, as well as to lower input resistance and provide a larger synaptic receiving area. Thinking fast we view as good, but how the time for thought is spent, we have little idea. We do appear to do a great deal of parallel processing at some levels, even if we feel it difficult to think about more than one thing at a time.

3. Command Neurons, Shared Circuitry, and Modulation of All-or-None Responses

In the Sherringtonian concept of the final common path, the motor neuron is shared; that is, converged on by many inputs, different subsets of which evoke its activity in different motor responses. (To define a common path requires an uncommon one.) For twitch of its particular motor unit as a behavior, the motoneuron is a command neuron. For whatever more complicated behaviors that involve its motor unit, the motoneuron is shared. In multiply innervated muscles of invertebrates, the motoneuron is not the final common path for contraction, although contractions initiated by different excitatory fibers generally are different. Moreover, responses may also be affected by inhibition. Thus, no final common path exists.

The question of how to define a command neuron has led to many thousands of words of discussion, to which this author has contributed his share. One must sympathize with the attempt of Kupferman and Weiss (1978) who defined a command neuron as fulfilling two criteria: sufficiency and necessity for evoking a "well-defined" behavior. Sufficiency has two aspects: activity of the neuron occurs during the naturally evoked behavior, and experimental stimulation of the cell evokes the behavior. Necessity means the behavior is prevented if the neuron's firing is blocked. It is obvious that the concept is closely related to that of the final common path. However, the common element(s) may be central to motoneurons shared by other behaviors.

The systems described in this volume are pertinent to the usefulness of the term command neuron as defined, and provide some of the classic examples. They also illustrate some of the difficulties and suggest that rarely is it in the interest of organisms to conform very precisely to the definition. One aspect is that behavior is often graded. Gradations may result from a change in frequency of activity in what might be called a command neuron; "might" because graded behavior is not necessarily well defined and may or may not be distinctly separable from other behaviors controlled by different neurons. Gradations also arise because motoneurons activated by the command system receive other inputs, and their excitability can thereby be affected. The sharing of motoneurons for different tailflip responses is well described in crayfish and presumably occurs in locust and *Drosophila* jumping motoneurons and in the mammalian startle reflex as well. In *Cnidaria* the same muscle fibers can be activated for both fast and slow swimming. The Mauthner system is not well studied in this respect. Does every axial motoneuron receive a Mauthner synapse or is there a subset innervating fast muscle fibers that is activated? Given the small number of Mauthner synaptic processes, innervation of a subset seems likely. Also, it seems likely that these motoneurons also receive other inputs. The pectoral fin motoneurons of the hatchet fish can be excited by other spinal fibers (Auerbach and Bennett, 1969), although the operation of these inputs in locomotion is undetermined, and the unlikely possibility remains that their role is solely a modulatory one for the Mauthner system (see below).

When one proceeds antidromically into the nervous system, similar questions of convergence and sharing arise. In relatively few instances is a final common path describable for a particular behavior. Electric organ systems provide a number of examples of chains of up to three neuronal levels (with a number of neurons at each level generally well synchronized by electrotonic coupling). The nuclei where the volley was presumed to arise, that is, the highest level of the final common path, were termed command nuclei by Albe-Fessard and her colleagues (1953) in the earliest use of the term command with respect to neural processing of which this author is aware.

The idea of a command neuron (or nucleus) that is a decision point (or nexus) for a given behavior is an attractive one, and the startle and escape systems appear to conform to the definition. Closer examination reveals, however, that the responses can be modulated. The situation is analogous to the all-or-none action potential that is either present or absent but variable in amplitude depending on refractoriness and, at the stimulation site, the shape of the initiating stimulus. The command element initiates the behavior, but the behavior is to varying degrees affected by activity in other neurons. To give a few examples, earthworm shortening is dependent on frequency of firing in the controlling giants, and action of the giants can be inhibited during crawling. In crayfish, tailflips evoked by giant fibers are completely inhibited when the animal is restrained by its carapace. Mauthner-evoked responses "habituate" to repeated stimulation of the Mauthner cell itself and are smoothly graded in amplitude. The evidence indicates that the gradation results from inhibition of the motoneurons (Aljure *et al.*, 1980). Thus, the amplitude of the responses to these putative command neurons is not fixed, but depends on other inputs to the neurons that they excite. There is a principle here that is likely to be of wide, if not general, application. Rapid escape movements involve a large amount of the body musculature, and there are likely to be circumstances under which the response should be controlled in amplitude or blocked completely. Complete block can be mediated by inhibition of a single command cell or nucleus, but amplitude gradation requires control at a lower level.

The locust jump also illustrates the action of other elements in a control system. The inhibitory command to jump can work only if the excitatory neurons have been previously activated. The inhibitor fulfills the criteria to be a command neuron, but the whole control system is much more interesting and complex.

The nice aspect of command neurons is that one doesn't have to think about patterns, which generally tends to be a somewhat vague enterprise. There is validity to the concept in that some nervous systems do have decision points, but obviously one does not know enough about the behavior when one knows the common command element. Command elements do exist, at least loosely defined, but patterns, also loosely defined, exist as well. Where there is a command element whose activity produces the behavior, one can make a more refined definition of the command. One can define the command as the activities of all subsets of neurons impinging on the command element that are adequate to excite it. Much of our motor activity at least appears to involve patterns or gradations of activity in many elements and is not readily fit into a rigid concept of neural command.

In Wiersma and Ikeda's (1964) original usage, the idea was of a single descending fiber that triggered complex swimmeret movements. The role was not unlike that of turning on a satellite computer to perform a local operation rather than having each step under central control. The biological usefulness of this approach is apparent, and many examples are now available including mammalian locomotion and peripheral visual processing. The systems described in this volume have some correspondance. They can involve a simple signal in few elements, the outcome of which depends on excitability in lower elements that can be affected by local inputs.

Rapid locomotion can have as its purpose getting away from somewhere or going somewhere. The former, as typified by escape, need not be oriented (and a random component could have a survival advantage); the latter requires orientation. A giant fiber system with few elements may not be as well adapted to oriented locomotion, and the crayfish swimming tailflip does not involve the giant fibers. Activity of the Mauthner system has not been studied in predatory behavior; its use in fast starts for prey capture is uncertain. One might predict that it would not be used for goal-oriented swimming because of lack of fine control of the initiating neural command, but, as noted above, the Mauthner impulse reaches motoneurons whose excitability can be preset. Casual observation suggests that the locust (or grasshopper) will jump to get places, and that the amplitude of the jump is controllable. Use of escape circuitry in non-escape responses may, in fact, be quite common.

A rapid movement in response to a sudden stimulus is a behavior we share with many, perhaps most animals (and perhaps even a few plants). To term the many forms of this behavior as startle simplified the naming of this volume, but may obscure the peculiar characteristic of the startle response as exhibited in higher vertebrates and about which we have introspective data. At first glance (introspectively or extrospectively), the startle response of mammals does not mediate escape and, if anything, appears insufficiently coordinated to have any locomotor value. It may be that the startle response either primes the nervous system for subsequent movement or itself acts on a previously primed system, one that is not involved in the ordinary testing of the reflex. If a "nervous" rat can jump so high as to bang his head on the top of his cage, we perceive a useful startle response, but one that is only exhibited against a suitable stimulus history.

Observationally, as well as introspectively, the startle response depends greatly on the state of the organism. As a child in third grade, one learns to inhibit at least the grosser aspects of the startle response—some readers may recall the penalty for flinching in response to a sudden visual stimulus. Thus, startle may produce useful actions only in the organism prepared to make the appropriate response in the sense that the PSPs produced by neurons of the startle system impinge on neurons whose excitability has been set at an appropriate level by other neural pathways. Alternatively, other pathways becoming active after the startle response may find the appropriate neurons more readily activated. Startle, then, may represent a way of using a single rapid warning system to trigger a variety of responses whose nature is based on prior experience. (Introspectively, and by casual observation, I take exception to the premise that startle requires a sudden external stimulus; visual recognition of an unexpected object can occur suddenly and evoke similar movements and identical sensations to the startle in response to stimuli of rapid onset. Startling revelations can be read.) This view of the startle response as an initiator or trigger of many behaviors, rather than as a command for a single behavior, is a simple extension of the concept of modulatability of escape responses as discussed above.

4. Some Conclusions and Some Questions

A phylogenetic tree is an historical fact; other classifications may give insight into function, but are arbitrary. The startle behavior of the title of this book is a good example of a functional as opposed to a phylogenetic class. One is not surprised that animals in the course of evolution have found more than one way to deal with the problem of rapid escape. Nor is it surprising that similar solutions have been arrived at independently.

Adaptations to increase speed are widely distributed, and one infers that certain kinds of changes are more readily evolved than others. Change or increase in relative size looks easy as Darcy Thompson's *On Growth and Form* suggested. Change from chemical to electrical transmission is easy for neurons, perhaps because it is an expression of an embryonic trait, early neuroblasts being generally coupled. Electrical transmission does not occur at neuromuscular junctions; perhaps in some way it cannot. Reduction in axonal capacity by multiple sheaths has not been invented very often, and the sheath anatomy differs in separate lines. Even Schwann cells and oligodendroglia do it differently, but are more like each other than like the invertebrate myelinating glia. How did the two vertebrate types arise? Was one first? And the fine structure of the Mauthner axon's myelin is still unclear.

Although the startle and escape responses have attracted study because of their stereotyped nature, they are generally not all-or-none. The giant fiber systems of earthworm, crayfish, and lower vertebrates do not give identical responses under all circumstances, and the common need for amplitude control, habituation, and sensory discrimination are likely to account for these differences. The mammalian startle response also is highly dependent on the state of the organism.

An important first step is taken in identifying the primary, rapidly conducting pathways by which the responses are mediated, and how far we have come is well illustrated in this volume. A great deal remains to be learned about how the responses are modulated and shaped to serve the organism's varying needs, and the known circuitries are becoming more elaborate. In this direction one is likely to find much more of the subtlety of neuronal processing that underlies the truly integrative action of the nervous system. Working with these fast systems may not turn out to be better for studying higher processes at the cellular level because the relevant cells may be no more accessible for study than in other systems. Nonetheless, the rapid response and large size of the primary elements give one a good starting point for further analysis.

ACKNOWLEDGMENTS. Work in the author's laboratory is supported in part by NIH grants NS-12627, NS-07512, and HD-04248.

5. References

Albe-Fessard, D., and Martins-Ferreira, H., 1953, Rôle de la commande nerveuse dans la synchronisation du functionnement des élément de l'organe électrique du Gymnote, *Electrophorus electricus* L. *J. Physiol. (Paris)* **45**:533–546.

Aljure, E., Day, J. W., Bennett, M. V. L., 1980, Postsynaptic depression of Mauthner cell-mediated startle reflex, a possible contributor to habituation, *Brain Res.* **188**:261–268.

Auerbach, A. A., and Bennett, M. V. L., 1969, A rectifying electrotonic synapse in the central nervous system of a vertebrate, *J. Gen. Physiol.* **53**:211–237.

Bennett, M. V. L., 1960, Comparative physiology of supramedullary neurons, *Biol. Bull.* **119**:303.

Bennett, M. V. L., 1971, Electric organs, in: *Fish Physiology*, Vol. 5 (W. S. Hoar and D. J. Randall, eds.), Academic Press, New York, pp. 347–491.

Bennett, M. V. L., 1977, Electrical transmission: A functional analysis and comparison to chemical transmission, in: *Cellular Biology of Neurons*, Vol. 1, Sec. 1, *Handbook of Physiology. The Nervous System*, (E. R. Kandel, ed.), Williams and Wilkins, Baltimore, pp. 357–416.

Bennett, M. V. L., Crain, S. M., and Grundfest, H., 1959, Electrophysiology of supramedullary neurons III. Organization of the supramedullary neurons, *J. Gen. Physiol.* **43**:221–250.

Ellisman, M. H., and Levinson, S. R., 1982, Immunocytochemical localization of sodium channel distributions in the excitable membranes of *Electrophorus electricus, Proc. Natl. Acad. Sci. U.S.A.* **79**:6707–6711.

Hamill, O. P., Marty, A., Neher, E., Sakmann, B., and Sigworth, F. J., 1981, Improved patch-clamp techniques for high-resolution current recording from cells and cell-free membrane patches, *Pflüg. Arch. Eur. J. Physiol.***391**:85–100.

Kupferman, I., and Weiss, K. R., 1978, The command neuron concept, *Behav. Brain Sci.* **1**:3–39.

Meszler, R. M., Pappas, G. D., and Bennett, M. V. L., 1974, Morphology of the electromotor system in the spinal cord of the electric eel, *Electrophorus electricus, J. Neurocytol.* **3**:251–261.

Pumphrey, R. J., and Young, J. Z., 1938, The rates of conduction of nerve fibres of various diameters in cephalopods, *J. Exp. Biol.* **15**:453–466.

Rayport, S. G., Ambron, R. T., and Babiarz, J., 1983, Identified cholinergic neurons R2 and LP_1 control mucous release in *Aplysia, J. Neurophysiol.* **49**:864–876.

Thompson, D. W., 1917, *On Growth and Form*. Cambridge University Press, Cambridge, England.

Wiersma, C. A. G., and Ikeda, K., 1964, Interneurons commanding swimmeret movements in the crayfish, *Procambarus clarkii* (Girard), *Comp. Biochem. Physiol.* **12**:509–524.

Index